D0434024

# Research and statistics

# Research and statistics

## A practical introduction for nurses

**Carolyn M. Hicks**
**B.A., M.A., Ph.D., Cert.Ed., A.B.Ps.S.,**
**C.Psychol.**

Senior Lecturer in Psychology
School of Continuing Studies
University of Birmingham
Birmingham

**Prentice Hall**
New York    London    Toronto    Sydney    Tokyo    Singapore

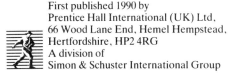

First published 1990 by
Prentice Hall International (UK) Ltd,
66 Wood Lane End, Hemel Hempstead,
Hertfordshire, HP2 4RG
A division of
Simon & Schuster International Group

Printed and bound in Great Britain by
Dotesios Printers Ltd, Trowbridge, Wiltshire

---

British Library Cataloguing in Publication Data

---

Hicks, Carolyn
  Research and statistics: a practical
  introduction for nurses.
  I. Title
  519.5

  ISBN 0-13-844077-8

---

1 2 3 4 5   94 93 92 91 90

*For Tom*

# Contents

*Preface*                                                                  xi

*Acknowledgements*                                                        xiii

**1** Research and statistics in nursing practice                           1
  1.1 Introduction                                                1
  1.2 The book's structure                                        3
  1.3 Some guidelines to using the book                           4

**2** Approaches to research: an introduction to inferential statistics     5
  2.1 Introduction                                                5
  2.2 Estimation                                                  8
  2.3 Hypothesis testing                                          9
  2.4 Key words and terms                                        10

**3** Some basic concepts involved in hypothesis testing                   12
  3.1 The hypothesis                                             13
  3.2 Designing a project to test your hypothesis                 16
  3.3 Key words and terms                                        32

**4** Additional aspects of research design                                33
  4.1 Selection of subjects                                      33
  4.2 Using the subjects in an experimental design               38
  4.3 Key words and terms                                        48

**5** Sources of bias and error in research                                49
  5.1 Order effects and counterbalancing                         49
  5.2 Experimenter bias effects                                  53
  5.3 Constant errors                                            54
  5.4 Random errors                                              56
  5.5 Probabilities                                              59
  5.6 Significance levels                                        60
  5.7 Key words and terms                                        62

**6** Collecting the data: types of measurement                            64
  6.1 Levels of measurement                                      64

6.2  The nominal level of measurement                                    65
6.3  The ordinal level of measurement                                    67
6.4  The interval level of measurement                                   69
6.5  The ratio level of measurement                                      70
6.6  Key words and terms                                                 72

**7** Selecting a statistical test to analyse your data                  73
7.1  Deciding which test to use                                          73
7.2  Parametric and non-parametric tests                                 77
7.3  Finding out whether your results support your hypothesis            79
7.4  One- and two-tailed tests or hypotheses                            80
7.5  Key words and terms                                                 85

**8** Statistical tests for same and matched subject designs with
two conditions only (experimental designs)                               86
8.1  Non-parametric tests                                                87
8.2  Parametric test                                                    100

**9** Statistical tests for same and matched subject designs with
three or more conditions (experimental designs)                         107
9.1  Non-parametric tests                                               109
9.2  Parametric tests                                                   126

**10** Statistical tests for different subject designs with two
conditions only (experimental designs)                                  141
10.1  Non-parametric tests                                              142
10.2  Parametric test                                                   154

**11** Statistical tests for different subject designs with three or
more conditions (experimental designs)                                  161
11.1  Non-parametric tests                                              162
11.2  Parametric tests                                                  181

**12** Statistical tests for correlational designs                      194
12.1  Introduction                                                      194
12.2  Non-parametric test for use with two sets of data                 197
12.3  Parametric test for use with two sets of data                     203
12.4  Non-parametric test for use with three or more sets of data       209
12.5  Making predictions from correlated data                           215

**13** Estimation                                                       225
13.1  An introduction to some basic issues and terms                    225
13.2  Point estimations and interval estimations                        226
13.3  Normal distribution                                               228
13.4  Standard deviation                                                230

13.5  The relationship between standard deviations and the normal
        distribution                                        232
13.6  Calculating interval estimates                        235
13.7  Key words and terms                                   240

**14**  Practical application of the theory: carrying out your research    241
14.1  Planning and preparing the research                   241
14.2  Writing a research proposal                           247
14.3  Obtaining funding                                     250
14.4  Working out the final details of the research procedure    252
14.5  Presenting your research findings                     253
14.6  Other useful sources of information                   254
14.7  Key words and terms                                   254

**15**  Presenting your research findings                   255
15.1  General guidelines                                    255
15.2  How to prepare an article for publication            257
15.3  Key words and terms                                   269

**16**  Reading research articles critically               270
16.1  Title                                                 270
16.2  Abstract or summary                                   271
16.3  Introduction                                          271
16.4  Method                                                272
16.5  Results                                               276
16.6  Discussion                                            277
16.7  References                                            278

*Appendix A*  A reminder of the basic mathematical principles    280
                A.1  Basic maths                            280
                A.2  Rank ordering data                     285

*Appendix B*  Some symbols commonly found in statistical formulae    289

*Appendix C*  Statistical probability tables               290

*Appendix D*  Answers to exercises in the text             311

*References*                                                321

*Index*                                                     322

# Preface

When writing this book, I had in mind one aim – to demystify the essential concepts and procedures of research design and analysis, so that the intending nurse researcher would have a reasonable knowledge-base from which to begin.

Research methodology and statistics are not most people's idea of fun, and yet the pressures currently operating within nursing, such as grading exercises and Project 2000, mean that research will become an increasingly important part of the nurse's job. But while many nurses acknowledge its potential value, they are unsure about how to embark on a research project – at whatever level – themselves. Enter this book.

There are numerous texts on the market which deal with paramedical research, so it would seem useful to justify the existence of this one. Firstly, unlike the majority of books on the subject, its main focus is *hypothesis testing* and *experimental design*. This approach to research involves testing ideas using scientific methods, analysing the resulting data with statistical tests and making inferences from these results.

Translated into an example, a nurse researcher might have observed that there seems to be a link between the wearing of surgical masks during wound dressing and the subsequent incidence of infection in the patients. This constitutes the initial idea or hypothesis. He or she might then design an experiment which tested the validity of this idea, by comparing the infection rates amongst the patients of nurses who do not wear masks with those who do. This constitutes the scientific methodology; if the infection rates for each group are analysed using statistical tests and the results indicate that there is a significant difference in infection according to the type of wound dressing practice, then this outcome might have more general implications for hospital policy on mask wearing among nursing personnel. This constitutes the statistical analysis and inference procedure. Clearly, then, hypothesis-testing is central to nursing research.

Thus, a primary aim of this book is to provide guidelines to formulating and testing hypotheses and analysing and interpreting the results.

A second feature of the book concerns its format and style. Because I am not a statistician, I hope that I can appreciate the difficulties some people have with figures and formulae. While this difficulty is typically emotional rather than intellectual in origin, it nonetheless leaves some students feeling punch-drunk in the face of a page of mathematical formulae. Being a one-time sufferer of this myself, I have tried to make all the explanations as straightforward as possible, presenting worked examples alongside the written instructions for each statistical test. I have also attempted to write informally, eliminating as much esoteric jargon as possible. There are numerous summary boxes, lists of key words and practical activities, all of which are intended to help the reader grasp the fundamental principles of experimental research.

In a continuing attempt to make this text as digestible as possible, very little statistical theory has been included – an omission which may arouse the ire of some statisticians. Although I recognise the value of a sound understanding of the background theory, I would justify its exclusion on the grounds that it puts many students off research and statistical analysis altogether. I take the view that the most important function of a text of this kind is to ensure that the would-be researcher knows how to construct a study along correct scientific principles and how to analyse and interpret the results appropriately. It should be remembered that statistics are simply a means of analysing data; a useful tool of research. And given that we do not need to understand how *other* tools and pieces of apparatus are constructed or work in order to use them appropriately, it seems logical that this should be true of statistics too.

And a final, though critical, difference is that this text relates research principles directly to nursing issues. It has long been established that the best way of imparting a difficult idea comprehensibly is by couching it in the terms and contexts with which the audience is familiar. I have run a great number of research methods courses for nurses and can confirm the wisdom and efficacy of an approach which takes account of the learner's existing expertise and knowledge.

A final word of caution. This text should be seen as a means to an end. While it is a guide to setting up, carrying out and analysing research, it also provides advice on how to get research published. Although the process of learning the techniques and principles may be dry and uninteresting, actually undertaking and publishing your own research is very exciting. (I know of no one who fails to get a 'buzz' out of appearing in print – even seasoned researchers who have published hundreds of articles before.) So do bear in mind that while the learning process may be dull, the end result is worthwhile. And remember, too, that even the smallest and most humble research project may have an important impact on the quality of patient care.

Lastly, some words of thanks are due.

I am grateful to Collette Clifford of the Queen Elizabeth School of Nursing, Birmingham, who advised me on many points concerning the nursing examples included in the text. Because I am not a nurse, this aid was invaluable and I am indebted to her. Secondly, my sincerest thanks are due to my wonderful secretary, Janet Francis, who received my totally unintelligible manuscript uncomplainingly, unravelled it and reproduced it as immaculate typescript. Thanks are also owed to Carol Nixon at whose vast fountain of statistical knowledge I frequently drank. And lastly, thanks to Tom, aged 1, without whose considerable contribution this book would have been completed months ago.

*Carolyn Hicks*
*Birmingham, 1989*

# Acknowledgements

I am grateful to the following sources for granting permission to reproduce the statistical tables in Appendix C of this book:

*Tables* C.1, C.3 and C.6 from Lindley, D.V. and Scott, W.F. (1984) *Cambridge Elementary Statistical Tables*, 10th edn, Cambridge: Cambridge University Press.

*Table* C.2 from Wilcoxon, F. and Wilcox, R.A. (1949) *Some Rapid Approximate Statistical Procedures*, American Cyanamid Company. Reproduced with the permission of the American Cyanamid Company.

*Table* C.4 from Friedman, M. (1937) 'The use of ranks to avoid the assumptions of normality implicit in the analysis of variance', *Journal of the American Statistical Association* **32**.

*Table* C.5 from Page, E.E. (1963) *Journal of the American Statistical Association* **58**.

*Table* C.7 from Runyon, R.P. and Haber, A. (1976) *Fundamentals of Behavioural Statistics*, 3rd edn, Reading, Mass: Addison Wesley.

*Table* C.8 from Kruskal W.H. and Wallis W.A. (1952) 'The use of ranks in one-criterion variance analysis', *Journal of the American Statistical Association* **47**.

*Table* C.9 from Jonckheere, A.R. (1954) 'A distribution-free $k$-sample test against ordered alternatives', *Biometrika* **14** (Biometrika Trustees).

*Table* C.10 from Olds, E.G. (1949) 'The 5% significance levels for sums of squares of rank differences and a correction', *Annals of Mathematical Statistics* **20** (The Institute of Mathematical Statistics).

*Table* C.11 adapted from Friedman, M. (1940) 'A comparison of alternative tests of significance for the problem of $m$ rankings, *Annals of Mathematical Statistics*.

*Table* C.12 from Fisher, R.A. and Yates, F. (1974) *Statistical Tables for Biological, Agricultural and Medical Research* Table VII (p. 63), London: Longman Group (previously published by Oliver and Boyd Ltd, Edinburgh). I am grateful to the Literary Executor of the late Sir Ronald Fisher, F.R.S., to Dr. Frank Yates and to Longman Group Ltd, London for permission to reprint Table VII from their book *Statistical Tables for Biological, Agricultural and Medical Research*, 6th edn, 1974.

# 1    Research and statistics in nursing practice

## 1.1  Introduction

Disraeli's comment that 'there are three kinds of lies: lies, damned lies, and statistics' seems to encapsulate most people's suspicion of numbers. At best the topic of statistics arouses indifference, at worst panic and distrust. While the origin of such attitudes may well reside in memories of a perpetual struggle with school maths, one fact remains inescapable: whatever our reactions to figures, they are an integral part of everyday life. And for no one is this more true than the nurse.

Lest anyone raises their hands to dismiss such an idea, it might be worth looking at the role of statistics and research in nursing. Nursing is, of course, principally concerned with caring and healing, but the process of achieving this depends very heavily on the use of statistics. For example, pulse monitoring, temperature taking and blood pressure reading all involve the collection of a set of numbers which are compared with the average for that function, in order to see whether the patient deviates from the normal limits as defined by previous research. Thus, in these most routine of nursing tasks, the nurse is involved in data collection and analysis.

Similarly, the nurse tutor in a school of nursing is unavoidably engaged in the same processes. Marking a set of exam papers for a group of second year students, calculating the average mark and comparing this with the performance of previous groups of second years also involves data collection, the use of averages and any deviations from these. And likewise, the nurse manager who is concerned with the ratio of trained staff to students might look at the off-duty rotas for a number of wards. From these, the relevant ratios could be calculated for each ward and compared, and trained nurses deployed to those areas where the ratio was inadequate. This process, too, inevitably involves data collection and statistical analysis.

So in these three major areas of nursing – clinical, educational and managerial – statistics are routinely used and interpreted.

However, while measuring a patient's blood pressure and comparing it with the norm, or marking an exam paper, are clearly acts of data collection and statistical analysis, they do not in themselves constitute *research*. Research is a mechanism for answering questions and addressing wider issues. It is a method by which information is collected using scientific principles and then summarised by statistical analyses. In this way, research can be construed as a process of problem-solving and statistics simply as a shorthand for presenting data in a way which can be universally understood.

In a nursing context, research should ultimately enhance the quality of patient care, either directly or indirectly, by forming the foundation on which nursing practice is based. At one level this means that the value of time-honoured practices such as routine four-hourly temperature taking, or the benefits of saline baths, has to be challenged. Undoubtedly, a lot of nursing practice has its foundations in history rather than science and has hitherto gone unquestioned. In the current climate of restricted resources and professional development, few would deny that it is unacceptable to continue with traditionally founded activities without first establishing their real efficacy through proper scientific investigation. Routine practices have to be justified in terms of their financial and therapeutic value if nursing is to operate effectively within reduced resources and an atmosphere of enhanced professionalism. Hence a major function of nursing research is one of scientifically demonstrating the medical and cost benefits of these everyday activities.

A second function of research in nursing is to extend and develop the profession's knowledge-base. This can be accomplished using readily available information; for instance, bed-state forms or patient dependency charts give a clear indication of manpower needs at given times of the day or on certain wards. If nursing resources are allocated appropriately on the basis of this information then the pay-off should be a better level of patient care, as well as comparable work-loads for the staff. Alternatively, nursing knowledge can be extended by looking at the efficacy of different treatment procedures. For example, does the provision of information about an operation reduce or increase a patient's pre-operative anxiety levels? Do salt-water baths have any *significant* effect on the healing of perineal wounds? Is treatment A more effective with eczema patients than treatment B? These questions can only be convincingly answered using formal research methodology and statistical analysis.

Research is now a central issue in nursing, as part of the projected plans for the profession's future, for training, clinical practice, grading assessments and even at the level of personal interest. In consequence, it is becoming the responsibility of every member of the profession to have some degree of familiarity with the essentials of research design and statistical analysis.

However, while these concepts may sound high-powered and awe-inspiring, they are easily conquerable even by those who have only the rudiments of GCSE maths. Carrying out research in nursing should not be confined to a select few who sit in dusty academic garrets crunching numbers; nor should it be the domain only of medics or of a limited number of specialist nurse researchers. Research methodology and statistical analysis should instead be part of the stock-in-trade of every nurse at whatever level, since even if they have no interest in conducting any research of their own, they should be in a position to evaluate the findings of others before implementing it in practice.

The main purpose of this book, then, is to provide intending nurse researchers with sufficient basic groundwork to enable them to carry out their own piece of scientific research or at least to have sufficient expertise to be able to assess the value of *existing* research before implementing the findings.

## 1.2  The book's structure

This text is primarily concerned with techniques of research which use inferential statistics. This is an essential tool in nursing research and simply means that information derived from one situation can be inferred to apply to another. A more detailed definition and explanation of this is presented in Chapter 2. Chapters 3–7 provide a general introduction on how to set up and carry out this sort of research, while Chapters 8–13 deal with the types of statistical analysis that are used to interpret and summarise the data collected. Appendix A provides an introduction to basic mathematical principles, including the rank ordering of data which is central to many statistical analyses. If you feel that you are rusty or ill-informed on these basic rules of maths you are strongly urged to read this appendix; in any event you should familiarise yourself with the ranking procedure outlined in Section A.2.

The remainder of the book presents advice on how to write research proposals and articles for publication, and obtain financial support. For those whose interest lies primarily in critically evaluating the research of others, Chapter 16 gives guidelines on how to do this.

Throughout the book I have tried to simplify the concepts and issues involved in a number of ways. Firstly, I have attempted to keep statistical jargon and theory to a minimum. Second, there are many summary boxes which contain the key points of the chapters. Third, there are glossaries of key words at the end of relevant chapters which can serve either as reminders of the terms and concepts covered or as a self-assessment exercise whereby the reader can test his or her own understanding of the major points. Fourth, there are exercises throughout the text, which are intended to encourage the reader to practise newly learned concepts and statistical procedures. Fifth, the statistical analyses are presented with worked examples alongside in order to simplify the process of calculating complex formulae. Also, within the chapters on statistical tests, there is a series of questions and answers that deal with commonly occurring problems.

Last, every example and idea has been related to nursing, drawing on the knowledge and experience of the reader in order to clarify the main concepts. I should say at this stage that I have made up every hypothesis, number and conclusion, so that, as far as I know, the examples bear no relationship to any actual studies or results in nursing.

I would advise the beginning researcher to read Chapters 2–7, since these provide an understanding of the basic principles involved in designing a research project. The chapters thereafter are devoted to specific statistical analyses which should be read when the need arises – in other words, when there is some actual research data to analyse.

Most of the chapters on statistical analysis stand relatively independently of each other. This, of course, means there is some overlap and repetition, but does allow the reader to turn directly to the sort of statistical analysis required without having the tedious and irritating task of reading several other chapters before understanding what needs to be done. Appendix B gives a list of the symbols used.

I hope this text will provide information, confidence and encouragement. Happy researching.

## 1.3  Some guidelines to using the book

1. Do not panic when you see a page of formulae and statistics. Take a deep breath, remember the problem is emotional not intellectual and just work through the stages systematically.
2. *Never* attempt to commit a statistical formula to memory – just know where to look it up.
3. Always use a calculator, unless you are a mathematical genius or a masochist. There is no virtue in sweating over pages of long division when a calculator will provide an answer in seconds.
4. Do not expect always to understand the concepts on first reading. Be patient with yourself, have a cup of coffee and try again later.
5. Remember that this book provides a means to an end – a route to embarking on research and publishing the findings. Even if the process of familiarising yourself with the basic principles of research methodology is dull, bear in mind the end result, which can be exciting and very valuable.
6. The ultimate aim of research is to improve patient care. However insignificant your research may seem to you, remember that its implications can be far-reaching and beyond any limits you might have considered.
7. All the calculations in this book have been worked to two decimal places; if your answer is marginally different from the one given, it could be that you were working to a different number of decimal places.
8. Try the exercises if you have the time and inclination. Unlikely as it may seem, statistical analysis can be very satisfying, if only to demonstrate your mastery over the dreaded numbers. This in turn will enable you to approach other statistical analyses more confidently.
9. Remember that research expertise develops with practice and that the more you do, the more automatic and skilful you become. Do not anticipate that you will push back the frontiers of science at the first attempt – leave that till later!
10. And finally, integrity is the hallmark of the good researcher: integrity not to use statistics to lie, or research to further personal ends; integrity to treat anyone who takes part in your research with dignity and respect; integrity of open mindedness and impartiality. Some researchers never let the data interfere with a good hypothesis – do not be one of them!

# 2 Approaches to research: an introduction to inferential statistics

## 2.1 Introduction

This chapter is an introduction to the particular approach to research and its analysis covered by this book. This approach is known as **inferential statistics**.

The research process is multi-faceted, in that there are a number of ways in which a topic can be studied. Whichever approach is adopted, however, the findings from the research have to be summarised and presented in a way which can be understood by other people. Research findings can be presented in two basic ways: they can be *described*, usually in terms of their most important features, or they can be used to make wider **inferences**. Let us look at some examples to illustrate these differences.

A survey might be conducted, for instance, which simply collected information on a given subject; this information is then presented in terms of its most interesting features. For example, a nurse manager might want to look at seasonal trends in the admissions to an acute psychiatric ward, so that more precise manpower planning can be achieved. The following sorts of data might be collected:

1. The total annual admission rate.
2. The average monthly admission rate.
3. The type of patient admitted (sex, age, psychiatric problem).
4. The average length of stay per patient.

From these data, a lot of information could be gathered which would tell the nurse manager how nursing resources should be deployed more accurately over the year.

Alternatively the nurse researcher might want to undertake a survey on the attitudes of qualified staff to the new grading structures. In this case, a questionnaire might be devised which would be sent to a sample of trained nurses. From the completed questionnaires, information could be gathered as to the number of nurses who were satisfied with the system, or who considered the principles of grading to be fair, etc.

Both these examples involve presenting the findings by *describing* the data in terms of its most interesting and relevant characteristics. Hence this approach to research involves **descriptive statistics**. While this technique of research and data analysis is suitable for a whole range of nursing-related topics, it limits the researcher to making statements just about the information that has been obtained, without being able to draw any wider inferences from it.

In other words, descriptive statistics *only* allows the nurse researcher concerned with attitudes to grading to make statements about the attitudes of the sample of nurses who returned the questionnaire. No statements can be made about the attitudes of nurses *in general*. While this may be acceptable for many topics in nursing research, there are clearly a whole host of occasions when the nurse wants to go beyond the data given, and make more general predictions and statements.

So, if we take again the example of attitudes to grading, the nurse researcher might want to estimate the percentage of *all* nurses who are dissatisfied with the scheme; or he or she might want to find out whether RMNs are, on the whole, more satisfied than midwives. The research techniques which allow the researcher to do this involve a form of data analysis called inferential statistics. It is inferential statistics that is the focus of this text, though if the reader is interested in techniques of descriptive statistics, the following books might be useful: Reid and Boore (1987), Jaeger (1983) and Clifford and Gough (1990).

Essentially, statistical inference allows us to go beyond the facts in front of us. In other words, we can *infer* things from established information. This process is one used in a multitude of everyday situations. For example, you may have been successful over the past five years in persuading Sister X to allow you to have Christmas off. On that basis, you contemplate booking yourself a skiing holiday this year because you assume she will grant you leave again. In other words, you have taken a set of facts (that Sister X has given you leave at Christmas for the previous five years) and inferred that these facts will also apply in other situations (i.e. that you will get leave again this year). Similarly, if you choose to drive to work today along a particular route because it has been less busy recently, you are inferring that today's traffic will be as light as that of the previous few days. You have gone beyond the established facts (the traffic density over the past days) to predictions or inferences about today's traffic. This is the basis of inferential statistics – that from established information, inferences can be made about how that information applies to other situations. Of course, the process of statistical inference is much more scientific and formal than this, but nonetheless, these are the principles on which it is founded.

In inferential statistics, a **sample** of people is selected for study. This sample is drawn from a parent **population**. The data collected from that sample are analysed and the results are inferred to apply to the wider parent population from which the sample was drawn. Let us illustrate this process using two examples. Imagine you are interested in looking at the side-effects of the MMR vaccine. You might take a *sample* of 100 1-year-old children who had received the injection and calculate what percentage of them suffered any side-effects. Supposing you found that 37 per cent suffered mild problems as a result. From this you could infer, using the procedures of inferential statistics, that 37 per cent of *all* 1 year olds receiving the MMR injection would experience mild side-effects, i.e. the results from your sample are assumed to apply to the whole population of 1 year olds who had received the injection.

The second example relates to the administration of insulin to diabetic patients. Suppose an insulin manufacturer has developed a new continuous-infusion insulin pump. You decide to compare the stability of blood sugar levels for twenty-five patients using this pump with those for twenty-five patients using the standard hypodermic syringe to see if there are any differences between the two methods. You analyse the results using techniques of inferential statistics and find that the new pump produces significantly greater stability. You recommend that all insulin-dependent diabetics consider changing over to the new pump. You make this re-commendation because you have predicted that blood sugar levels will be more stable with this pump for all insulin-dependent diabetics. In other words, you have inferred that the results derived from your sample of insulin-dependent diabetics also apply to the whole population of insulin-dependent diabetics.

Two important terms have emerged here – population and sample. Population can be defined as a collection of people or objects who have *at least one* characteristic in common, e.g. hysterectomy patients, third-year student nurses, schizophrenics – in other words, the group of interest. The sample is a small section of that population.

Now, there are two vital points here. While it *is* possible to study the entire parent population in your research project, it is usually far too costly and time-consuming to do so. (Consider the last example, where the population of insulin-dependent diabetics must run into hundreds of thousands.) In consequence, a small sample of that populaton is selected instead, and the results from the sample are assumed to apply to the parent population from which it was drawn. This assumption is only made possible by the correct use of research procedures and of inferential statistics when designing your research project and analysing the results.

While these procedures will be described in detail throughout the rest of the book, one issue will be emphasised here. The accuracy of the inferences made depend on the nature of the sample selected. Let me explain what I mean.

Imagine you want to establish the average cholesterol levels for the adult popu-lation. You select a sample of 20–30-year-old women, who have low stress jobs, are vegetarian, eat a high fibre, low fat diet and do aerobics three times a week. You might find (not surprisingly) that their cholesterol levels are at the low end of the scale. You would be most unwise to infer that the rest of the adult population also had low cholesterol levels, because the sample you selected was not **repre-sentative** of that population. So, the sample selected to take part in your research must represent the parent population from which it was drawn. If it does not, then erroneous inferences and conclusions may be drawn. So accurate inferences depend very heavily on the quality of the sampling procedure: the more representative the sample, the more accurate the inferences. More details on representative sampling in theory and practice are given in Chapter 4.

Now, referring back to the examples concerning the MMR vaccine and the con-tinuous infusion insulin pump, you might notice that while both are concerned with making inferences about a population from just a small sample of that population, they differ in what they actually infer and how they infer it. These illustrations represent the two separate branches of inferential statistics. In the example of the MMR vaccine, a single statistic was calculated for the percentage of the sample of

100 1-year olds who had suffered side-effects from the injection. From this figure a 'guess' or estimate was made about the percentage of the whole population of 1-year olds who would also experience side-effects. This is a branch of inferential statistics known as **estimation**.

The second example, in contrast, tested the idea that the continuous infusion insulin pump and the traditional hypodermic syringe differed in their ability to stabilise the blood sugar levels of insulin-dependent diabetics. The two methods of administering insulin were compared. From the results, predictions or inferences were made about which technique would be better for stabilising blood sugar levels in *other* insulin-dependent diabetics. This branch of inferential statistics is known as **hypothesis testing**. Both aspects will be described in more detail below.

## 2.2 Estimation

Estimation is a very useful technique in nursing research. It allows the researcher to infer the value of a given **characteristic** (or **parameter**) of a population from the data derived from a sample of that population. Imagine you are interested in finding out the percentage of trained nurses who leave the NHS within five years of qualifying. You might select a sample of 300 trained nurses and from their employment records, calculate that 61% leave the NHS within that period. Providing you have selected your sample properly (see Chapter 4) you can infer that 61% of *all* trained nurses leave the NHS within this time period. Therefore, you have estimated the population parameter (percentage of all trained nurses leaving the NHS within a given period) from the data derived from a sample of 300 nurses. This would clearly be a useful piece of information to have when planning training, recruitment drives, etc.

However, it is probably obvious to you that a *single* figure estimate (which is known as **point estimation**) may not be a good basis on which to plan detailed financial or manpower resources, because it stands a good chance of being wrong. After all, an estimate is only as good as the quality of sampling procedure which was involved, and however sound this was, it is still highly likely that the estimate will be incorrect – if only marginally.

When a researcher needs to feel more confident about an estimate, a second form of estimation is typically used: **interval estimation**. Here, two figures, called **confidence limits**, are calculated for the sample's characteristic, using a special formula. From this, the researcher can estimate, with a fair degree of confidence, that the population characteristic will fall within these limits. So, if in the previous example, the researcher was not happy about making a single figure guess at the proportion of all nurses leaving the NHS, it might be possible to calculate from the sample an upper and lower figure (or confidence limit) for this parameter. Say these figures were 54 and 69%. The researcher could then say, with a high degree of confidence, that between 54 and 69% of all nurses will leave the NHS within five years of qualifying. The mechanism for calculating these confidence limits is outlined in Chapter 13.

Thus, there are two types of estimation. Point estimation calculates a single number or percentage for the sample's characteristic. This single figure is then inferred to apply to the population from which the sample was selected. The second type of estimation is called interval estimation. This involves the calculation of two figures for the sample's characteristic. The researcher can they say, with some degree of confidence, that the parent population's characteristic also falls within these two figures. More information on the methods of making both types of estimate is given in Chapter 13.

## 2.3 Hypothesis testing

The second branch of inferential statistics is called hypothesis testing. This involves stating a hypothesis or belief that something is the case, and then testing this belief on a sample of people, using scientific methods, to see if it is correct. The data from the research are then analysed using statistical tests and the conclusions derived from the sample are then inferred to apply to the population from which the sample was drawn. Let us take an example. Suppose you have spent a long time as a casualty nurse and have observed that two sorts of dressing – Fibrinex and Melanocyl* – seem to be differentially effective in the treatment of surface wounds. You decide to test out this belief by selecting a group of patients with surface wounds, and allocating half of them to treatment with Fibrinex dressing and the remainder to Melanocyl dressing. You assess how many days it takes the patients in each group to achieve total healing. The results are then compared using a statistical test. The statistical test tells you that Fibrinex patients recover more quickly; in consequence, you decide to use this preparation in future. In other words, you have selected a sample of patients on whom you tested your hypothesis. From their results, you inferred that *other* surface wound patients would respond similarly, and this led you to make recommendations for future treatment.

This approach to research, then, involves formulating a hypothesis, selecting a sample of people on which to test the hypothesis (using correct scientific principles) and from the data analysis inferring that the results would also apply to the population from which the sample came.

It will undoubtedly be evident to you that testing hypotheses in this way is central to nursing research, in that it allows the nurse researcher to compare the effectiveness of different treatment procedures or the responses of patients to particular types of nursing care, or the value of various nurse training programmes, etc. It is this sort of empirical research that forms the knowledge-base of sound nursing practice and hence is an essential technique in paramedical research.

---

* These are fictitious names.

There are two basic types of scientific method which can be used to test hypotheses: **experimental designs** and **correlational designs**. Both have some different assumptions and approaches and will be described in detail in the next chapter. Both, however, allow the researcher to make inferences from a sample about the population from which the sample was drawn. It should be emphasised, however, that despite the title of the book, correlational designs are not a branch of experimental research. However, they do employ a number of principles of experimental research and have been included in this text because of their usefulness in paramedical research.

---

**Summary of key points**

1  There are a number of ways in which research can be carried out and the findings presented. One approach is called inferential statistics.
2  Inferential statistics involve collecting information from a sample of people and then inferring that the information derived from the sample also applies to the population from which that sample was drawn.
3  In order to make these inferences, the sample must be representative of the parent population.
4  There are two branches of inferential statistics: estimation and hypothesis testing.
5  Estimation involves calculating a value for a characteristic of the sample and then inferring that this value also applies to the population's characteristic. There are two types of estimate: point estimates and interval estimates.
6  Hypothesis testing involves testing an idea to see if it is correct, using scientific principles. The hypothesis is tested on a sample of people and the data from the study are analysed using statistical tests. The results from the tests are inferred to apply to the parent population from which the sample comes.
7  There are two basic approaches in hypothesis testing: experimental designs and correlational designs.

---

## 2.4  Key words and terms

Confidence limits
Correlational design
Descriptive statistics
Estimation
Experimental design
Hypothesis testing

Inferences
Inferential statistics
Interval estimation
Parameter/characteristic
Point estimation
Population
Representative
Sample
Statistical tests

# 3 Some basic concepts involved in hypothesis testing

This book's focus is **inferential statistics**. This approach, as was mentioned in the last chapter, has two branches: hypothesis testing and estimation. This chapter, and Chapters 4–7 will deal with some of the concepts fundamental to hypothesis testing, while Chapter 13 deals with estimation.

Hypothesis testing is central to nursing research, in that it is usually concerned with experiments. I do not mean by this that you sit in a laboratory with rows of chemicals and jars, but, instead, that you have a prediction about whether or not something is the case (for example, that treatment A is more effective than treatment B), and you want to test this idea to see whether you are right. In order to test an idea in this way you need to do the following:

1. Set out your idea clearly, in the form of a *hypothesis*.
2. Design a study or *experiment* to test the hypothesis.
3. Analyse the data you get from your experiment, using a *statistical test* to see whether it supports your hypothesis.

As was explained in the last chapter, inferential statistics allow you to infer that the outcome or results of your study do not just apply to that study, but would also apply to a broader situation. An example might be useful to refresh your memory.

Supposing you had hypothesised that one hour of pre-operation counselling aided the subsequent psychological adjustment of mastectomy patients; you tested your hypothesis and the analysis of your results had supported your idea. How would you go on to use these results? You would probably want to implement one hour of counselling for *all* mastectomy patients coming into hospital. However, there would be no point in doing this if the only people who benefited from the procedure were the small sample of women in your original study. It is essential that you are able to assume that other mastectomy patients would also profit. In other words, the results from your study should not just apply to the patients who participated in it, but to other mastectomy patients as well.

Two important terms introduced in the last chapter should be stressed here. The people you use in your study are called a **sample** and the group to which they belong is called the **population**. So, in this example, you would select a sample of, say, twenty mastectomy patients from all the available population of mastectomy patients.

This is the role of inferential statistics: using an appropriate research design and statistical test to infer that the results you obtain do not just relate to the particular sample of people used in your study, but to the wider population from which the sample was drawn. And if your study is to have any wider implications for nursing practice and policy, this inference is essential.

However, if you are to use research techniques which involve inferential statistics, you must design your study carefully, because any oversights or errors may lead you to make incorrect inferences from your results. This chapter and Chapters 4–7 are concerned with the principles of sound research design so that this particular pitfall can be avoided.

Let us look now at some of the first steps you need to take when setting up a piece of experimental research – which, if you refer to Section 2.3, begins with stating a hypothesis.

## 3.1 The hypothesis

### 3.1.1 The experimental hypothesis

It was mentioned just now that inferential statistics are used to analyse data which had been derived from an experiment or study which tested out an idea.

This idea is the first stage in any research of this type and is called the **experimental hypothesis**. You will see it referred to as $H_1$. Typically, the experimental hypothesis emerges from some observations you have made during the course of your work or perhaps during discussions with colleagues or after reading a piece of research which makes you wonder what results would emerge if the procedure was modified slightly. The following are all examples of experimental hypotheses:

1. Episiotomies are more likely to be performed following epidural anaesthesia than following inhalation anaesthesia.
2. Superficial burns heal more quickly if they are immediately immersed in ice-cold water rather than being covered up.
3. Diabetics are more likely to comply with their medical regimen if they are educated about the condition as in-patients rather than as out-patients.
4. Agoraphobic patients make quicker recoveries if they set their own goals in therapy, rather than having the goals imposed by the nurses.

Essentially what the experimental hypothesis does is to predict a relationship between two things which are called **variables**. (In more complex research, the experimental hypothesis can predict a relationship between more than two variables, but that is beyond the scope of this text. We shall only be looking at hypotheses which predict a relationship between two variables. Readers interested in more advanced research are referred to Jaeger (1983), Greene and D'Oliveira (1982) and Ferguson (1976).)

This relationship between the two variables is assumed to be reliable and consistent and not just a result of chance factors.

If we look back at the examples just given, the first hypothesis predicts a relationship between type of anaesthesia given (epidural or inhalation) and likelihood of an episiotomy. The two variables here, then, are type of anaesthesia and likelihood of an episiotomy. The second hypothesis predicts a relationship between type of immediate treatment for superficial burns (ice-cold water or being covered up) and speed of recovery. The two variables here are type of immediate treatment and speed of recovery. The third hypothesis predicts a relationship between how the diabetic was educated (as an in- or out-patient) and the degree of compliance with the medical regimen. The two variables here are how the diabetic was educated and compliance with the medical regimen. The fourth hypothesis predicts a relationship between who set the therapy goals (patients or nurses) and the speed of recovery. The two variables here are the goal-setter and speed of recovery.

You may have a number of ideas you want to test out. It would be a good idea to jot them down, making sure you have predicted a clear relationship between two variables. By each hypothesis, state what the variables are. If you have any difficulties with this, just ask yourself the following question: 'What is the relationship predicted here – between what and what?'

The next clear stage to your research is to design a project which tests whether the relationship predicted in your experimental hypothesis does, in fact, exist. If you find it does, your hypothesis is said to be *supported*. (Please note that the hypothesis is not proven – we cannot prove hypotheses in psychology, nursing or other branches of paramedicine. We can only ever *support* a hypothesis. This point is particularly important if you wish to publish your research and will be referred to later.)

### 3.1.2 The null hypothesis

However, before we can proceed to the next stage of designing a project, one vital point must be raised. In order to make it worth your while setting up an experiment which tests your hypothesis, it must first be logically possible for your hypothesis to be wrong, i.e. that the predicted relationship does not exist. If there is no chance that your hypothesis can possibly ever be incorrect, then you are wasting your time testing it.

Let us take an example. Suppose you have come up with the experimental hypothesis: 'Women are more likely to have hysterectomy operations than men'. (What is the relationship which is being predicted here? Between what and what?) You then take yourself off to the hospital record books and after several days of work, you find that, lo and behold, your hypothesis has been supported! I am sure you can see that you would have been wasting your time and energy on this project, because there was no chance you could be wrong. So, to justify your research, it must be logically possible that the relationship predicted in your hypothesis does not exist.

Now rather than just assuming this, it is necessary to state this possibility clearly each time you construct an experimental hypothesis. In order to do this, an alternative hypothesis must be formulated. This is called the **null hypothesis** which is referred to as $H_0$. For every experimental hypothesis, you should state the corresponding null hypothesis.

Let us see how this works in practice. The experimental hypothesis predicts the existence of a relationship between two variables, and this relationship is assumed to be a consistent and reliable one. The null hypothesis predicts that *no* relationship exists between the two variables, or if, on some occasions, a relationship looks as though it exists, it is only due to chance fluctuations and not to a reliable and real association between the two variables. In doing this, the null hypothesis states the possibility that the experimental hypothesis is wrong. If we go back to the four experimental hypotheses given as examples, we can state the null (no relationship) hypothesis for each:

1. The first experimental hypothesis predicts a relationship between type of anaesthesia and the likelihood of an episiotomy. The corresponding null hypothesis would be: 'There is no relationship between the type of anaesthesia given and the likelihood of an episiotomy'.
2. The second experimental hypothesis predicts a relationship between the type of immediate treatment for superficial burns and speed of recovery. The corresponding null hypothesis would be: 'There is no relationship between type of immediate treatment for superficial burns and speed of recovery'.
3. The third experimental hypothesis predicts a relationship between how a diabetic was educated and degree of compliance with the medical regimen. The corresponding null hypothesis would be: 'There is no relationship between how a diabetic is educated and the degree of compliance with the medical regimen'.
4. The fourth experimental hypothesis predicts a relationship between who sets the therapy goals and speed of recovery of agoraphobic patients. The corresponding null hypothesis would be: 'There is no relationship between who sets the goals of therapy and the speed of recovery of agoraphobic patients'.

The easiest way to state the null hypothesis is a two-stage process:

1. Identify the relationship which has been predicted in your experimental hypothesis by saying: 'There is a relationship between variable A and variable B'.
2. State the null hypothesis by stating: 'There is no relationship between variable A and variable B'.

One point must be emphasised here. Some students get confused over the null hypothesis because they (wrongly) think it predicts the *opposite* relationship to that predicted in the experimental hypothesis. So, if we look back to the first experimental hypothesis given as an example: 'Episiotomies are more likely to be performed following epidural anaesthesia than following inhalation anaesthesia' these students would state the null hypothesis as: 'Episiotomies are more likely to be performed following *inhalation* anaesthesia than following epidural anaesthesia'.

In other words, they think the null hypothesis predicts the *opposite* relationship. But if we look at their null hypothesis, it is still predicting a relationship between episiotomies and type of anaesthesia. This is incorrect, because the null hypothesis predicts that no relationship exists of any sort.

Bearing this in mind, have a look at the experimental hypothesis you constructed earlier, and for each one, state the corresponding null hypothesis.

Why is it necessary to go to all the trouble of stating the null hypothesis, when we could just infer it from the experimental one? To answer this fully, we would need to trace the origins of the philosophy of scientific method, which are really too extensive to cover in a short text. Suffice it to say that when we carry out any research, we should not set out to find support for our experimental hypothesis directly, but that, instead, we should support it indirectly, by finding evidence which suggests that the null hypothesis is incorrect. The end result is the same of course, but convention has it that we should claim support for the experimental hypothesis only by the rejection or falsification of the null hypothesis. Do not worry about this. These points will be made clearer when examples of how to write up your research are given later on.

---

**Summary of key points**
1 *Inferential statistics* are used when the research involves hypothesis testing.
2 The starting point of this type of approach is the experimental hypothesis which predicts a relationship between two variables.
3 However, to justify your study, you must acknowledge the possibility that your experimental hypothesis is incorrect by stating the corresponding null hypothesis.
4 The null hypothesis states that the relationship predicted in the experimental hypothesis does not exist.

---

## 3.2 Designing a project to test your hypothesis

The next stage in your research is to design an experiment or project which tests your hypothesis in the most appropriate way. The phrase 'in the most appropriate way' is used because, while sometimes there is an obvious decision about the best design for your hypothesis, on other occasions, two or three designs might be feasible, and you will have to make a reasoned choice. The remainder of this chapter will outline the basic points of two designs – **experimental designs** and **correlational designs** – because the first stage in your decision-making process is which of these two designs would be a more appropriate test of your hypothesis.

### 3.2.1 Experimental designs

Experimental designs start off with the experimental hypothesis which predicts a relationship between two variables. To test whether this relationship actually exists using this type of design, one of these variables is altered or **manipulated** to see what effect it has on the other variable. While this sounds very complicated, it is in fact something you do unwittingly every day.

Imagine you are vacuum cleaning your house and after a while the vacuum cleaner stops picking up the bits. You would probably open it up, and see whether the dust-bag was full. If it was, then you would empty it and start again. What you have done here is to formulate an experimental hypothesis: 'There is a relationship between the state of the dust-bag and the effectiveness of the vacuum cleaner' and you have altered or manipulated the state of the dust-bag (variable 1) to see what effect it has on the vacuum cleaner's efficacy (variable 2). If your vacuum cleaner now works properly, you conclude that the full dust-bag was the *cause* of the appliance's failure to collect dust; and altering this has the *effect* of improving its performance. In other words, you have carried out a simple experiment which determines *cause and effect* of an everyday event. Should your vacuum cleaner still not work, you would perhaps check the plug or the rubber belt – in other words you go on formulating hypotheses until you find the *cause* of the problem.

There are many other everyday examples of hypothesis-testing: you may switch on a lamp and find that no light comes on. You might change the bulb to see what effect that has. You have formulated the hypothesis that there is a relationship between the state of the bulb and the presence of light, and have manipulated the state of the bulb to see what effect it has on the production of light. Perhaps you can think of other examples.

Essentially, though, the same principles apply to more formal experimental designs in research. Once you have formulated your hypothesis, you **manipulate one of the variables** to see whether it has an effect on the other. If it does (and you have designed your experiment properly!) you can conclude that altering that variable is the cause of the observed effect on the other variable.

Experimental designs, therefore, if properly carried out, allow you to infer the cause and effect of events – and this is an extremely important distinction from correlational designs, which do not allow such an assumption, as we shall see later.

**Independent and dependent variables**
The two variables in the hypothesis are called by different names. The variable which is manipulated is called the **independent variable** (or IV), while the one which is monitored for any effects is called the **dependent variable** (or DV)*, because its state is *dependent* upon what is done to the other variable. The independent variable then is the cause and the dependent variable, the effect. The effects on the DV form your *data* in an experiment. If we look back to the two

---

* Do ensure that you do not confuse the abbreviations DV and IV as defined here with those commonly used in nursing.

domestic examples, we can see that the IV in the first case is the state of the dust-bag and the DV is the effectiveness of the vacuum cleaner. In the second case, the IV is the state of the light bulb and the DV the production of light.

Some students get worried about deciding which variable should be manipulated, but if we look at this a bit more closely, there is only one logical choice. You cannot, for example, alter the production of light to see what effect it has on the bulb because the state of the bulb does not depend on whether or not there is light. The reverse only is true – that the production of light *depends* on the state of the bulb – so the production of light must be the *dependent* variable. There is a logical and temporal sequence to these events – altering the bulb comes first, followed by the production of light. The IV is always the first variable in the logical sequence and the DV is always the second. Many students find this problem easier to deal with if they ask themselves which variable depends on which. Once they have decided this, they can easily attach the correct names to the variables.

Let us look back at the four hypotheses, and identify which variable is which in each case:

1. *Hypothesis 1*: the two variables here are likelihood of episiotomy and type of anaesthesia given during labour. The DV is the likelihood of episiotomy because it *depends* on the type of anaesthesia given (which is the IV). Whether or not an epidural is performed would be the *data* in this experiment.
2. *Hypothesis 2*: the two variables here are type of immediate treatment for burns and speed of recovery. The DV is speed of recovery because it *depends* on the type of treatment (the IV). Speed of recovery would form the data in this study.
3. *Hypothesis 3*: the two variables here are how a diabetic was educated and degree of compliance with the medical regimen. Degree of compliance is the DV because it *depends* on how the diabetic was educated (the IV), and degree of compliance would be the data in this study.
4. *Hypothesis 4*: the two variables here are who the goal setter is and speed of recovery. Speed of recovery is the DV because it *depends* on who sets the goals (which is the IV). Speed of recovery would be the data in this study.

If you are still not clear, just reverse the variables to see if their sequence is logical. For instance, in the last example, who sets the goals of therapy does not depend on the speed of recovery and therefore the sequence of events is not logical.

Using the term 'manipulate' implies that you, as the researcher, have total control over the IV. Obviously, this is not necessarily the case. It would be unlikely, for example, that the researcher would be in a position to control which patient has inhalation anaesthesia, because the decision is usually taken either by the patient or by the patient in conjunction with the midwife, obstetrician and other relevant personnel. What would be meant by manipulation here is simply the selection of one group of patients who had already chosen an epidural and another group who had opted for inhalation anaesthesia. The incidence of episiotomies for each group would then be compared. There will be many occasions when you do not have complete control over how to alter the IV and in such cases you would simply select a sample of people from already existing groups.

Look back now at the hypotheses you constructed and identify the IV and DV in each case.

**Some considerations when constructing an experimental design**
Once you have stated your hypothesis and (having read the rest of this chapter) decided to test it with an experimental rather than correlational design, there are a number of issues you should take into account, and these are best illustrated by an example.

Imagine that you are working on a busy surgical ward and a common problem that you have noticed is raised blood pressure amongst the pre-operative patients. You wonder whether this is the direct result of fear of the unknown, i.e. that lack of information about the operation and not knowing what to expect is causing the high blood pressure. You decide that you would like to see whether talking to post-operative patients with the same condition alleviates the problem in these patients. After checking with the relevant personnel about permission and ethics relating to this study, you go about testing the hypothesis: 'Information about an operation reduces raised blood pressure among pre-operative patients'.

The IV is the information about the surgical procedures and the DV is the level of blood pressure. The data here would be the blood pressure readings.

You select ten pre-operative patients from your ward and encourage them to talk to patients who have already had the same operation. You then measure the blood pressure of the pre-operative patients. Your design looks like Table 3.1.

After you have measured these patients' blood pressures, could you conclude that information reduced it? You have probably come up with an emphatic 'no', on the basis that we did not know what their blood pressure was *before* they received the information – and unless we know what their blood pressure was to start with, we cannot make any statement about whether or not information altered it. So, a vital ingredient of an experimental design is measuring the DV *before* administering the IV and then measuring it again afterwards. We call this a **pre-test measure**. (It should be pointed out that for all sorts of reasons, it may not always be possible to take a pre-test measure of the DV. Where this is the case it will be obvious to you – see page 22. However, if you can take a pre-test measure, you should.)

If we incorporate the pre-test measure, our design now looks like Table 3.2 (overleaf). Because you now have a 'before' and 'after' measure of the DV you can compare the two sets of results to see if they had changed at post-test. If blood pressure had reduced on the post-test, could you now conclude that your hypothesis was correct and that information does relieve raised blood pressure?

Well, it is certainly *one* possible explanation, but there may also be others. For

**Table 3.1** Initial experimental design

| Independent variable | Dependent variable |
| --- | --- |
| Information about operation | Blood pressure |

**Table 3.2** Revised experimental design with pre-test measure

| Pre-test measure of dependent variable | Independent variable | Post-test measure of dependent variable |
|---|---|---|
| Blood pressure | Information about operation | Blood pressure |

**Table 3.3** Revised experimental design with control condition

| | Pre-test measure of DV | IV | Post-test measure of DV |
|---|---|---|---|
| 1. Experimental condition | Blood pressure | Information | Blood pressure |
| 2. Control condition | Blood pressure | No information | Blood pressure |

example, just talking to someone (not necessarily about the operation) might take a pre-operative patient's mind off the operation and so have a calming effect; or as the patients get used to being in hospital, they may become less tense, so their blood pressure might have gone down anyway with the passage of time. You may have thought of other explanations. The point is, though, how can we ascertain whether the change in the DV (blood pressure) is due to the IV (information) as we predicted?

The answer to this lies in the selection of another group of patients who have similar experiences to the first group, but who do *not* receive the IV, which in this case is information. We call the group which receives the IV the **experimental condition** or group and the group which does *not* receive the IV, the **control condition** or group. If we incorporate this revision, our design looks like Table 3.3.

The IV has been manipulated here by giving one group of patients information and the other group no information. If the groups are identical in every other way, then any difference between them in terms of reduction in blood pressure from pre- to post-test measures must be attributable to whether or not they received the IV, i.e. information about the operation.

Obviously here you would compare the blood pressure changes from pre- to post-test for each group and see whether there are any *differences* between the groups. This is an essential feature of experimental designs – they look for differences between the conditions. It is a salient distinction from correlational designs, as we shall see later.

There are still many more modifications which need to be made to this design in order for it to be acceptable and watertight, and these will be covered in the next four chapters. The foregoing principles will, however, serve as an introduction to some basic issues of experimental design and so will allow you to compare them with correlational designs.

At this point, ethical issues start raising their heads. The sort of design just outlined using an experimental condition (which receives the IV) and a control condition (which does not) may be quite acceptable for the hypothesis we were testing, where there may be no particular moral dilemma as to whether or not someone receives information about an operation. Nevertheless, there will be occasions when it is simply ethically unacceptable *not* to administer some form of the IV to patients. Let us take a hypothetical situation (in which you are highly unlikely to find yourself because it is not within the domain of nursing research) by way of illustrating this moral issue.

Supposing you were working with renal transplant patients and heard about a new drug (drug X) which, if taken for one month following transplant, removed the need for permanent anti-rejection drugs. You would like to see whether this drug is effective in suppressing rejection and so you construct the hypothesis:

**H$_1$:** Treatment by drug X is effective in the post-operative treatment of tissue rejection following renal transplants.

Here the IV is drug treatment and the DV is the effectiveness of treatment of tissue rejection. If you were to adopt the experimental design just outlined, you would use an experimental condition where, say, twelve patients received treatment by drug X following renal transplant and a control condition where a further twelve renal transplant patients received *no* drug treatment to protect against rejection. As it takes little imagination to predict what the outcome for the patients in the control group would be, you could not, in all conscience, carry out this particular study using this design.

You have probably already guessed what would be done in such a situation. Rather than comparing treatment by drug X with *no* treatment (the standard control procedure), you would compare it with the *traditional* treatment, i.e. the usual anti-rejection drugs. This means that instead of using *one* experimental condition and one control condition, you use *two* experimental conditions as in Table 3.4.

At the end of the study, you would compare the differences in progress of both groups to see which treatment was more effective. Because both groups are receiving some form or *level* of IV (treatment) they are both experimental conditions. (Remember that the control condition is defined as a *no* IV, i.e. no treatment, procedure.) Such an approach is commonly used in medical research in order to offset ethical problems.

An experimental design which uses two conditions, whether it be two experi-

**Table 3.4** Experimental design using two experimental conditions

|  | *Pre-test measure of DV* | *IV* | *Post-test measure of DV* |
|---|---|---|---|
| Experimental condition 1 |  | Drug X |  |
| Experimental condition 2 |  | Traditional anti-rejection treatment |  |

**Table 3.5**  A two-condition experimental design

|  | Pre-test measure of DV | IV | Post-test measure of DV |
|---|---|---|---|
| Experimental condition | (No pre-test measure of DV can be carried out here for obvious reasons!) | Epidural anaesthesia | Apgar score |
| Control condition | None | No anaesthesia | Apgar score |

**Table 3.6**  A three-condition experimental design

|  | Pre-test measure of DV | IV | Post-test measure of DV |
|---|---|---|---|
| Experimental condition 1 | None | Epidural anaesthesia | Apgar score |
| Experimental condition 2 | None | Inhalation anaesthesia (Trilene) | Apgar score |
| Control condition | None | No anaesthesia | Apgar score |

mental conditions, or one experimental condition and one control condition, is the simplest design of all, and may well satisfy many of your research needs. However, there will undoubtedly be occasions when a two-condition design does not meet your requirements.

Imagine you were working in obstetrics and are interested in the effects of anaesthesia during labour on the baby's Apgar score. Your hypothesis might be: 'Anaesthesia during labour affects the neonate's Apgar score'. Here the IV is anaesthesia and the DV is the baby's Apgar score. You might decide to test this hypothesis by comparing the Apgar scores of babies whose mothers received epidural anaesthesia with those whose mothers received *no* anaesthesia – a traditional one experimental and one control condition procedure; Table 3.5. You would compare the Apgar scores for each condition to see if there were any *differences* between them.

But there are other types of anaesthesia besides epidurals, for example inhalation anaesthesias (e.g. Entonox, Trilene), 'tens' (transcutaneous nerve stimulation), acupuncture, hypnosis, narcotics, sedatives and tranquillisers. Rather than carrying out separate experiments to compare the effects of each one with a control condition, you can be much more economical and combine them. Therefore, without needing to alter your hypothesis, you might add another experimental condition to your original design; Table 3.6. Again, you would compare the conditions for *differences* in the Apgar scores.

You now have three conditions: two experimental conditions (i.e. two **levels of**

**Table 3.7**  A five-condition experimental design

|  | Pre-test measure of DV | IV | Post-test measure of DV |
|---|---|---|---|
| Experimental condition 1 | None | Epidural anaesthesia | Apgar score |
| Experimental condition 2 | None | Inhalation anaesthesia (Trilene) | Apgar score |
| Experimental condition 3 | None | Narcotics | Apgar score |
| Experimental condition 4 | None | Tens | Apgar score |
| Control condition | None | No anaesthesia | Apgar score |

**the IV**) plus a control (no anaesthesia) condition. You are still testing the original hypothesis, but using a more sophisticated design. You could take this even further and include narcotics and tens in your design; Table 3.7.

You now have five conditions in total: four experimental conditions (or four levels of the IV) and one control condition. You could, of course, omit the control condition if you felt your study warranted it, or you could go on adding further experimental conditions. However, it should be stressed that in each of the above examples, we are still testing the same hypothesis, with type of anaesthesia as the IV and Apgar score as the DV. All that is changing with each design is the number of *types* of anaesthesia (or levels of IV) being compared. Do ensure, if you ever need to carry out this sort of design, that you do not change the actual IV in any of the experimental conditions, as this would complicate the issue considerably and require design considerations and forms of statistical analysis which are beyond the scope of this book.

Knowing whether your design has two conditions or three or more conditions is essential since each one necessitates a different sort of statistical analysis.

## Summary of key points
1 You can test a hypothesis using an experimental design.
2 This involves manipulating one variable to see what effect it has on the other.
3 The variable which is manipulated is called the independent variable (IV). The variable which is monitored for any changes or effects is called the dependent variable (DV).
4 The changes in the dependent variable constitute the data in the study.
5 As long as the study is carried out properly, the IV can be assumed to be the cause of the effects on the DV. Thus, experimental designs can determine cause and effect.
6 To carry out an experimental design, two or more conditions are set up,

and the data from each are compared at the end of the study for any differences between them. Thus, an experimental design is said to look for differences between conditions.

7 These conditions can be of two types: *experimental conditions* which involve the application of some form or level of the IV and *control conditions* which involve no application of the IV. As long as the IV does not alter, several levels of the same IV can be incorporated in the same experiment.

### 3.2.2 Correlational designs

Your other option of design when testing hypotheses is the *correlational design*. When adopting this approach, the researcher starts off in exactly the same way as with experimental designs – that is, by formulating an experimental hypothesis which predicts a relationship between two variables.

However, the assumptions underlying correlational designs are somewhat different from those underlying experimental designs. Correlational designs are concerned with the *interrelationships* between the two variables in the hypothesis, noting how much of a change in one variable is associated with an alteration in the other. In order to assess these interrelationships, the researcher has to collect data on both variables, but neither variable is manipulated or altered intentionally in any way by the researcher. So what typically happens instead, is that the researcher takes a whole *range* of scores on both of the variables to see if they are related to each other in some way. (Do note the word 'typically'. It is possible to carry out a correlational design without collecting a range of data, but these designs are not often used because they are less informative. As a result the correlational designs that are referred to in this text assume that a range of data has been collected for each variable.)

To illustrate the differences in the two designs, let us imagine you have formulated the hypothesis:

$H_1$: High alcohol consumption is related to frequent urinary tract infections.

If you were to test this hypothesis using an experimental design, you would compare a group of regular drinkers with a group of teetotallers on their incidence of urinary tract infections. Your design would look like Table 3.8. You would compare the incidence of infection for each group to see whether there was any *difference* between them.

If there was, and the experiment had been carried out properly, you could conclude that alcohol consumption is the cause of urinary tract infections.

However, you could test the same hypothesis using a correlational design. Here you would take a range of, say, twelve people, each of whose alcohol intake varied from teetotal to regular heavy drinking; Table 3.9.

**Table 3.8**

| Condition | Subjects | Assessed on |
|---|---|---|
| Experimental condition | Regular drinkers | Incidence of urinary tract infections |
| Control condition | Teetotallers | Incidence of urinary tract infections |

**Table 3.9** Range of scores on one variable using a correlational design

| Subject | Variable 1: average alcohol intake per week |
|---|---|
| 1 | Teetotal |
| 2 | One unit |
| 3 | Three units |
| 4 | Five units |
| . | . |
| . | . |
| . | . |
| 12 | Thirty-five units |

**Table 3.10**

| Subject | Variable 1: average alcohol intake per week in units | Variable 2: incidence of urinary tract infections over the past five years |
|---|---|---|
| 1 | Teetotal | 0 |
| 2 | 1 | 1 |
| 3 | 3 | 2 |
| 4 | 5 | 5 |
| 5 | 6 | 7 |
| 6 | 9 | 8 |
| 7 | 10 | 10 |
| 8 | 12 | 12 |
| 9 | 19 | 13 |
| 10 | 24 | 14 |
| 11 | 30 | 16 |
| 12 | 35 | 18 |

You would then collect data about these people in terms of the incidence of their urinary tract infections over the previous, say, five years. If your hypothesis is correct you would expect that the heaviest drinkers experienced the highest frequency of infection, and the teetotallers the lowest frequency, with the others ranged in between accordingly. Your results might look like Table 3.10.

I have deliberately made these results very clear cut, with the higher the alcohol intake, the higher the incidence of urinary tract infection, in order to illustrate the point. It is highly unlikely that you would ever get such a clear pattern emerging – what you are more likely to get is a trend or tendency, such that people with a high alcohol intake *tend* to experience a high incidence of infection, while a person with a low alcohol intake *tends* to have a low incidence of infection.

A number of points emerge from this. Firstly, the researcher does not manipulate any variable, but simply selects data which cover a range of scores on one variable. Data on the other variable is then collected to see whether there is any link or *association* between the two sets of data.

The collection of data can take the form of a *retrospective* or *current* study. A retrospective study simply means that the data for each variable are collected from existing medical records, biographical information, health authority facts and figures, etc., without the researcher having to come into contact with the subjects. On the other hand, in a current study, the researcher collects the data on each variable from the subject directly, for instance, taking blood pressure, temperature, height, amount of analgesia taken daily, etc.

From this, you might have realised that a correlational study can combine both retrospective *and* current collection of data, such that data on one variable can be collected from medical records, while data on the other can be collected directly.

However the data are collected, though, the purpose of a correlational design is to find out whether there is any association or pattern between the two sets of data. This is a fundamental distinction between experimental and correlational designs. Experimental designs look for *differences* in the data deriving from the conditions; correlational designs look for a link or association between the data collected for each variable.

Because no variable is manipulated, the concepts of independent and dependent variable are inappropriate in a correlational design. No cause or effect can therefore be ascertained, unlike experimental designs which allow the researcher to infer cause and effect. If we look at this in relation to the example just quoted, while it is tempting to make an educated guess that alcohol causes urinary tract infection, the correlational design does *not* allow us to conclude this definitively. All that we know from a correlational design is that the data on one variable are associated with the data on the other. So in this case while it is possible that alcohol causes infection, it is also conceivable that urinary tract infection leads to excess alcohol consumption (perhaps through the individual's wish to drown their discomfort in an alcoholic haze – particularly if they have a long history of urinary tract infections). It is also possible that both alcohol intake and urinary tract infection are caused by a third, unidentified variable, such as a particular biochemical in some individuals which predisposes them both to drink *and* develop infection. While these two suggestions are unlikely, they cannot be ruled out by a correlational design. So, in the present example, the following are possible:

1. Infection is the result of alcohol intake.
2. Alcohol intake is the result of infection.
3. Alcohol intake and infection are both the result of some third variable.

This is an important issue and it might be worth clarifying with one further example. Imagine your hypothesis is:

**H₁:** High cholesterol diets are related to high blood pressure.

Your prediction here might be that with increasing cholesterol in the diet, blood pressure gets higher. If you carried out a correlational test of this hypothesis, you might take a group of people on diets containing varying amounts of cholesterol, such that there was a range of cholesterol data from low to high, and then measure their blood pressure.

If your prediction was correct, the people with diets heaviest in cholesterol would have the highest blood pressure, and those with diets containing least cholesterol the lowest, with the others ranged in between accordingly. However, you could not conclude that cholesterol in the diet caused high blood pressure (even though you may be convinced of its culpability) because the following are also possible:

1. High blood pressure causes some need to take in cholesterol.
2. High blood pressure and high cholesterol diets are the result of a third factor (obesity, job stress, etc.).

You may find that being unable to make a categorical statement about cause and effect is highly unsatisfactory and puts you off using a correlational design. Indeed, many researchers avoid this design for exactly that reason. However, it should be pointed out that what the correlational design lacks in certainty it compensates for in ethics, since because the researcher is not deliberately altering variables to see what effect this has, there is less interference with patients and procedures. This means that a correlational design may be much more acceptable in many situations involving medical research.

Table 3.11 recaps briefly on the difference between experimental and correlational designs.

**Positive and negative correlations**
Correlational designs are a bit more complex than this, however, because scores on the two variables may be linked in one of two ways.

Firstly *high* scores on variable 1 may be linked with *high* scores on variable 2 (and, inevitably, *low* scores on variable 1 with *low* scores on variable 2). In other words, the data may vary in the same direction for each variable.

For example, if we took the hypothesis: 'High weight gain during pregnancy is related to high blood pressure' we would be predicting that the link between the data on each variable would be one where people with *high* weight gain scores would also have *high* scores on the blood pressure variable, and similarly, people with *low* scores on weight gain would also have *low* scores on blood pressure. Another way of looking at this is that the *higher* the score on variable 1, the *higher* the score on variable 2, and of course, the *lower* the score on variable 1, the *lower* the score on variable 2. This is called a **positive correlation**.

Other examples of hypotheses which predict a positive correlation are as follows:

**Table 3.11** A comparison of experimental and correlational designs

| Experimental designs | Correlational designs |
| --- | --- |
| 1. Start with an experimental hypothesis which predicts a relationship between two variables. | 1. Start with an experimental hypothesis which predicts a relationship between two variables. |
| 2. Manipulate one variable (the IV) to see what effect it has on the other (the DV). | 2. Do not manipulate either variable, but simply select a whole range of data on one variable and then collect data on the other to ascertain whether there is a *link* between the data. |
| 3. Experimental designs look for *differences* between the data deriving from the conditions in the experiment. | 3. Correlational designs do not look for differences, but for *associations or patterns* between the sets of data deriving from each variable. |
| 4. Experimental designs can ascertain causes of events. | 4. Correlational designs cannot ascertain causes of events. |
| 5. Experimental designs, in consequence, provide conclusive results. | 5. Correlational designs cannot provide conclusive results. |
| 6. Experimental designs may be ethically dubious because of the deliberate manipulation of variables. | 6. Correlational designs are much more acceptable ethically because they involve no deliberate manipulation. |

1. *High* anxiety among patients is associated with *high* levels of reported pain.
2. *Low* haemoglobin counts are associated with *low* energy levels.
3. The *longer* it takes for patients to travel to hospital, the *higher* the incidence of missed out-patients appointments.

In each case, it is predicted that the higher the scores on one variable, the higher the corresponding score on the other.

Alternatively, the data from the two variables may be linked in an inverse way, such that *high* scores on variable 1 are associated with *low* scores on variable 2, and conversely, *low* scores on variable 1 are associated with *high* scores on variable 2. In this way, the data on each variable varies in the *opposite* direction.

For example, if we took the hypothesis: 'High absenteeism among student nurses is related to low job satisfaction' we would expect that student nurses with *high* scores on the absenteeism variable would also have *low* scores on the job satisfaction variable and similarly that student nurses with *low* scores on the absenteeism variable would also have *high* scores on the job satisfaction variable. This can also be stated as the *higher* the scores on variable 1, the *lower* the scores on variable 2, and likewise, the *lower* the scores on variable 1, the higher the scores on variable 2. This is called a **negative correlation**.

Examples of hypotheses which predict a negative correlation are as follows:

1. As age *increases*, so memory for recent events *decreases*.
2. The *higher* the patient's social class, the *shorter* the recovery time from open heart surgery.
3. As age *increases*, so fertility *decreases*.

All the above hypotheses predict that the *higher* the score on one variable, the *lower* the corresponding score on the other.

### The correlation coefficient

Why is this distinction important? When you carry out a correlational design you will analyse the data deriving from both variables using an appropriate statistical test to see whether the two sets of data are asociated in some way (see Chapter 12).

The results of this test will give you a number somewhere between −1.0 and +1.0:

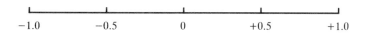

| | | | | |
|---|---|---|---|---|
| −1.0 | −0.5 | 0 | +0.5 | +1.0 |

This figure may be, for example, −0.72 or +0.49 and is called a **correlation coefficient**. The correlation coefficient is an index of the degree of association between the two variables, and the closer it is to +1.0 or −1.0, the stronger the association. Therefore, a correlation coefficient close to −1.0 suggests a *negative* correlation (because of the minus sign) and the closer it is to −1.0 the stronger the negative correlation. In this way −0.92 indicates a stronger negative association than −0.79. Similarly, a correlation coefficient close to +1.0 suggests a positive correlation (because of the plus sign) and the closer the figure is to +1.0, the stronger the positive association between the two sets of data. Thus, a figure of +0.65 represents a closer positive association than a figure of +0.54.

If, however, the correlation coefficient is around 0, this suggests a weak link between the two sets of data and is indicative of no correlation between the data. The closer the figure is to 0, the weaker the link between the two sets of data. Therefore, a correlation coefficient of +0.119 indicates a weaker relationship than one of −0.231. Do not confuse negative and no correlations: negative correlation suggests that high scores on one variable are related to low scores on the other, while no correlation means there is no relationship between the two sets of data.

Therefore, the correlation coefficient which emerges from your statistical analysis tells you the following:

1. Whether you have a positive or negative correlation (by the plus or minus in front of the figure).
2. How strong this correlation is (by its proximity to +1.0 or −1.0).
3. Whether any direction to the two sets of data as predicted by your hypothesis has been confirmed. (For instance, if you predicted a *positive* correlation in your hypothesis but ended up with a *negative* correlation coefficient, the predictions would not be supported.)

One other point must be stressed here. How close must your correlation coefficient be to $-1.0$ before you can claim that there is a negative association between the two sets of data? Similarly, how close must the correlation coefficient be to $+1.0$ before you can claim that there is a positive association between the two sets of data? The decision about this is not arbitrary and you cannot draw your own cut-off points. You need to use a set of statistical probability tables to tell you whether your correlation coefficient represents a significant association between the two sets of data. Do not worry about this, because it will be discussed in much greater detail in Chapters 5 and 12.

If, however, you do find that you have a significant relationship between the data, it allows you subsequently to make some predictions about one variable if you have scores for the other. Let me explain what is meant by this. Supposing you found that people who smoked a high number of cigarettes had a high incidence of respiratory disease, i.e. a positive correlation. You then encounter a patient who smokes sixty a day and another who smokes ten a day. You could predict on the basis of your study that the first patient would experience a high incidence of respiratory disease, while the second would have a much lower incidence. Therefore, having established that a significant positive or negative correlation exists between two variables, you are then in a position to make predictions about the scores on one variable, from knowledge of the scores on the other. The way in which this is done is covered in Chapter 12.

---

### Summary of key points

1 Correlational designs start off with an experimental hypothesis which predicts a relationship between two variables.

2 To see whether the relationship exists using this design, a whole range of scores on one variable is usually selected and the corresponding scores on the other variable collected, to see if there is some link between the two sets of data.

3 As no variable is manipulated, there is no IV or DV and no causal association between the variables can be ascertained.

4 The type of link between the data can be one of two sorts:
   (a) A positive correlation whereby high scores on one variable are associated with high scores on the other (and accordingly, low scores on one variable associated with low scores on the other). This is indicated by a correlation coefficient of around $+1.0$.
   (b) A negative correlation, whereby high scores on one variable are associated with low scores on the other. This is indicated by a correlation coefficient of around $-1.0$.

5 The closer the correlation coefficient is to $+1$ *or* $-1$, the stronger the link between the variables.

6 Sometimes you will find there is *no* correlation between your two sets of data and this is indicated by a correlation coefficient of around 0.

7 Once you have established that a significant correlation exists between two variables, you are in a position to make predictions about the score on one variable if you know *only* the scores on the other.

---

*Exercises (answers on page 311)*

1 Look at the following hypotheses and for each one state: (a) what the variables are; (b) what the null hypothesis is; (c) which variable is the IV and which the DV (assuming you were going to test each hypothesis using an experimental design).
   (i) Satisfaction with nursing care is greater when patients are nursed in side-wards than when they are nursed in main wards.
   (ii) Asian patients and West Indian patients differ in their responsiveness to pain.
   (iii) Rate of healing of varicose ulcers is greater on partial bed-rest than on total bed-rest.
   (iv) Student nurses trained on traditional RGN courses are more likely to stay in ward-based jobs on completion of training than are student nurses trained on degree courses.
2 Look at the following hypothesis and outline both an experimental *and* a correlational design to test it: 'Women who have taken oral contraceptives have difficulty in conceiving'.
3 Look at the following hypotheses and for each state whether the predicted relationship suggests: (a) a negative or positive correlation; (b) where you would expect the corresponding correlation coefficient to be on a scale from $-1.0$ to $+1.0$ (i.e. towards the $+1.0$ end or towards the $-1.0$ end).
   (i) As temperature decreases so the number of elderly people suffering from hypothermia increases.
   (ii) The greater the AFP levels in maternal blood samples, the greater the likely severity of spina bifida.
   (iii) The longer patients have to wait in out-patient departments, the less the reported satisfaction with care.
   (iv) The greater the amount of jogging carried out on hard surfaces, the greater the damage to the leg joints.
4 Look at the following correlation coefficients and state which one indicates the strongest relationship and which the weakest.

   $+0.521$    $-0.741$    $-0.803$    $+0.113$    $+0.605$    $-0.194$    $-0.301$

   (Note that the strongest correlation is that number closest to either $+1.0$ or $-1.0$, while the weakest correlation is that number closest to 0.)

## 3.3  Key words and terms

Control condition
Correlation coefficient
Correlational designs
Dependent variable
Experimental condition
Experimental designs
Experimental hypothesis
Independent variable
Inferential statistics
Levels of the independent variable
Manipulation of variables
Negative correlation
Null hypothesis
Population
Positive correlation
Post-test measures
Pre-test measures
Sample
Variables

# 4    Additional aspects of research design

## 4.1  Selection of subjects

Whenever you carry out any research in nursing you will collect facts and figures about something – for instance, blood pressure, urine output, responses to a drug, healing rate or whatever. These facts and figures are called **data** and can take a multitude of forms. Typically, the source of the data is other people: patients, colleagues, student nurses, etc., and they are called your **subjects** (often referred to as Ss). So, for instance, you might be interested in testing the hypothesis: 'Breast cancer is more common in women who have only one child, than in multi-parity women'. Here your data would be the incidence of breast cancer and you would collect these data from a number of subjects who would be women who have either one child or many children.

There are a number of vitally important issues which you must bear in mind when you select your subjects. These are best illustrated by examples. Supposing you carried out a (well-designed!) piece of research to test the hypothesis just given and you found that your hypothesis was supported, i.e. that there *is* a link between breast cancer and having only one child. To make your study worth while, you have to put the results to some use, rather than keeping them to yourself, since this is pointless and probably unethical. One way in which you might do this is to publish these results in a professional journal, where they may be read by other health care workers. The ultimate outcome of your study might be to direct the available screening devices to women with one child (in preference to those who have had more than one) since you have shown single-child women to be at greatest risk. Thus, your study would have far-reaching implications for health education programmes and the availability of mammography for this particular group of women as an aid to early detection of carcinoma of the breast. This may all sound a bit grand, but any good research which has interesting results should influence health care practice at some level, however minimal.

Underlying all this is a fundamental assumption in inferential statistics – that the results you obtained do not just apply to the subjects you used in your study, but to women in general who have only one child. If we can only assume that the results apply to your subject sample alone, then the research has not been very valuable.

The essential point here, then, is that the results from any piece of research should be **generalisable** from your small sample to the larger population from which the subjects were drawn. A second point which derives from this is the notion of **prediction**.

If we are in a position to assume that the results from your study are generalisable – that is, that women in general with only one child are more likely to develop carcinoma of the breast – then we could *predict* that *any* woman who has a single child is going to be more vulnerable to developing this form of cancer. These women might then be targeted by GPs, health visitors, district nurses, etc., for closer scrutiny for lumps and other signs, or for health education leaflets on self-examination, because it has been *predicted* on the basis of your results, that they will be more likely to get cancer of the breast.

So, the results from your study must be generalisable to other similar subjects, in order that predictions can be made.

In order for results to be generalisable and for consequent predictions to be made it is essential that your subjects are selected properly, according to certain principles. Let us go back to our hypothesis. Imagine you had selected the following subjects to take part in your study.

*Group 1*: Five women who had only one child and also had the following characteristics:
1. Were over 35 when they had the child.
2. Have a family history of breast cancer.
3. Have a history of benign breast lumps.

*Group 2*: Five women who had more than one child but less than five and also had the following characteristics:
1. Were under 30 when they had the children.
2. Have no family history of breast cancer.
3. Have no history of breast problems.

How confident would you be to make any generalisations or predictions from your results? If your answer is 'not very confident', why not? You might give two reasons for your diffidence:

1. You have only five subjects in each group.
2. The subjects in the first group have a number of other factors besides having one child which would predispose them to developing cancer. Therefore, the sample is *biased*.

In order that any results from a study can be generalised to other similar subjects, the subject sample selected from the study must be sufficient in number and representative of the population from which they were drawn. This latter requirement can be achieved by **random selection** of the subjects. Each of these points will be discussed in turn.

## 4.1.1  Sufficent numbers of subjects

If you select only a small sample of subjects to take part in your study, it is less likely that you will get a set of results which affords generalisation or predictions.

Let us take the illustration of political opinion polls. It is obviously impossible to ask every member of the voting public who they will be supporting in a general election, so the poll companies select a **sample** of the electorate to question. From their responses, predictions are made about the outcome of the election. Now, if one opinion poll company, prior to an election, only asked half a dozen people about their voting intentions and another company asked a thousand people, whose results would you set more store by? Your answer, almost undoubtedly, is the second company, because of the size of their sample. A large sample, while in itself is no guarantee of representativeness, is more likely to produce a less biased set of results.

The same principle applies to nursing research. You could not, for a whole host of practical and financial reasons, select every woman in the country with just one child and compare them for breast cancer with every woman in the country with more than one child. What you have to do is to select a small sample of the population of women with one child and a small sample of the population of women with more than one and compare the groups on their incidence of breast cancer. But how big should these samples be in order for the researcher to be reasonably confident of detecting reliable support for the experimental hypothesis? Generally speaking, the bigger the sample, the more chance there is of finding support, although in reality, the number of subjects recruited depends mainly on the amount of time and money available to the researcher.

What does all this mean in practical terms for nursing research? Obviously it is costly and time-consuming to select huge subject samples even if the subjects you required belonged to a large and easily accessible populaton, like blood group O+ve. While there are many opinions on the matter, it is generally considered that twelve subjects per group or condition is a minimum figure to aim for while twenty to twenty-five is considered acceptable for most studies. Try not to select fewer than twelve, though if your subjects are people with congenital methaemoglobinaemia you might find it extremely difficult to find even this number!

If you are using subjects who belong to a rare population and you consequently cannot achieve a minimum of twelve, then use the available subjects, but be aware of the limitations restricted numbers place on the reliability and generalisability of your results. Do not worry too much about this, though, because the statistical test you use to analyse your results takes into account the number of subjects you have used and therefore how far they can be considered generalisable.

On the other hand, it is unnecessary to use dozens of subjects even if they are readily available. Aim for an optimum number of around twenty-five subjects per group or condition and a minimum of twelve. Two points should be made here. Firstly, some statistical tests which you will be using to analyse the data from your research set their own minimum and maximum figures. Where this is the case, it will be noted at the appropriate point. Second, throughout the chapters on statistical tests, I have frequently used fewer than twelve subjects in the worked examples. The only reason for this is to avoid huge tables of data in the text and the lengthy calculations which would inevitably result. (Further information on appropriate size of samples in research can be found in Reid and Boore 1987.)

## 4.1.2 Random selection of subjects

If you select your subjects randomly then you will be ensuring, to some degree, that the results you obtain are fairly representative and unbiased. If we go back to the opinion poll companies, we can see this concept more clearly. Supposing company A selected their sample from the boat owners at Henley Regatta and asked how they would vote; company B selected theirs by putting an advert in *The Guardian* to solicit replies; and company C selected their sample by asking every second person who entered Euston railway station about their voting intention. Each poll, because of the sample it has chosen to question, will come up with very different results.

If you were to bet your life savings on one of these results being the most accurate predictor of the general election's outcome, which poll would you choose? The chances are you would select company C, because companies A and B have only polled a very biased sample of the electorate, whereas company C has selected people from a less biased cross section of the general public. In other words, we would be more likely to place our faith in predictions deriving from the results of a representative and unbiased sample of subjects.

While a deliberately distorted example has been given here, it does, nonetheless, illustrate an important point. Each of the three opinion polls would have produced rather different results because of the nature of the sample they selected for questioning. This difference in results (which is to some degree inevitable) is called **sampling error** and simply means that if a number of subject samples are selected from the population, the results from each sample will vary slightly, because of the inherent differences in the samples. While *some* differences between subject samples are inevitable, very biased results are unacceptable, but are less likely if the subjects are selected randomly. Proper random selection, although ensuring that the sample is representative of the population from which it comes, is very difficult to achieve. However, what it usually means in practice is the selection of a subject sample in one of the following ways:

1. Putting all the names of the available subjects in a hat and pulling out, say, twenty. So in the previous example, you might have identified forty-eight women who have only one child. You would put their names in a hat and pull out the number you wanted. You repeat the procedure for the multi-parity women.
2. Selecting every second (or third, fourth, fifth, etc.) subject who presents. So, again in this example, you might decide to select every third woman you come

across in the course of your work who has one child, until you have your required number. This procedure would be repeated for the multi-parity women.
3. Assigning a number to each of the available subjects and using random number
   tables to select the sample size you want. (Random number tables can be found
   in a variety of texts, e.g. Robson 1974 and Fisher and Yates 1963.)

Random number tables simply consist of large sets of computer-generated
figures, in which the numbers 0–9 all have the same probability of occurring at each
position in the table. If you wanted to select a random sample of fifteen multi-parity
women from the population of fifty-two available to you, you would firstly need to
number the women in the population from 0 to 51; then you should enter the table
at a random point (shutting your eyes and sticking a pin in will suffice). From the
number pin-pointed, work either to the left, to the right, upwards or downwards,
and write down the first fifteen numbers you come to. Pick the subjects whose
numbers correspond with these.

Should you wish to repeat the process, remember to enter the table randomly
again (*not* at the same place) and take a different direction from the one previously
chosen. If you always start at the same point in the table and move in the same
direction, you will always be using the same set of numbers.

Many statisticians would say that these methods of random sampling will not
guarantee a pure representative sample. While this is undoubtedly the case, these
methods should ensure that each member of the population has an equal chance of
being selected and in consequence, the sample will not have any overwhelming
bias. For example, any one of the above methods would be highly unlikely to select
twenty women with one child who all have histories of benign breast lumps, family
histories of breast cancer and were over 35 when they had their child. Certainly
some of thee factors would be present among the sample, but providing you choose
your multi-parity women according to the same process, these factors are just as
likely to be present among *them* and so would balance each other out. It should be
noted that in practice it is extremely difficult to obtain a perfect random sample.
However, as long as the sample shows no obvious bias, it is usually acceptable to
treat it as random.

One final point about subject selection. The manner in which you select your
subjects influences to some degree their performance during the research study. It
has been extensively documented that subjects who *volunteer* to take part react
differently from those who are paid or who are asked or coerced into participation.
There is no 'right' way of selecting subjects – the practical considerations of your
study will determine the method chosen. However, you should always make a note
of how you obtained your subject sample in any research report you ultimately
compile.

It should also be emphasised that as well as selecting your subject sample randomly, it is often possible (and highly desirable) to allocate the sample randomly to
the experimental and control conditions. If your design allows this random allocation then you should endeavour to do so. (Note that the methods of obtaining a
random sample described above can also be used to allocate subjects to conditions

– for instance, the names of all the subjects in the sample could be put into a hat, and the number required for each of the conditions could be drawn out.)

## 4.2  Using the subjects in an experimental design

Once you have decided how many subjects you will need and how you are going to select them, you then have to make a very important decision: how should they best be used as a test for your hypothesis. If you are using a correlational design, the answer is simple and is explained in detail in Chapter 12. Briefly, however, selection of subjects for a correlational design means taking a sample of people who cover a range of scores on one of the variables and then collecting their scores on the second variable to see if the sets of data are linked. The selection of subjects should be random as far as possible, while still meeting the need for a range of scores on one of the variables.

If, however, you are using an experimental design, you have a choice of three ways in which to use your subjects. (The following discussion refers to experimental designs only.) An illustration would be useful to clarify what is meant by this concept, although the following outline is by way of introduction only. More details will be given later in this chapter. Imagine you have been asked to work with a plastic surgeon on a topic which has been getting a lot of publicity recently – the use of leeches on wounds. The hypothesis you are to test is: 'Leeches produce a lower risk of wound infection than conventional wound care procedures'.

The hypothesis is to be tested on patients receiving plastic surgery following burns injuries. Obviously this calls for an experimental design since you are looking for *differences* between the treatments. There are two levels of the IV here, conventional treatment and leeches, so you would have to have two experimental conditions (see Section 3.2.1). There are three basic options as to how you are going to use the patients to test this hypothesis:

1. Select *one* group of, say, twenty-four burns patients; on some of their plastic surgery wounds you could use conventional wound care techniques, while on the remaining areas you could use leeches. The degree of infection for each process could be monitored and compared. You therefore have *one* group of subjects who receive *both* of the two experimental conditions (conventional treatment and leeches). You have the design shown in Figure 4.1.

    This is called a **same subject design** (sometimes referred to as a related or within subject design). The example provided involves testing *one* group of subjects on *two* occasions, but a single group can be tested on *more* than two occasions if required.

2. Alternatively you could select twenty-four burns patients and randomly allocate twelve to the conventional wound care treatment and the remaining twelve to the leech treatment. The infection levels of both groups would be compared. You would therefore have two separate groups of patients, each of which would be receiving one form of treatment only; Figure 4.2.

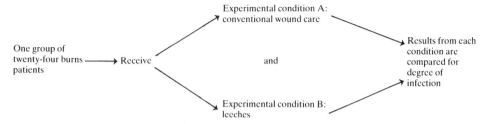

**Fig. 4.1**    An experimental design to compare two different treatments using one group of subjects.

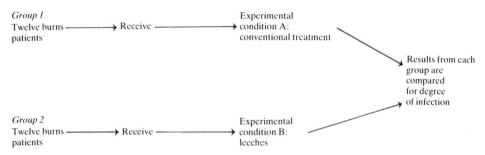

**Fig. 4.2**    An experimental design to compare two different treatments using two groups of subjects.

This is called a **different subject design** (sometimes referred to as an unrelated or between subject design), and can involve two or more groups of subjects.

3. However, you might feel that there may be so many individual differences between the groups using the above design, which might bias the results (in terms of nature of surgery, healing rate, position and cause of burn, etc.) that this method of testing the hypothesis is unacceptable. You might therefore decide that the patients in the two groups should be matched with each other on each of the factors which might bias the results, such that any possible causes of bias to the results are evened out. In other words, causes of burn, site, age of patient, etc., should be the same in each group. You would therefore have *two* groups of patients, *matched* for certain critical factors which may influence the results, and each group would receive one form of treatment only. The degree of infection for each group would be compared for differences. You would have the design shown in Figure 4.3 (overleaf).

This is known as a **matched subject design**. The illustration uses two matched groups of subjects, but more than two matched groups can be compared if the experiment warrants it.

In other words, you have three basic options of how to use a sample of subjects to test a hypothesis when using an experimental design. In more complex research

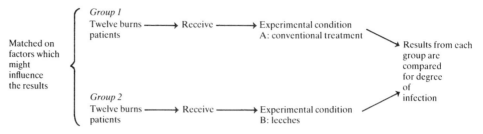

**Fig. 4.3**    An experimental design to compare two different treatments using two groups of matched subjects.

there are other possible designs, but for the purposes of an introductory text, we shall confine ourselves to these three basic designs. Which design you select is dependent on the hypothesis under investigation; sometimes – as in the example – you could use any one of the options, but as long as your final choice can be justified and is an adequate test of the hypothesis, then it would be acceptable. On the other hand, for some hypotheses, there is really only one choice of design. The points will be made clearer as each design is considered in more detail.

### 4.2.1 Different subject designs

This design raises a number of important points about hypothesis testing. As you will recall, it involves randomly selecting two (or more) separate groups of subjects and exposing each group to one condition only (experimental *or* control). The reactions of the groups are compared for differences.

The sort of hypothesis which of necessity calls for this design is one where groups of people with *inherent existing* differences between them are being compared, e.g. males v. females, young v. old, staff nurse v. nursing officer, atopic asthma patients v. allergic asthma patients, Asian patients v. Caucasian patients.

Let us take an example. Anti-tumour drugs can be given in many ways. Suppose you are working on an oncology ward and you want to compare the side-effects in two different types of cancer patient during the administration of high dose inter-mittent combination therapy. Your hypothesis is: 'Patients with carcinoma of the larynx and patients with carcinoma of the bowel differ in terms of the severity of side-effects experienced during administration of high dose intermittent combination therapy'. (The IV is site of carcinoma and the DV is the severity of side-effects.) The relationship you are anticipating is that different severity of side-effects results from the *site* of the tumour.

You need to select randomly one group of patients with carcinoma of the larynx and a second group of patients with carcinoma of the bowel and compare the side-effects of the anti-tumour drug. You would have the design shown in Figure 4.4. With this hypothesis you have no option but to use two groups of subjects, because

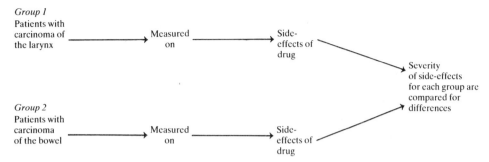

*Group 1*
Patients with
carcinoma of ——————————▶ Measured —————————— Side-
the larynx                        on                     effects of
                                                         drug
                                                                        Severity
                                                                        of side-effects
                                                                        for each group are
                                                                        compared for
*Group 2*                                                               differences
Patients with
carcinoma   ——————————▶ Measured —————————— Side-
of the bowel                     on                     effects of
                                                         drug

**Fig. 4.4**  An experimental design which necessitates the use of different subject groups.

you are comparing two levels of the IV (site of carcinoma) which constitute *inherent existing* differences between patients. In other words you are comparing patients with *either* carcinoma of the larynx *or* carcinoma of the bowel. It would make no sense here to select *one* group of patients who had both levels of the IV (i.e. carcinoma of the bowel *and* larynx) and look at the side-effects of the anti-tumour drug because you would not be able to determine which site of the tumour was producing the greater severity of side-effects; consequently you could not test the hypothesis adequately.

Any hypothesis which involves the comparison of inherently different groups of subjects necessitates a different subject design and it may be good practice for you to jot down two or three hypotheses to which this design applies. However, while this design is often essential, it does have a major flaw. In the present example, we have selected two groups of carcinoma patients, whom we are comparing for side-effects. What we are hoping to find is a relationship between the site of the tumour and severity of these side-effects. Suppose we found that carcinoma of the larynx produced more severe side-effects. However, because different patients are involved in each group, it is quite conceivable (if the subjects have not been selected properly) that the carcinoma of the larynx patients were older and had a history of gastric and urinary problems. On this basis, then, it is possible that they would be more prone to side-effects such as nausea, diarrhoea, bone marrow depression, nephrotoxicity and chemical cystitis, simply because of their age and previous medical predispositions, and *not* because of the site of the tumour. In other words there could be alternative explanations for your results *besides* the site of the tumour. These alternative explanations are called **constant errors** and are discussed more fully in Chapter 5. Even if you ensured that the age and medical history of the patients in each group were the same, there are still individual differences that cannot be eliminated, because they cannot necessarily be identified, e.g. idiosyncratic biochemical responses to the anti-tumour drug. These are called **random errors** and will also be discussed fully in Chapter 5. This means that alternative explanations for the results remain besides the one predicted in the hypothesis. While it is quite likely that individual differences such as biochemical reactions will be equally distributed between the two groups of patients, thereby cancelling out their biasing effects on the results, it is also likely that they all occur within one group. Because of their unidentifiable and unmonitorable nature, you will be unaware of this. Consequently, any results you obtain may be the product of these individual differences and not of the IV.

While it is impossible to get rid of these individual differences totally, random selection and allocation of subjects to conditions goes some way towards overcoming the problem. Thus, here you would randomly select a sample of patients with carcinoma of the larynx from the population of such patients available to you and you would do the same when selecting the group with carcinoma of the bowel. Random allocation of subjects to conditions cannot be carried out in this example, because you cannot select patients and allocate them to having carcinoma of the larynx *or* of the bowel. However, if we look back to the example with burns patients, we could take a randomly selected sample of twenty-four burns patients and then randomly allocate half to conventional treatment and the remainder to leech treatment. You *must* randomly allocate your subjects to conditions if your study permits it.

Despite this criticism of different subject designs, it does have an important asset. Sometimes you may want to carry out a study which looks, for instance, at the effectiveness of two different treatments. Let us take an example concerning the prevention of pressure sores, comparing sheepskins and water beds. If we were to take a same subject design, we would select *one* group of patients of a similar type (e.g. elderly bed-bound) and give them firstly sheepskins to lie on for a fortnight, followed by a water bed for a fortnight (or you could do it the other way round).

You have the design shown in Figure 4.5. Supposing we found that towards the middle of week 4 patients were beginning to develop pressure sores; could we conclude that water beds were less effective in their prevention? Your answer is probably 'no', because there may be other reasons for their emergence at this point, the most obvious of which is the fact that the patient had already been bed-bound for two weeks when they were placed on the water bed. Therefore, it could be that length of time spent bed-bound was the responsible factor and not the surface on which the patients were lying. In other words, the *order* in which the

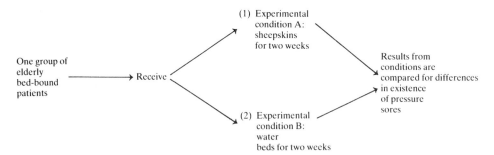

**Fig. 4.5** An experimental design which uses one group of subjects where order effects may be a problem.

treatment was received was responsible for the pressure sores and *not* the surface. **Order effects** are discussed in detail in Chapter 5.

A different subject design gets round this problem. Rather than having *one* group of patients receiving *both* treatments, you could have two different groups, one of which lies on sheepskin and the other of which lies on a water bed. Order of presentation of the treatments is therefore not a relevant problem here.

So, in brief, different subject designs have a major problem of individual difference biasing the results, but they are not beset by the problems of order effects.

### 4.2.2 Same subject designs

If you look back, you will see that this design involves one group of subjects which receives all the conditions – be they experimental or control conditions. The group's performance or reaction to each condition is compared for differences. It is the design typically used in 'before and after' washing powder type experiments where the efficacy of a particular product or procedure is under consideration. Let us take an example. Supposing you were working on a ward with a number of Crohn's disease patients, with chronic and unremitting pain, for whom mild analgesia was ineffective and strong analgesia unsuitable because of its potential addictive qualities. Because of the intensity of pain experienced by some of these patients, they are in need of some form of treatment and you decide to assess the value of local heat.

Your hypothesis is: 'Local heat is effective in reducing the pain experienced by patients with Crohn's disease'. (The IV is local heat and the DV is amount of pain experienced.) Such a hypothesis is best tested by a same subject design because it calls for a before and after approach. Let us look at this more carefully.

Using a same subject design, you might randomly select fifteen Crohn's disease patients over a period of, say, three months. You might then ask them to rate their pain *before* local heat is applied (control condition because no IV is administered); give them local heat (experimental condition, because the IV is given); ask them immediately *after* the application of local heat to rate their pain again. You have the design shown in Figure 4.6. If local heat is effective in reducing pain, you should find the reported pain levels lower in the experimental condition.

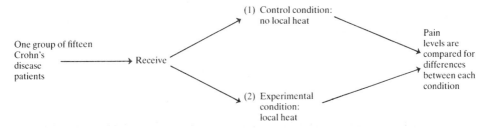

**Fig. 4.6**   An experimental design which requires the use of one group of subjects.

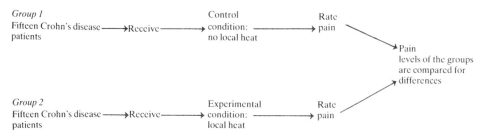

**Fig. 4.7**   The inappropriate use of a different subject design.

This design is a better test of this hypothesis than a different subject design; let us see why. If we were to use a different subject design, we would need to select randomly *two* groups of Crohn's disease patients, give one group *no* local heat (control condition) and the other group local heat (experimental condition). The reported pain experienced by each group would be compared. We would have the design shown in Figure 4.7.

However, we know that a number of factors might influence a patient's pain level; for example, the *actual* degree of pain, tendency to complain verbally, personality factors and a host of other sources of illness behaviour. These differences may be difficult to identify and to eliminate and so constitute random errors (see Chapter 5). In other words they are factors which might bias the results and are difficult to control because of their unpredictable nature. Thus, it may be these individual differences between the patients that produces the results and not the effect of local heat. Given that what we want to demonstrate is the efficacy of local heat, to have other possible explanations infiltrating is unacceptable, because it means that no definitive conclusions can be drawn about the hypothesis. Even though there will be individual differences among the group of fifteen patients in the same subject design, these differences will be carried through both conditions, and so will at least be a *constant* factor each time the patient rates his or her pain level.

A same subject design has a number of advantages. Firstly, it eliminates the individual difference problem inherent in different subject designs, and so eradicates this source of potential bias to the results. Second, fewer subjects are required to take part in the study, which, if they belong to an unusual population, is an important consideration.

However, its major disadvantage is one which has already been referred to – that of order effects. If one group is to take part in two or more conditions, then by definition, there has to be some sequence to the presentation of conditions. Because of this, it is conceivable that any results obtained are due to the order of presentation and *not* to the IV. If we go back to the example about pressure sores, it can be seen that any development of pressure sores could be due to one of two factors:

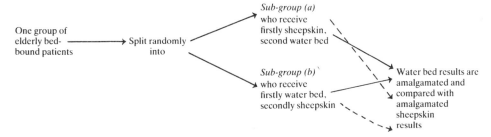

**Fig. 4.8**  A same subject design which uses counterbalancing to overcome order effects.

1. The nature of the surface on which the patient was lying.
2. The time spent in bed. In other words whatever surface the patient was lying on *second* would have coincided with the development of pressure sores, not because of the surface itself, but simply because it was second.

There are, then, two possible explanations for the results – order of presentation and the IV. Unless the order effects can be eliminated, we cannot claim that the results support the hypothesis. The only way round these order effects is a technique called **counterbalancing** (see Chapter 5). This involves dividing our group of patients into two sub-groups, by randomly allocating half the patients to the sheepskin surface first and the water bed second, and the remaining patients to the water bed first and the sheepskin second. Following this procedure, the results from *both* sub-groups are amalgamated to compare the effects of sheepskin and the water bed. We have the design shown in Figure 4.8.

The other disadvantage of a same subject design is that it obviously cannot be used when manipulation of the IV simply means comparing two or more groups of subjects of a fixed, existing nature, e.g. male v. female, West Indian v. Asian, premature babies v. full-term babies, people living on their own v. people living with families, social class V v. social class II, colostomy patients v. ileostomy patients.

### 4.2.3 Matched subject designs

Matched subject designs overcome all the problems inherent in both same and different designs, in that they eliminate individual differences, eliminate order effects and allow comparisons of 'fixed' groups while maintaining all the advantages of each design.

Given that such designs sound like the answer to a researcher's prayer, why do we not just forget about the other designs and always use this one? The answer lies simply in the fact that they are extremely difficult to carry out properly. An example will clarify this. Supposing you wanted to look at the efficacy of tar baths for patients with chronic psoriasis and patients with erythrodermic psoriasis. Your hypothesis might be: The reaction to tar baths of patients with chronic psoriasis is different from that of patients with erythrodermic psoriasis'. The IV is type of psoriasis (chronic or erythrodermic) while the DV is reaction to tar baths.

Obviously a same subject design cannot be used here because patients cannot have (or are very unlikely to have) both chronic and erythrodermic psoriasis. But if a different subject design is used then there are all the problems of individual differences. For example, the chronic psoriasis patients may all be younger, fitter, have never had any previous treatment for their condition and have no family history of the condition. On the other hand, the erythrodermic psoriasis patients may be over 60, in poor health, had extensive previous treatment and have a family history of psoriasis. All these factors may influence the way in which each group responds to tar baths, and so distort the results and conclusions from the study. One way of getting round these potentially biasing factors is by *matching* the subjects in each group on all the factors which could influence the results. The way in which this is done is quite difficult and time-consuming. Firstly, you must identify all the factors which could potentially bias your results. Here we have pin-ponted four (although you may have thought of others). These are age, general health, previous medication and family history.

Then a subject from one category is selected – *either* the chronic *or* the erythrodermic group, and is assessed on these four factors. We might find, then, that subject 1 from the chronic group is 42 years of age, rates at 8 on a 10-point scale of general health, has used Dithranol ointment for the previous five months and has no family history of the condition.

You now have to find a subject in the erythrodermic category who corresponds exactly to this subject on each of the above factors, i.e. is *also* 42 years of age, rates at 8 on a 10-point scale of general health, has used Dithranol ointment for the previous five months and has no family history of the condition.

This first pair of matched subjects may take a very long time to find, and you still have at least another eleven pairs to go! (And, of course, if you were wanting to compare more than two matched groups it will take even longer to find an appropriate number of matched 'triplets', 'quads', etc.)

The next pair of subjects only needs to be identical to *each other* on these characteristics, i.e. they do not *all* have to be 42 years of age, etc. In this way we would end up with pairs of 'twins', identical to each other on all the salient factors which might influence the results, but one 'twin' would have chronic psoriasis and the other erythrodermic psoriasis.

The ultimate in matching, of course, would be *real* identical twins, one of whom had chronic psoriasis and the other erythrodermic psoriasis. However, the chances of finding at least a dozen such sets of identical twins are virtually nil. Nonetheless, the theory behind matching subjects is that if the matching is done properly, then the pairs of subjects will, for the purposes of the experiment, be equivalent to identical twins. For this reason, matched subject designs and same subject designs are treated in the same way when it comes to statistical analysis. We would therefore have the design shown in Figure 4.9.

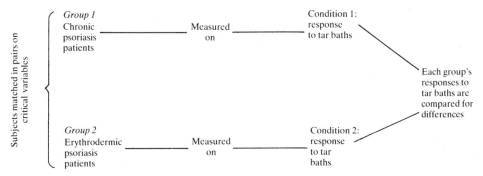

**Fig. 4.9**    An experimental design which uses matched subjects.

Matched subject designs, although very time-consuming and problematic to set up, do overcome problems of order effects (because there are still two or more groups), and of individual differences (because the differences have been matched across groups), while also allowing 'fixed' groups to be compared. However, there is a very important problem underlying all this. While I have just said that individual differences can be eliminated by matching up the subjects, this only applies to those individual differences which are identifiable and measurable. There may still be many other distinctive characteristics of the subjects which could influence the results and which we cannot detect, e.g. skin type, healing capacity and biochemical reaction to the tar. Even if we could identify such idiosyncrasies it is usually impossible to measure and match them. So, basically there may be many factors which may affect the results, but which we do not know about or cannot measure. For this reason, it is usually recommended that a same subjects design is used in preference to a matched subject design.

It should be pointed out that many researchers do not take the trouble to match their subjects properly. What they do, instead, is to identify the salient factors, and select subjects in each group *all* of whom are between, say, 40 and 50 years of age and are above average in general health, etc. In other words, they set upper and lower limits on each factor and simply ensure that *all* the subjects fall within these limits. This is *not* matching. Matching involves the selection of identical matched *pairs* (or 'triplets', etc.) of subjects, such that every subject in one group has a 'twin' (or 'triplet', etc.) in the other.

---

**Summary of key points**
A hypothesis can be tested by an experimental design using the subjects in one of three ways.

1 Different subject designs, whereby two (or more) different groups of subjects each receive one condition (experimental or control) and their performances on these conditions are compared. Such a design is

essential when 'fixed' existing groups are to be compared. However, individual differences between the groups may distort the results.

2 Same subject designs, whereby one group of subjects takes part in all the conditions and their performance in each is compared. This is a valuable design for 'before and after' type experiments where a single treatment procedure is being investigated. It can be used to evaluate more than one treatment, i.e. the same group of subjects experiences all the treatments, but complications occur here due to order effects. However, these designs overcome bias due to individual differences, but cannot be used to compare 'fixed' groups.

3 Match subject designs, whereby pairs (or more) of subjects are selected who are matched on all the factors which could influence the results. One 'twin' ('triplet', etc.) is allocated to one group and the other 'twin' ('triplet', etc.) to the second group. While this design overcomes the problems inherent in the previous two designs, it is very difficult to carry out properly, for two reasons: (a) it is unlikely that adequate pairs (or more) of subjects can be found who are sufficiently similar and (b) it is impossible to identify and match all the factors which could affect the experiment's outcome.

## 4.3  Key words and terms

Constant errors
Counterbalancing
Different subject design
Generalisability
Matched subject design
Order effects
Prediction
Random errors
Random selection
Same subject design
Sample
Sampling error
Subjects

# 5     Sources of bias and error in research

The last two chapters outlined a number of important guidelines which you should follow when designing your study. However, although there seem to be a lot of things to take into account, these guidelines are still not exhaustive; there are other factors which may influence the outcome of your study which must be considered if your research is to be of use. These factors are called **sources of error** and, as the name implies, may bias and obscure your results if they are not controlled, so that you could end up with a totally erroneous set of conclusions from your study.

Let me clarify this. If you recall, I said that when you prepare a piece of research which uses an experimental design, you manipulate the independent variable (IV) to see what effect this has on the dependent variable (DV). If you have followed the procedures outlined in Chapters 3 and 4, you could conclude that any change in the DV is the result of altering the IV. Since this would support your experimental hypothesis, this is exactly what you would hope for. However, if your experimental design does not also take account of the possible sources of error, it is conceivable that any change in the DV may be the result of the influence of these sources of error and not of the IV. Unless you can establish, without doubt, that the IV is the *only* possible explanation for the changes in the DV then your research will be wasted.

Even if you are using a correlational design, which cannot establish causal connections, you should still try to eliminate as many sources of error as possible. The points covered by the remainder of this chapter apply to both correlational and experimental designs.

There are several major sources of error and each will be discussed in turn.

## 5.1   Order effects and counterbalancing

### 5.1.1   Experimental designs and order effects

These can be a major source of error and are best explained by an example. Suppose that you have hypothesised that psychiatric out-patients suffering from anxiety neuroses make more improvement when treated by relaxation and biofeedback techniques than when treated by tranquillisers. Your hypothesis predicts a relationship between type of treatment (IV) and degree of improvement (DV). Because

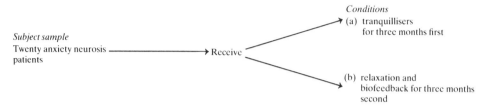

**Fig. 5.1**    A same subject design where order effects could bias the results.

you know that each patient has a very different type of personality, neurosis, psychiatric history, etc., you realise it would be impossible to use a matched subject design, so you decide on a same subject design. This will involve selecting a number of anxiety neurosis patients (say twenty), giving every patient both treatments and comparing the progress made on each.

You decide on the design shown in Figure 5.1. You compare the patient's psychological state when on tranquillers with that when on relaxation and biofeedback.

Let us again imagine that you have carried out your research, and analysed the data. You find that the patients do, in fact, seem to be in a better psychological state when being treated by relaxation and biofeedback, and you conclude triumphantly that your hypothesis has been supported.

However, after a little while, some doubts start to creep in, and you begin to wonder whether the treatment technique is the *only* possible explanation for your results. Since *all* patients had drug therapy first, and relaxation and biofeedback second, it is conceivable that patients did better on the second treatment simply because it was second and not because of the inherent benefits of the technique. In other words, the patients could have improved on *any* second treatment, irrespective of what it was, simply because they were further on in their psychological condition and therefore over the crisis, or because they had, by definition, received more treatment and the cumulative effect of this was producing the improvement. If patients are subjected to two or more treatments in a fixed sequence in this way, it is possible that it is the *order* of the treatments and nothing whatever to do with the *type* of treatment, that produces the results. This is known as an **order effect**, and there are two types of these – **practice effects** and **fatigue effects**. When subjects do *better* on the last treatment, as in the last example, this is called a practice effect.

It is, of course, just as likely that you thought that the patients in the previous example would do consistently worse on the relaxation and biofeedback treatment, perhaps because they were suffering withdrawal symptoms after being taken off the tranquillisers, or because they were reaching the peak of their problem, or for a host of other reasons. When subjects perform *worse* on the last treatment of a fixed order design of this type, it is known as a fatigue effect. (Of course, if you have more than two conditions, practice and fatigue effects may start operating before the final condition and will therefore have a progressive impact.)

**Table 5.1** How to avoid order effects when three groups of subjects are used

| Subject group | Order of conditions |
|---|---|
| (1) One-third of the original subject group receive | A then B then C |
| (2) One-third of the original subject group receive | B then C then A |
| (3) One-third of the original subject group receive | C then A then B |

It does not really matter whether you argued that a practice or fatigue effect was more likely to occur in this example, since the point is that both are examples of a potent source of error – the order effect.

Despite the fact that order effects are a potential source of error when you have *one* group of subjects undertaking two or more conditions they are relatively easy to overcome, if you use a technique called **counterbalancing**.

Counterbalancing is, as the name suggests, a method by which order effects can be balanced out. If we go back to our example, all we need to do to even out this source of error is to split the patient group randomly into two. Half the patients would then receive the drug treatment first, followed by relaxation and biofeedback, while the other half would receive relaxation and biofeedback first, followed by drugs. When the study is complete you simply amalgamate *all* the drug results from both halves of the group and *all* the relaxation and biofeedback results from both halves and compare them. In this way, the order effects are balanced out.

Obviously, if you have one subject group and three or more conditions, you need to divide your subjects into the appropriate number of groups, and ensure that the order of presentation of the conditions is totally altered for each group. So, supposing we had three treatment conditions (A, B and C), we would need to divide our subject group randomly into three and could present the conditions in the order given in Table 5.1.

In this way, each treatment condition has been received in first, second and third place. You would then, of course, just combine the results from all the A treatments, all the B treatments and all the C treatments, and compare them. Remember, though, that you must divide your subject group randomly (see Chapter 4), or you may get other types of error creeping in.

### 5.1.2 Correlational designs and order effects

Order effects in correlational designs are somewhat easier to deal with as they are not such a prominent problem. To illustrate this point, it might be useful to recap on some of the essential points of these designs. In a correlational design, the researcher simply collects a range of data for each variable before analysing it to see if there are any interrelationships or patterns between them. Often these data already exist in medical records, biographical information, etc., and so the study is *retrospective*. For example, if you were interested in testing the hypothesis 'Amount of insulin required daily is related to the weight of the patient at the onset of diabetes (the higher the weight, the more insulin)', you are predicting that weight and amount of insulin are associated. To test this hypothesis, you might select a number of subjects who varied in weight and then note the amount of insulin they needed. Both sets of data could be derived from existing medical records, and so would be a retrospective study. The researcher in this case would probably not come into any contact with the subjects.

However, not all correlational designs have to be retrospective; it would be quite possible in the example just given to carry out a *current* study. In other words, you could select patients as they attended a diabetic clinic for the first time and ask them their weight at onset and insulin requirements. The researcher would therefore collect data on both variables at that point in face-to-face contact. In some studies, however, data for one variable may be available from existing records, while for the other it may be collected directly from the subject. This sort of data collection is therefore both current and retrospective. In all three types of correlational studies you are simply collecting data which is not the result of any experimenter manipulation, but which is an inherent feature of the subjects selected.

However, the crucial difference between these approaches, from the viewpoint of order effects, is that while it is irrelevant in a retrospective and retrospective/current study which variable is dealt with first in terms of data collection, it may make a difference in a *current* study. Here the researcher will be involved in collecting data on *both* variables *directly* from the subjects, and the order in which it is collected may bias the subject's response. In other words, the fact that information had been elicited on one variable first may have the effect of sensitising the subject to the purpose of the study. This may result in a biased response to the second variable particularly if the study is concerned with a subject's opinion, attitude or report of an issue, i.e. subjective data. Typically a distorted reply of this kind is not necessarily an attempt to deceive the researcher or to wreck the study, but rather to present the person in a more favourable or acceptable light, particularly if the issues involved are sensitive ones. The simple expedient of alternating the order in which the data on each variable are collected for each subject should ensure that any bias in the responses due to order effects is counterbalanced.

---

**Summary of key points**

1 (a) *Experimental designs and order effects*. When you use just one group of subjects in all the conditions, the order in which the conditions are undertaken may distort the results. If all the subjects are given the conditions in an identical sequence it is possible for one of two types of order effect to occur. These two types of order effect are called practice effects, where the subjects do better on the last

condition, and fatigue effects, where the subjects do worse on the last condition.

(b) *Correlational designs and order effects*. When a current data collection approach is used, the order in which the data are elicited can bias the subjects' responses.

2 Order effects can be controlled by a technique called counterbalancing which for an experimental design involves randomly dividing the subject group into two. One half then carries out condition 1 first, followed by condition 2, while for the remainder of the group the sequence is reversed. If there are three or more conditions in the experiment, the subject group is divided into the appropriate number and each sub-group receives the conditions in a different order.

For a correlational design, the order in which the data for each variable are collected is alternated for each subject.

---

Although it is quite easy to control order effects by counterbalancing them, this only deals with one source of error. There are several others which are just as powerful.

## 5.2  Experimenter bias effects

If you have a good idea for a piece of research and you go to some lengths to design it properly, set it up and carry it out, it is inevitable that you will be very concerned that the results work out as you predicted. This is an entirely natural commitment arising out of your personal investment in the project. However, this degree of commitment may provide another source of error, since it is conceivable that in your enthusiasm to support your hypothesis you subconsciously bias the results. I am not suggesting here that you will do anything devious or dishonest, but that in your keenness you will unwittingly influence the process of the experiment. It is a well-known fact that people see what they want to see or hear what they want to hear, and so it is in research. For instance, in a previous example, because you had predicted that patients would do better on relaxation/biofeedback techniques, it is possible that when you evaluated their performance on this condition, you subconsciously modified the assessment criteria, or smiled a bit more in encouragement when a patient did something that accorded with your predictions. Such behaviours are called **experimenter bias effects**.

Many students become very indignant at such a suggestion, because it suggests a lack of professionalism at the very least. But because they are carried out subconsciously, we are all guilty of such actions and they have been well documented in the literature. Other examples of experimenter bias effects include the more fixed (rather than behavioural) characteristics of the experimenter. For example, age, race, sex, social class and attractiveness can all influence the subject's performance, e.g. it is known that black children do better on IQ tests when tested by a black experimenter rather than a white one.

If these experimenter bias behaviours are subconscious, how can we do anything about them?

The most common solution, especially in medical-type research, is the **double-blind technique**. This procedure involves recruiting someone who is totally ignorant of your hypothesis to collect the data for you. In double-blind procedures the subjects are also unaware of the purpose of the research. As no one knows what results you are predicting they cannot influence the data one way or the other and so the data collection should be fairly objective and impartial. (You might feel that in nursing research it is unethical to keep patients in the dark about the aims of the study they are participating in. This is a valid objection in many cases, and you might consider informing the patients but *not* the person collecting the data.)

Obviously the individual you recruit to collect your data would have to be competent and it would be your responsibility as the experimenter-in-charge to ensure that this was so.

## 5.3 Constant errors

### 5.3.1 Experimental designs

The term **constant errors** is given to any source of error which will distort your data in a consistent, predictable way – hence the name. Let us take an example. Suppose you have hypothesised that male premature babies are more likely to suffer respiratory complications than female premature babies. You are therefore predicting a relationship between the sex of the premature baby and the incidence of respiratory complication. In order to find out whether the predicted relationship exists you select fifteen male premature babies and fifteen female premature babies and compare their respiratory condition. Imagine that you find more respiratory difficulties in the males, which supports your hypothesis. However, unless you have selected your babies and carried out your study properly, there may well be a considerable number of constant errors which might have biased your data and produced the results.

Suppose that the male babies you selected for study were all born between 30 and 32 weeks' gestation, weighed between 2 and 3 lb at birth, had mothers who smoked heavily throughout pregnancy and are less than a week post-partum. On the other hand, the female babies in the study were all born between 34 and 36 weeks' gestation, of non-smoking mothers, weighed between 4 and 5 lb at birth and were at least 2 weeks post-partum. Given such discrepancies between the subject sample, it would be expected that the male babies would have more respiratory complications anyway because of the maternal history, gestational immaturity, lower birth weight and age and not necessarily because they were male. In other words there are five possible explanations for your results:

1. Gestational stage.
2. Birth weight. ⎫ constant
3. Maternal history. ⎬ errors
4. Post-partum age of baby. ⎭
5. Sex, i.e. the independent variable.

Unless you can eliminate the first four explanations you cannot claim that your hypothesis has been supported.

These other possible explanations are called constant errors, because they will affect your results in a constant and predictable way. For example, we know that the earlier the gestational stage of a baby, the more immature and vulnerable the lungs and so the infant would be *more* likely to suffer respiratory difficulties. Similarly, lower birth weights are associated with greater immaturity of vital organs and so the lighter babies will be more likely to suffer respiratory complications. In the same way, the younger the post-partum age of the baby and the more the mother smoked, the more likely the baby is to have problems. So all the constant errors present in this study would have distorted the results in the same manner and would have biased the outcome in a constant and predictable way, such that on completion of your study you would not know which factor, or combination of factors, was accounting for your results.

This is obviously a hopeless situation to be in, since you can establish nothing at all in the way of a causal link between sex and respiratory complications. It should be one of your main concerns as a researcher to track down all the sources of constant error and either control them or eliminate them in some way. How is this done?

If we take the previous example, we can deal with the sources of constant error in the following manner:

1. Differences in gestational stage: deal with it by ensuring male and female babies are comparable on gestational stage, e.g. between 32 and 34 weeks.
2. Birth weight: deal with it by ensuring that male and female babies are similar in birth weight, e.g. all between 3 and 4 lb.
3. Maternal history of smoking: deal with it by ensuring that *none* of the mothers of either group smoked during pregnancy.
4. Post-partum age of baby: deal with it by ensuring that all the babies, both male and female are of similar post-partum ages, e.g. 1–2 weeks.

If you control the constant errors in this manner, then any differences in results you get from the experiment have to be attributable to the sex of the child, because all the other possible explanations have been excluded.

So whenever you carry out an experiment, you *must* identify all the factors which will bias your results in a constant and predictable way and get rid of them by ensuring that your subject groups are comparable and the experimental procedures used are standardised.

## 5.3.2 Correlational designs

Obviously in correlational designs, you are not concerned with establishing causal links, so the concept of constant errors accounting for alternative explanations for your results is not appropriate. However, it is still advisable in correlational designs to take account of those factors which could bias the conclusions of the study. Let us take the example that the greater the amount of saturated fat in the diet, the earlier the onset of coronary heart disease. To test this hypothesis, you might select a number of subjects whose intake of saturated fats ranged from none to very high, and would then note the age at which coronary heart disease was first manifested. If you were unscrupulous, you might select two or three subjects with a high intake of saturated fats but who also had a family history of heart problems, were over-weight, smoked, ate a considerable amount of salt, etc. In addition, you might include in your sample a number of subjects who besides eating very little or no saturated fat, took a lot of exercise, did not smoke, were of normal weight, etc. The presence of just a few subjects of this type would have the effect of biasing your results in favour of your hypothesis. While this would not affect any assumptions of cause and effect, as in experimental designs, it is nonetheless unethical and poor research practice. Therefore, in correlational designs, it is still important to identify and eliminate the constant errors as far as possible. Once this has been done, the subject sample should be selected randomly from the remaining available subjects.

## 5.4 Random errors

**Random errors** are less obliging than constant errors, order effects or experimenter bias, in that they are unpredictable in their effects and very difficult to deal with properly.

Let us explain this further. All experiments have random errors, but it is not always easy to identify what they are; even if we can identify them, it is impossible to forecast what their effects will be, since random errors operate in a totally variable manner. This makes them impossible to control. All we know is that random errors are present in all experiments and that they affect the results in a random or *chance* way.

These insidious random errors include many transitory individual differences like personality, moods, current state of health, biochemical predisposition, attitudes, motivation and so on all of which may influence the way in which a subject responds, etc.

## 5.4.1 Experimental designs

Let us go back to the earlier illustration in Section 5.1.1 comparing relaxation/ biofeedback techniques with drug therapy for patients with anxiety neuroses.

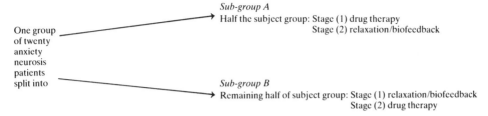

One group
of twenty
anxiety
neurosis
patients
split into

*Sub-group A*
Half the subject group: Stage (1) drug therapy
                          Stage (2) relaxation/biofeedback

*Sub-group B*
Remaining half of subject group: Stage (1) relaxation/biofeedback
                                  Stage (2) drug therapy

**Fig. 5.2**   A same subject design which uses counterbalancing to overcome order effects.

Imagine that we have designed this experiment properly, controlling for order effects, experimenter bias and constant errors and that we are about to make the final assessments of the patients' progress. Remember that we divided the group of twenty anxiety neurotics into two, and gave one half drug therapy followed by relaxation and biofeedback, while for the other half the order was reversed; Figure 5.2.

We are about to assess the patients' progress after completing the second stage of their therapy. There will be a large number of random errors here which could bias our results in a totally chance way. For example, some of the patients might be in a bad mood on the day of assessment, because they missed the bus, had a row at home, broke a cup, etc. Other patients might be really motivated to do well on their assessment because they want to please the experimenter, while others might want to do badly for the opposite reason. The experimenter may be overtired and irritable and so is less patient with some of the subject group. Other patients may have an unidentifiable biochemical reaction to the drugs they have been on, which makes them short-tempered. Obviously, all these things will affect the patients' performance in some way. The list of possible random errors is endless, and it would be impossible to identify them all. And if we could list them, we would not be able to find out which patients they were affecting (or to what degree) with any accuracy.

Now, although it is highly improbable, suppose that *all* the random errors which affected the patients' performance in an adverse way suddenly occurred in the group which received drug therapy second, just at the time they were being assessed. This would have the effect of depressing their performance. So, when all the drug therapy scores are combined in order to be compared with relaxation and biofeedback, the drug scores will be lower overall. This will have the effect of creating a large difference between the two therapies, which might lead you to conclude that relaxation and biofeedback is better and this would support your hypothesis. Now as you can see from this, it is conceivable that relaxation and biofeedback is more effective and therefore is responsible for your results, but it is also possible that random errors could be the cause. How can you disentangle which of the two explanations is responsible? The problems which random errors can cause are potentially immense, and yet because of their transitory and un-predictable nature, we obviously have very little control over them. So is there *anything* we can do?

The only answer to this is a fairly unsatisfactory one, in that it will not take away random errors completely. What is required is *random sampling* (see Section 4.1.2). If we select our subjects randomly out of the population from which they come, then it is highly likely that they will be *typical* and *representative* of that population. In this case, if we required 20 anxiety neurosis patients for our study and there is a pool of 153 anxiety neurosis patients from which we can select them, as long as the selection is done randomly, then the patients in our study should be fairly typical of anxiety neurosis patients as a whole. They are unlikely to be *particularly* awkward, irascible, compliant, dependent, moody, extreme in attitudes, unhealthy or whatever. Furthermore, if we then *randomly* allocate 10 of these patients to receive drug therapy first, and the remaining 10 to receive relaxation and biofeedback first, any random errors should be relatively evenly divided across the two groups. If the random errors are equally distributed, then their effects should theoretically be comparable for each group. If the effects are comparable for each group, then any differences arising between them must be the result of the independent variable.

You will notice that all we have done here is to attempt to distribute the random errors equally; we have not been able to get rid of them totally. They will still be present in our study, affecting the results in some obscure way.

From this, you may have concluded that any results we get from an experiment could be explained *either* by the influence of the independent variable (which is what we would like) *or* by the influence of random errors (which we would not like). How can we ever decide whether the IV is responsible for our results, if random errors are always around? The answer to this lies in the use of statistical tests. Sections 5.5 and 5.6 explain their importance.

## 5.4.2 Correlational designs

Clearly random errors will not influence causal inference here since this concept is not relevant in correlational designs. However, random errors such as the individual differences already noted, may influence the subject's response on one or both the variables being measured and this could artificially distort the results. So, returning to the earlier example which predicted a correlation between the intake of saturated fats and the age of onset of coronary heart disease, it is conceivable (though unlikely) that all the subjects who had a high intake of saturated fats also had a host of individual differences which predisposed them to heart disease. These would artificially distort any correlation found and would be an unfair reflection of the interrelationship between the two variables in the hypothesis. Similarly, a subject's response when being assessed on something may be altered by individual differences such as mood, personality, temporary health state, etc.

Again, the solution to the problem lies in the random selection of the subject sample. Once a pool of potential subjects has been identified, and the constant errors eliminated, a sample of subjects from this pool should be randomly selected whose responses on one variable cover a range of scores. If this process accords with the guidelines given for random selection, then any distorting random errors should theoretically be evenly distributed across all the subjects selected, thereby cancelling their impact.

Again, any results from a correlational study could be explained by the relationship predicted in the hypothesis or by the presence of random errors, since these cannot be totally eliminated. Given this, how can we ascertain whether the results support our hypothesis or are simply the outcome of random error? Again, the answer to this lies in the use of statistical tests to analyse the data.

---

**Summary of key points**
1 It is possible for experimenters to influence the outcome of their research subconsciously, by the way they behave or appear to the subjects. This is called the experimenter bias effect and can be controlled by using the double-blind procedure. This involves recruiting someone who is totally ignorant of your hypothesis to collect your data for you. At the same time the subjects are also unaware of the hypothesis.
2 Constant errors bias the results of a study in a constant and predictable way. They can be controlled by ensuring that the procedures and conditions in the study are identical for all subjects.
3 Random errors obscure the results of a study in a random or chance way. They cannot be excluded, but they can be partially controlled by ensuring that subjects for an experiment are randomly selected and allocated to conditions.

---

## 5.5 Probabilities

In order to decide how far the results from a research project can be explained by the relationship predicted in the hypothesis and how far by random error, we have to use statistical tests to analyse our data.

When we calculate any statistical test we end up with a number, which is then looked up in a set of probability tables (each test has its own probability tables); this gives us a **probability or *p* value** for that number. What this *p* value actually tells us is how probable it is that our results could be explained by random errors. This concept is fundamental to statistics and you should make sure you understand it fully.

Now, as you know, the experimenter wants the results to be explained by the experimental hypothesis (and *not* by random errors) since this will support the original prediction. So it is obvious from this that the experimenter wants to obtain the *smallest p* value possible for results, since the smaller the *p* value, the less the likelihood that random errors could explain the results. (It might be worth reading this bit again, just to make sure you are clear about it.)

If we move on one step further, the less the likelihood that random errors could explain the results, the greater the likelihood that it is the relationship predicted in the hypothesis that is responsible and so the greater the support for your hypothesis. What we are saying here, in crude terms, is that the smaller the $p$ value you obtain, the better it is for your hypothesis.

Values of $p$ are always expressed as a decimal of less than 1.0 or as a percentage, e.g. 0.05 or 5%. So, if you ended up with a $p$ value of 3%, this means that there is a 3/100 probability that chance or random error could explain your results.

Your $p$ value will *always* be greater than 0, which means that random error will have affected your results to some degree. However, if the $p$ value is sufficiently low we can say that our experimental hypothesis has been supported and therefore our null hypothesis can be rejected. But how low is sufficiently low? The answer to this lies in the next section.

## 5.6  Significance levels

I posed the question in the last section of how low should your $p$ value be before you can say that your results support your hypothesis. This is not an easy question to answer, since it depends on the nature of your experiment.

Let us imagine you are comparing the healing rates of a superficial wound when dressed by a plaster or left open. You predict that exposed wounds heal quicker. You carry out your experiment, collect your data and analyse your results. You end up with a $p$ value of 10%. This means that there is a 10/100 or 1 in 10 chance that random errors could account for the results. Because no one would suffer greatly if your results were the product of random errors, you would probably recommend leaving all superficial wounds undressed, since the vast majority healed more quickly with this treatment and those few that did not seemed to suffer no adverse side-effects.

However, let us suppose instead, that leaving wounds exposed produced nasty side-effects – gangrene which necessitated the amputation of the affected limb. This means that 10% of all patients who followed your recommendations would end up with amputations. You would now be very much less likely to recommend leaving wounds exposed on the basis of the 10% error margin derived from your previous set of results because the effects of the random errors are so appalling. You probably would only suggest exposing wounds if your results had yielded a much smaller $p$ value, say 1/1000, or 1/10 000. This would mean that the possibility of error creeping in would be infinitely reduced, and so the disastrous effects on the patients much less likely to occur.

What all this means is that the effect of the random errors in your experiment will dictate how small your $p$ value must be before you can claim that your hypothesis has been supported. If the effects are minimal, the $p$ value can be fairly large; if the effects are adverse, then the *p* value must be much smaller. You should select your cut-off *p* value before starting your experiment.

This all seems to be very unhelpful and rather vague. However, a good guideline for any research which does not have any dangerous side-effects is a cut-off *p* value of 5%. What this means is that if your results have a probability of 5% or less, you can claim that your hypothesis has been supported, since there is only a 5/100 (or less) chance that random error could account for your results. We call this cut-off *p* value the **significance level**, because if your results have a probability which is equal to or less than the selected cut-off point, we say that *the results are significant*. This means that we can reject the null hypothesis and say that the experimental hypothesis has been supported. If your results have a probability which is larger than the selected cut-off point, then the results are said to be not significant. In such a case the null hypothesis has to be accepted and not rejected, because your results do not support the prediction in your experimental hypothesis.

---

**Summary of key points**
1 When you have analysed your results using a statistical test, you will end up with a *p* value which tells you how likely it is that your results are due to chance or random errors. The smaller the *p* value, the greater the support for your hypothesis. If your *p* value is very small then you can say the experimental hypothesis has been supported.
2 The experimenter should decide how small the *p* value must be before this is concluded. This decision is based on the potential effects of the random errors. If the effects are likely to be adverse, the cut-off *p* value should be very small; if they are likely to be minimal, then the cut-off *p* value can be larger. The cut-off *p* value is called the significance level, because if the *p* value obtained in your study is equal to or smaller than the selected cut-off point, your results are said to be significant. If the obtained *p* value is larger, the results are said to be not significant.
3 The standard cut-off *p* value selected for most studies is 5%.

---

*Exercises (answers on page 312)*

1 Imagine you have decided to test the hypothesis: 'Burns caused by dry heat take longer to heal than those caused by wet heat'.
  (a) Identify and list the sources of constant error and state how you would control them.
  (b) Try to list as many possible sources of random error and say how you would deal with these.
2 You are going to carry out some research to test the hypothesis: 'Patients in

side-wards are more likely to suffer depression in the immediate post-operative period, than patients in main wards'.
(a)  Identify the sources of constant errors and how they could be controlled.
(b)  List as many sources of random error as you can think of and state how you would try to reduce their effects.

3  Which of the following $p$ values implies stronger support for your hypothesis?
(a)  $p = 5\%$ or $3\%$
(b)  $p = 7\%$ or $9\%$
(c)  $p = 0.1\%$ or $1\%$

4  It has already been said that $p$ values can be expressed either as decimals or as percentages. Convert the following $p$ values to percentages:
(a)  $p = 0.01$
(b)  $p = 0.05$
(c)  $p = 0.001$
(d)  $p = 0.5$
(e)  $p = 0.1$
Convert the following $p$ values to decimals:
(f)  $p = 5\%$
(g)  $p = 3\%$
(h)  $p = 1\%$
(i)  $p = 15\%$
(j)  $p = 2\%$

5  Imagine you saw the following $p$ values in some nursing research articles. What would they mean in terms of the percentage probability of random error accounting for the results?
(a)  $p = 0.01$
(b)  $p = 0.07$
(c)  $p = 0.03$
(d)  $p = 0.05$
(e)  $p = 0.1$

6  Using the standard significance level, which of the above $p$ values would indicate that the results were significant?

7  Look at the $p$ values in Exercise 5 and put them in order of greatest support for the experimental hypothesis (greatest support at the start, least support at the end).

## 5.7  Key words and terms

Constant errors
Counterbalancing
Double-blind technique

Experimenter bias
Fatigue effects
Order effects
Practice effects
Probabilities or $p$ values
Random errors
Significance level
Sources of error

# 6    Collecting the data: types of measurement

It was mentioned in Chapter 4 that when you carry out any research, you will be collecting facts – or **data**. For example, you might want to know how many right hip replacements are carried out as opposed to left hip replacements. The numbers you end up with for each hip are your data. Or you might want to know whether the time taken to discharge from hospital is longer if the patient recovers from by-pass surgery in a side ward or a main ward. The number of days (or weeks) to discharge is your data. Or you may want to compare the weight of newborn babies whose mothers have had iron and vitamin supplements during pregnancy with those whose mothers have received no such supplement. The weights are your data.

Have a look at the hypotheses you have devised and decide what is your data. Now, if you look at both your own and the above examples, you will probably see that the type of data you have in each case is quite different:

1. Number of left v. right hip replacements.
2. Number of days to discharge.
3. Weight (in lb and oz or kg).

You might have added to this, from your own list, data such as the following:

4. Height (in feet and inches or metres).
5. Distance walked.
6. Numbers of nurses.
7. Types of patients.
8. Professional status.
9. Milligrams of a drug.

## 6.1  Levels of measurement

You can see from this that metres, kilograms, milligrams and number of days to discharge are all different forms of measurement. However, while the actual types of data probably run into their hundreds, they all belong to one of four categories or **levels of measurement** (sometimes called *scales* of measurement).

Whatever data you collect, you must be sure which level of measurement it belongs to, because you may wish to analyse your data using a statistical test, and some statistical tests can only be used with certain levels of measurement.

Each level of measurement implies that a different amount of information can be obtained from your data and this can be outlined in the following way:

1. The **nominal level** of measurement implies *least* information can be obtained from your data.
2. The **ordinal level** of measurement gives you all the information provided by the nominal level plus a bit more.
3. The **interval level** of measurement gives you all the information provided by the nominal and ordinal levels, plus a bit more.
4. The **ratio level** of measurement gives you most information – all the information of the nominal, ordinal and interval levels, plus a bit more.

So there is an increasing amount of information provided by each level:

1. Nominal | least information
2. Ordinal
3. Interval
4. Ratio     ↓ most information

However, for the purposes of statistical tests, the interval and ratio levels are usually amalgamated to form a single category. After describing what these levels mean, they will be dealt with as one level for the rest of the book.

Let us look at each level of measurement more closely to find out what is meant by 'obtaining information from the data'.

## 6.2  The nominal level of measurement

This category, as you will recall, derives *least* information from your data, and as you have probably deduced from the title, is a naming category. What this means in reality, is that a nominal level of measurement allows you to allocate your data into named categories, *only*. Once the data are in a named category, there is no implication about size, worth, value or anything else. An everyday example might clarify this.

During the run-up to the 1987 General Election the public was bombarded with the results of opinion polls, which typically provided information about the percentage of people expected to vote Conservative, Labour or SDP/Liberal Alliance. In order to achieve these results, the pollsters usually asked a sample of, say, 1000 people which party they intended to support. According to their reply, each person's response was allocated to a category named Conservative, Labour or SDP/Liberal Alliance. In other words, the data obtained during an opinion poll were of a nominal level, because they were simply placed in named categories.

Now there are a number of reasons why this level of measurement does not allow us to infer much information about our data. These are as follows. Firstly (despite personal opinions!) there is no *absolute* or commonly understood level of worth implied by the categories. That is, conservatism is not inherently better than socialism, which in turn is not inherently better than the Alliance. This can be contrasted with data measured, for example, in feet and inches, where six feet is *always* and *absolutely* bigger than five feet.

Because there is no inherent and commonly agreed worth to the category labels, it means that the labels could be altered and yet the same results would *still* be obtained. So Conservative could be relabelled capitalist, Thatcher, right of centre, Party A; Labour could be relabelled socialist, Kinnock, left of centre, Party B; etc. and the outcome would be the same.

If we look back at the comparison I have just made with measurement in feet and inches, this point quite clearly does not apply. If we relabel feet and inches as metres, then despite the fact they both apply to length, the same results would not occur: 3 feet and 3 inches would become 1 metre.

Secondly, there is no grading of *degree* in the responses in each category. Of the sample, 42% may have said they would vote Tory, but from that alone, we have absolutely no notion of how far to the political right they are, or whether they are just bordering on voting Alliance instead; how long they have been a Tory; how firm their commitment is. All we know from this type of data is that the respondents belong to a particular category and that is all. Therefore, no numerical value is attached to the nominal categories.

One important point derives from this. The respondent can only be allocated to *one* category – once they have been put in, say, the Labour category, they cannot be put in the Conservative category as well. In other words, nominal categories are mutually exclusive.

Let us apply this concept a bit more closely to nursing. Imagine you are working on a cardiology ward and you decide you would like to look at the recovery rates from myocardial infarction of patients who smoke as opposed to those who do not. You decide to take a cut-off point of 10 days from admission and you simply ask those patients who are discharged within this period: Do you smoke? Yes/No.

All you are doing here is allocating the patient's response to a Smoke/Don't smoke category. From the data you have very little information. You do not know, for example: how many cigarettes the smokers smoke; how long they have been smoking; whether they smoke low, medium or high tar cigarettes; whether the non-smokers have ever smoked, and if so, how long they have refrained from smoking. All you know from this level of measurement is that a certain number of patients can be labelled 'smoker', and a certain number 'non-smoker'.

The following examples are all nominal measures:

1. Number of males v. females with carcinoma of the bowel. In this instance, you might take fifty patients admitted with carcinoma of the bowel, and just allocate them to either the male or female category.
2. Number of successful v. unsuccessful liver transplants. Here you might look at all the liver transplants carried out in a given hospital over the past twelve

months and allocate those who survived past the first three months to the 'successful' category, and those who failed to survive this period to the 'unsuccessful' category.
3. The type of anaesthesia given to women in labour in a particular maternity hospital. Here you might allocate patients to epidural anaesthesia, inhalation anaesthesia or 'tens'.

## 6.3  The ordinal level of measurement

This level of measurement provides a bit more information from our data, because it allows us to *order* the data according to a *dimension*. For example, we might be interested in assessing the clinical competence of nurses trained on an RGN course with those trained on a polytechnic degree course. Here we would be interested in a dimension of

most competent ⟶ least competent

Or we might be interested in the degree of healing of ulcerated legs when treated by one of two methods. Here the dimension we would be measuring would be

most healed ⟶ least healed

Or we might be carrying out some research on job satisfaction among staff nurses on two different wards. Here we would be assessing their responses on a dimension of

most satisfied ⟶ least satisfied

In other words, the ordinal level of measurement allows you to grade your responses on a dimension of *magnitude*.

The most usual way of operating an ordinal scale is by applying a *point scale*. For instance, we might compare the clinical competence of the RGN trainees with the polytechnic trainees by rating each student on the following scale:

|            1             |             2           |            3            |           4           |           5            |
| Completely incompetent | Fairly incompetent | Averagely competent | Fairly competent | Very competent |

We have here a dimension of magnitude from completely incompetent to very competent. However, it is essential to note that the differences between each point on the scale are not necessarily equal and must *not* be assumed to be so. In other words, the *actual* difference in competence between points 1 and 2 may not be the same as the difference in competence between points 3 and 4, even though the numerical difference in each case is the same (i.e. one point). All an ordinal scale of this type tells us is that someone who scores at 2 is more competent than someone who scores at 1. Similarly, someone who scores at 3 is more competent than someone who scores at 2 or 1. What we do not know is exactly how much more competent each student is. All we have is a relative assessment or measure. On this basis then, we might find that student A on the RGN course is fairly competent and so she would score 4. On the other hand, we may find that student 1 on the degree course was fairly incompetent and so would score at 2. This does *not* mean that the RGN student is *twice* as competent as the polytechnic student (a score of 4 as opposed to 2). All it means, is that she is *more* competent. Just to reiterate, an ordinal scale allows a measure of *relative* size, competence, healing ability or whatever and not an absolute measure.

It should be noted that usually an ordinal point scale has up to seven points on it. If it has more than seven points researchers tend to treat it as an interval scale. This will be explained in the next section.

Another, though less common, way of using an ordinal scale is by rank ordering your data according to the dimension you are interested in, without using a point scale. A typical example of this is the school report which states the pupil's position in class. For example, a pupil may be placed ninth in the class, which means that on a dimension of scholastic achievement, he has achieved less than the pupil placed eighth (but we do not know how much less) and more than the pupil placed tenth (although we do not know how much more). In this way, then, we might look at all the ulcerated legs in our sample, and rather than classifying them on a point scale of most–least healed, we might give Mrs A's leg a score of 1, because it is most healed, Mr B's leg a score of 2 because it is next most healed and so on.

How is it that the ordinal scale gives us more information than the nominal scale? Let us return to the example of patients with myocardial infarction. The nominal scale simply allocated the patients to a smoker/non-smoker category. As an alternative to this, we could use an *ordinal* scale, which employs, say, a five point scale in the following way:

| 1 | 2 | 3 | 4 | 5 |
|---|---|---|---|---|
| Do not smoke | Light smoker | Moderate smoker | Fairly heavy smoker | Very heavy smoker |

Each patient could classify himself or herself according to this scale. Now from these results we could derive *all* the information of the nominal scale by simply adding up: all the scores of 1 (non-smoker) and all the scores of 2, 3, 4 and 5 (smokers). This would give us the number of smokers v. non-smokers, i.e. all the information of the nominal scale. However, the ordinal scale has taken us a stage further, because it has allowed us to rank order out data along a dimension of smoking as well, i.e. light to very heavy. In other words we can grade the patients. We now know not just that *x* number of patients smoke, but also the number who are light, moderate, fairly heavy and very smokers. (Remember! Scoring at 2 (light smoker) does *not* mean this patient smokes *half* the number of the patient who scores at 4 (fairly heavy). It just means they smoke less.)

The following are examples of ordinal scales of measurement:

1. Least – most disabled multiple sclerosis patients.
2. Most – least pain using a new analgesic drug for pain following a hysterectomy.
3. Least – most respiratory problems among premature babies.

## 6.4  The interval level of measurement

The interval level of measurement allows the researcher to measure data along a scale on which the intervals between each point of measurement are assumed to be exactly the same. This is in contrast to the ordinal scale. However, the interval level is still only a *relative* measure and not an absolute one, in that although it enables the data to be compared directly, it does not provide any knowledge of the *absolute* magnitude of the data in question.

Let me simplify this. Suppposing you were interested in comparing the final theory examination performance in physiology of a group of students trained on a traditional RGN course, with those trained on one of the new RGN schemes. Your data are the *percentage* marks achieved by the students. Percentages are typical of an interval scale of measurement in that the difference between 10 and 25% is the same as the difference between 50 and 65%, i.e. 15%. Because of this, we could assume that the gap in performance of a student who achieved 62% and one who achieved 67% is exactly the same as the gap between the student with 67% and one with 72%. We can conclude this, because each 1% interval is assumed to be identical. As a result of this equivalence, we can say that the difference in performance of Student A with 50% and Student B with 70% is *twice* the difference of that between Student C with 65% and Student D with 55%.

In this way, we can make direct comparisons of the differences in scores. If you look back at the ordinal level of measurement, we cannot do this, because the intervals are *not* equal. Therefore the difference in smoking behaviour of a patient who scores 1 and a patient who scores 3 on an ordinal scale is *not* the same as the difference in smoking behaviour of a patient who scores 3 and one who scores 5 (i.e. two points in each case). In this way, the interval scale provides more information than the ordinal scale, because it allows us to rank order the students' performance according to percentage marks obtained, but *also* allows us to make comparisons of performance in this way.

It was said earlier that the interval scale does not imply any absolute measure. What this means, essentially, is that although we might have a measure of the students' performances on a physiology examination, this does not imply any absolute and exhaustive measure of their total physiology knowledge; it only provides a measure of what they were tested on in the exam. In order to have an *absolute* measure of physiology knowledge (if, indeed, it were possible) the question paper would have to cover every fact ever known in physiology. This obviously is not feasible, and so the physiology exam can only make a relative comparison of students' performance on a *sample* of physiology. Therefore, the measurement is still relative and not absolute. Even if a student got 0%, we cannot assume he or she has no knowledge of physiology – only that there were no questions about the bits he or she did know. Therefore, an interval scale does not have a true and meaningful zero point.

Just to clarify how the interval level provides more information than the nominal and ordinal scales, let us continue with the example of the physiology exam. From the percentages obtained we could classify students according to whether they passed or failed (nominal category), and in terms of their position in the group by rank ordering their performance (ordinal scale). But in addition, we could also make sensible comparisons of the differences between students' performances on the exam.

Sometimes researchers using a point scale (usually of more than seven intervals) assume equal distances between the points and therefore use it as an interval rather than ordinal scale. Sometimes this is quite acceptable, but if you wish to do this, you must make it clear that the underlying assumption is one of equal distances between the points.

The following are examples of an interval level of measurement:

1. Temperature – even if the temperature drops to 0° on a centigrade thermo-meter, this would still register as 32° on a Fahrenheit thermometer. This is because the 0° point has been arbitrarily imposed and does not mean there is *no* temperature. If you think about the weather at $-15°C$, this point becomes clearer. So, because temperature on a centigrade or Fahrenheit scale has no *absolute* zero point, and because the intervals between degrees are assumed to be equal, this is classified as an interval level.
2. IQ test results assume equal intervals between scores, such that the difference between an IQ score of 90 and 100 is the same as that between 100 and 110. Again there is no absolute zero – simply because someone does not score on an IQ test does not mean they have zero intelligence, but just that they did not register on an arbitrary test of ability. There is no *absolute* meaning to the term 'average intelligence'.

## 6.5  The ratio level of measurement

The ratio level of measurement, however, is an *absolute* measure and includes measurements such as feet and inches, metres, weight, time, volume, etc. Because of its absolute nature, it does have a true zero point and so we can meaningfully say that someone who can walk zero metres after a hip replacement operation has no mobility.

We can also say that someone who walks four metres after the operation has walked twice the distance of someone who only managed two metres. In contrast, it makes no sense to say that a student who got 0% in a physiology exam has no physiology knowledge or that someone who scored 60% on the exam has twice the knowledge of physiology of a student who achieved 30%.

If we go back to the smoking behaviour of myocardial infarction patients, we could use a ratio measure of their smoking, simply by asking how many cigarettes they smoked daily. From the responses, we could (a) identify the non-smokers and smokers (nominal level of measurement); (b) rank order the smoking on a dimension of none through light, moderate to heavy (ordinal); and (c) make *absolute* comparisons of the number of cigarettes smoked, such that the patient who smokes sixty a day smokes twice as many as someone who smokes thirty, and three times the quantity of someone who smokes twenty (ratio level of measurement). Therefore, the ratio level of measurement tells us a lot more about the smoking behaviour of these patients.

Do not be overly concerned with the distinction between interval and ratio scales of measurement, because for the purposes of statistical analysis, they are treated as similar and are combined to form a single category.

One last point is that many nurse researchers want to use *questionnaires* to collect their project data. Questionnaire data can be of nominal, ordinal or interval/ratio type. However, it must be stressed that designing a questionnaire is not simply a matter of dreaming up a few questions and administering these to your subject sample – it is an art in itself. There is no room in this text to provide details on questionnaire construction, but the reader is referred to Oppenheim (1966).

---

**Summary of key points**
When you carry out any research you will be involved in collecting data. Data can be categorised into one of four levels of measurement. Each level of measurement indicates that a different amount of information is provided by the data.

(i) *Nominal level* provides least information, and only allows the data to be placed into named categories.
(ii) *Ordinal level* provides a bit more information and allows the researcher to rank order the data according to the dimension the researcher is interested in.
(iii) *Interval level* provides yet more information in that it assumes equal intervals on the measurement scale and so allows the researcher to make direct comparisons of differences between pairs of scores. No absolute and true zero applies to this level because it is not an absolute measure in itself.
(iv) *Ratio level* provides most information in that it does give an absolute measure of the attribute we are interested in, and so has a meaningful zero point.

For the purposes of statistical analysis, the interval and ratio scales are combined to form a single category.

---

*Exercises (answers on page 313)*

1 Identify what level of measurement each of the following belongs to (i.e. nominal, ordinal or interval/ratio).
   (a) Volume of blood transfused in litres.
   (b) Types of cancer patient on an oncology ward.
   (c) Degree of job satisfaction on a five point scale.
   (d) Male v. female applicants for places at a school of nursing.
   (e) Vital capacity of a emphysema patient in ml.
2 Imagine you are carrying out a piece of research which involves monitoring the food intake of gastrectomy patients in the twenty-four hours post-operation. Devise ways of measuring the food intake using:
   (a) a nominal level of measurement;
   (b) an ordinal level of measurement;
   (c) an interval/ratio level of measurement.
3 You are involved in measuring the urinary incontinence of patients following prostatectomy. Again, devise a method of measuring this using:
   (a) a nominal level of measurement;
   (b) an ordinal level of measurement;
   (c) an interval/ratio level of measurement.

## 6.6  Key words and terms

Data
Interval level
Levels of measurement
Nominal level
Ordinal level
Ratio level

# 7    Selecting a statistical test to analyse your data

## 7.1  Deciding which test to use

So far we have looked at how to design a study to test your experimental hypothesis. Once you have designed your research and carried it out, you need to analyse the results to see whether or not they support your hypothesis. To do this, you will need to use statistical tests. Unfortunately there is no single test to analyse the results of every type of experiment – in fact there are dozens of statistical tests and how you have designed your research will determine which one of these you should select. Each of the research designs we have looked at so far has its own statistical test. It is your job as a researcher to decide which test to use, on the basis of the main design features of your experiment. A word of caution, though. If you select the wrong test to analyse the data from a particular design then your results will be null and void. So matching up the research design with the appropriate statistical test is critical if your study is to have any value.

This is a cardinal rule of statistics – each design has its own statistical test and you must select the correct test to analyse your data, otherwise the conclusions will be completely incorrect.

Many researchers have considerable difficulty with this matching task and there is evidence of this in several journal articles, where the incorrect test has been used. Do not worry unduly about this aspect of research, because as long as you ask yourself some key questions (which will be outlined shortly), you should arrive at the correct answer.

One further word of caution: do ensure, when you are designing your research that you know in advance how to analyse the data. Do not wait until you have collected your data before deciding which test to use. The reasons for this warning are twofold. Firstly, if you answer the key questions about the design as you are *planning* your study then there is less chance that you will select the wrong test than if you try to answer the questions retrospectively. Second, many researchers (and I confess to having been one of them) plan and carry out complex studies and then find, after months of hard work and data collection, that there is no available statistical test to analyse the results. While such a problem is unlikely if you stay within the designs outlined in this book, it becomes more of a possibility if you decide to carry out more advanced research. So decide (at the planning stage) which statistical test you will be using.

When deciding which test fits your design, you need to ask yourself four key questions:

1. Do I have a correlational or experimental design?
2. How many conditions or sets of data do I have?
3. Have I got a different, same or matched subject design? (experimental designs only)?
4. What level of measurement have I used?

While these issues have already been covered in some detail, it may be valuable to look a bit more closely at each of the questions in order to refresh your memory.

### 7.1.1  Do I have a correlational or experimental design?

This question refers to Chapter 3. If you recall, it was said that a correlational design did the following:

1. Looked for a link or association between the scores on the two variables in the experimental hypothesis.
2. Tested one of two predictions – either (a) that as scores on one variable went up, so scores on the second variable went down, i.e. *high* scores on variable 1 are linked with *low* scores on variable 2 (negative correlation) or (b) that as scores on one variable went up, so scores on the second variable also went up, i.e. *high* scores on variable 1 are associated with *high* scores on variable 2 (and similarly *low* scores on variable 1 are associated with *low* scores on variable 2). This is a positive correlation.

An experimental design, on the other hand did the following:

1. Manipulated one of the variables in the hypothesis (the IV) to see whether this had any effect on the other variable (DV).
2. Involved manipulating the IV by setting up two (or more) conditions all of which could be experimental, or alternatively could incorporate a control condition, and comparing the results from each to see if there were any *differences* between the conditions.

If you are still unclear about this distinction, re-read Chapter 3. When you have done this, look at the hypothesis you want to test and see whether it requires a correlational or experimental design. One of the easiest ways of deciding this, if it is not immediately clear from your hypothesis, is to identify the two variables in the hypothesis and ask yourself whether you are collecting a range of scores on *both* variables to see whether they are associated. If you are, then you have a correlational design. If not, you have an experimental design. This may make more sense if you look back at the sample hypotheses in Section 3.2.2.

If you have a correlational design, you need only ask yourself two further questions. Firstly, how many sets of data do I have? If you have three or more sets of data then you will use a different test from the one used if you have only two sets of data. You then have to ask which level of measurement has been used. If you turn to Section 7.1.4 you can refresh your memory on this. Alternatively, you can move to Chapter 12, where you will find information on statistical tests which should be used with various correlational designs. These tests are also given in Figure 7.1.

### 7.1.2  How many conditions or sets of data do I have?

If you are using an experimental design to test your hypothesis, you now have to decide how many conditions or sets of data you have in your experiment. If you look back to Section 3.2.1 you will note that these can be *just* experimental conditions (where each condition receives one form or level of the IV) or experimental and control conditions (where the control condition receives no IV at all). As each condition yields one set of data, you can think of this issue in terms of the number of sets of data you have. It was pointed out in Section 3.2.1 that you could compare the data from two conditions only, or more than two conditions. At this stage of deciding which test to use, you have to identify how many conditions you set up in your experiment – whether you used either two conditions only or more than two conditions. This is the only decision you have to make at this point, since experimental designs which employ two conditions are analysed by one form of test, while those which used more than two conditions need another form of test.

### 7.1.3  Have I got a different, same or matched subject design?

This question (for experimental designs only) refers to Chapter 4. Did you use just one group of subjects who took part in all the conditions (same subject design)? Or did you use two (or more) separate groups of subjects each of which took part in one condition only (different subject design)? Or did you use two (or more) groups of subjects, each of which did one condition only, but who were *matched* on the key factors that would influence the results (matched subject design?). Once again, you only have to decide between different subject design and same and matched subject design since it was pointed out in Chapter 4 that for the purposes of statistical analysis, same and matched subject designs were treated as one and the same thing.

### 7.1.4  What level of measurement have I used?

This, of course, refers to Chapter 6 and the nature of your data. If you recall, there are four levels of measurement which are important to you:

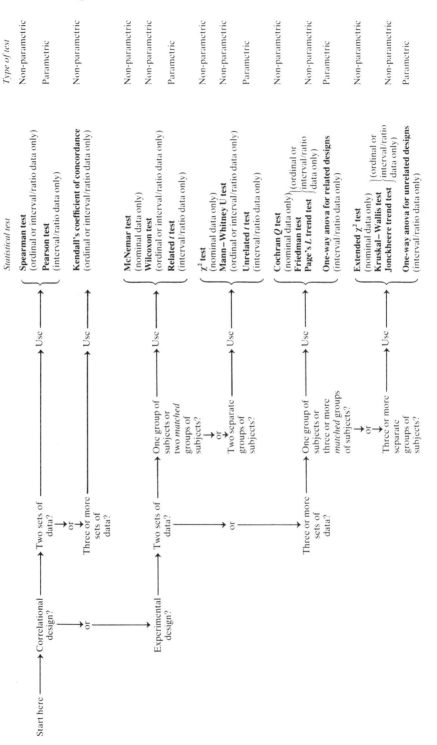

**Fig. 7.1**  Decision chart for selecting a statistical test.

1. Nominal – this allows you to allocate your subjects to named categories such as out-patient/in-patient, lumpectomy/radical mastectomy, etc.
2. Ordinal – this allows you to rank order your data according to the dimension in which you are interested, e.g. most–least, fastest–slowest, most satisfactory–least satisfactory.
3. Interval/ratio – this assumes equal intervals in your measurement scale and so allows you to make relative or absolute comparisons of your data, e.g. percentages, blood pressure, heart rate.

Knowing what type of measurement you have used is essential, since some statistical tests can only be used with certain levels of measurement.

Once you have answered these four questions you will be able to identify which statistical test is appropriate for your design, from Figure 7.1. Two points should first be made. Firstly, you will see that in most cases it appears that there are usually at least two tests that could be used to analyse the data from your research. Some of these are labelled 'parametric tests' and the rest 'non-parametric'. The significance of these terms will be explained in the next section. You will also see that, even within the headings of 'parametric' and 'non-parametric', there is occasionally more than one test option. This does *not* mean that you can select any of the tests indicated, but that each test requires a slightly different set of criteria before it can be used. These criteria are referred to briefly in the decision chart and more fully in the chapter which deals with those particular tests.

It might be useful at this point to look at the design you have chosen to test your own hypothesis, and using these questions as guidelines, select the test(s) that are appropriate from Figure 7.1. You will not be able to make a final choice until you have read the next section.

## 7.2  Parametric and non-parametric tests

If you look at Figure 7.1, you will notice that for any particular design, you usually have a choice of a **parametric test** or a **non-parametric test**. For example, if you were using an experimental design, two conditions, a same subject design and interval/ratio data you could use either a Wilcoxon test (non-parametric) or a related *t* test (parametric).

On what basis should you use the Wilcoxon rather than the related *t* test, and vice versa? To answer this question, the differences between parametric and non-parametric tests must be outlined. Basically, parametric tests are much more sensitive tools of analysis than non-parametric tests, in that if there is any support for your hypothesis in the data, the parametric test is more likely to pick it up. Thus, with the sort of design described above, the related *t* test would be more likely to highlight the differences in the results from your two conditions than would the Wilcoxon test. Two simple nursing analogies may confirm the difference between the two types of test. A patient's temperature can be monitored by touch or by clinical thermometer. If you wanted to find out whether a patient was suffering from pyrexia, you could either put your hand on his or her forehead or use a clinical thermometer in the axilla.

The first method is the nursing equivalent of the non-parametric test – it is a crude method of measurement which will give you a rough guide as to temperature change, but it is not as accurate as the clinical thermometer. The clinical thermometer will give you a much more precise index of the patient's temperature and so is the equivalent of the parametric test.

Similarly, if you had to check a patient for anaemia, you might either look at the inside of the lower eyelids and note the colour or, alternatively, take a haemoglobin count. The second method will give you a much more accurate reading and so is the equivalent of a parametric test. The eyelid technique will allow you to record changes, but is cruder and so is the equivalent of the non-parametric test.

In both these examples, the more sensitive nursing technique would usually be selected in preference, but they obviously require more time, skill and equipment – in other words, they are only used if certain conditions can be met. The same applies to parametric tests in statistics. They are preferable to non-parametric tests, but can only be used if four conditions or *parameters* can be fulfilled – hence their name. These four conditions are as follows:

1. The data from your study *must* be on an interval/ratio scale. Parametric tests cannot, under any circumstances, be used with nominal or ordinal data. This condition is vital and cannot be waived. The remaining three conditions are not quite as critical and may be interpreted more flexibly.
2. The data from your study should be (roughly) normally distributed. If you look at Chapter 13, you will see that the **normal distribution** curve is an inverted U-shape and is a concept which is fundamental to the assumptions underlying many statistical tests. To ascertain whether your data meet this requirement, you could plot a graph. If the data are approximately normally distributed, this condition or parameter will be met.
3. Your subjects should be randomly selected from the population from which they are derived. This point was covered in Chapter 4, and is an important factor in subject selection for research.
4. The variation in the results from each group or condition in your study should be more or less similar.

This concept of variation in results needs some clarification. If you look at the data from each condition in your study and note the lowest and highest scores in each case, the *difference* between these two figures should be similar for each. If it is not, then you should not use a parametric test. For example, if you were comparing the birth weight of cot-death babies with the birth weight of babies who did not die of cot-death, you might find the lowest and highest weights in each group as in Table 7.1.

In each case you take the lowest weight from the highest, to give you the *range*. In the first case (cot-death infants), the range is 3 lb 11 oz while in the second it is 3 lb 2 oz. The ranges are fairly similar and would allow you to use a parametric test.

**Table 7.1**

| Birth weight of babies who died from cot-death | | Birth weight of babies who did not die from cot-death | |
|---|---|---|---|
| Lowest | Highest | Lowest | Highest |
| 6 lb 14 oz | 10 lb 9 oz | 5 lb 10 oz | 8 lb 12 oz |

Had the ranges been 6 lb and 2 lb instead then they would have been too different to fulfil this condition.

Two important points should be raised here. Firstly, although the first condition (level of measurement) is absolutely essential to using a parametric test, the remaining three can be violated to some extent, without repercussions. The reason for the relative flexibility of these conditions is the 'robust' nature of parametric tests. In other words, as long as the data are interval/ratio and there are no *major* violations of the other three criteria, a parametric test may be used to analyse your results. So when choosing between a parametric and non-parametric test, concentrate primarily on the type of data you have.

Second, while parametric tests require stringency as to the type of measurement used, non-parametric tests are less fussy. They can normally be used with *any* level of data. This point will be clarified in the chapters on statistical tests. A second cardinal rule of statistics should be emphasised here. If you are not sure whether you can use a parametric test on your results, always choose the non-parametric equivalent since this is an error to caution.

## 7.3  Finding out whether your results support your hypothesis

When you calculate a statistical test on your results you will end up with a number. This number is looked up in a set of probability tables to find out whether or not random error could account for your results (see Section 5.4). Every statistical test has its own set of probability tables which can be found in Appendix C. The way you do this is outlined for each statistical test in the relevant chapter. Some tests require additional information, such as the number of subjects used or the number of **degrees of freedom** (also referred to as the d.f.). This is a fairly complex concept deriving from statistical theory and refers to the extent to which the scores can vary once certain restrictions have been imposed on them. These restrictions are derived from the nature of the distribution of the scores. Do not worry about all this, because although the concept is complex, the computation of the d.f. is very easy and you do not need to understand the theory in order to calculate it. If you are interested in the theory, Nunnally (1975) and Siegel (1956) provide a more detailed discussion.

Whatever additional information is required by a particular test, details will be provided as to how it should be computed in the relevant section.

However, one further piece of information about your hypothesis is required before you can look up the results of your statistical calculations. If you turn to Table C.2, for example, you will see at the top, two lines, which says 'Levels of significance for a one-tailed test' and 'Levels of significance for a two-tailed test'. This notion of **one- and two-tailed hypotheses** or tests is relevant here and will be discussed in some detail.

## 7.4  One- and two-tailed tests or hypotheses

There are thousands of hypotheses which can be formulated in nursing research; however, they can all be classified into one of two categories: one-tailed hypotheses or two-tailed hypotheses (sometimes referred to as one- and two-tailed *tests*). How a hypothesis is classified depends on how specific the prediction is in the experimental hypothesis. Some hypotheses are very precise in how they expect the results to turn out, while other hypotheses are rather vague. Let us clarify this.

If you formulated the hypothesis: 'Autoclaving is a more effective method of sterilising equipment than gamma irradiation' you are making a very precise prediction that *autoclaving* is *more* effective than gamma irradiation in equipment sterilisation. This, therefore, is a one-tailed hypothesis because you have predicted only *one* outcome of your study, namely that autoclaving is better than gamma irradiation sterilisation.

If you had not been as confident of the outcome of your study, you might have decided on a vaguer hypothesis which did not predict which method of sterilisation would be better. In this case you might have stated: 'Autoclaving and gamma irradiation are differentially effective in the sterilisation of equipment'. Here you have not predicted which method is better and so your results could show either that autoclaving is better than gamma irradiation or that gamma irradiation is better than autoclaving.

In this hypothesis, you have left your options open about the final results of your study – they could show either of the two outcomes indicated. Because you have predicted in your hypothesis that there could be either of *two* outcomes of your study, this hypothesis is *two-tailed*. The following hypotheses are all one-tailed:

1. Malignant tumours of the lung are more likely to metastasise than are malignant tumours of the rectum.
2. Ulcerative colitis is more likely to occur in people who have experienced prolonged severe stress than in people who have experienced minimal stress.
3. Acute renal failure is more likely to be reversible if it follows a period of prolonged vomiting as opposed to severe diarrhoea.
4. Women who have had genital herpes are more likely to develop carcinoma of the cervix than women who have not had genital herpes.

5. Women who smoke are more likely to develop carcinoma of the bladder than are non-smokers.

In each case the hypothesis predicts only *one* possible outcome for the results:

1. Lung tumours are *more* likely to metastasise than rectal tumours.
2. Ulcerative colitis is a *more* likely occurrence in people who suffer stress than in those who do not.
3. Acute renal failure is *more* likely to be reversible when vomiting is the cause rather than when diarrhoea is the cause.
4. Carcinoma of the cervix is *more* likely to be associated with a history of genital herpes than with no such history.
5. Carcinoma of the bladder is *more* likely to occur in smokers than in non-smokers.

Had you not felt sufficiently confident to make such precise predictions, you could keep your options open by making these hypotheses two-tailed. If we make them two-tailed, the hypotheses become:

1. Malignant tumours of the lung and malignant tumours of the rectum differ in their tendency to metastasise.
2. People who experience prolonged severe stress and those who experience minimal stress differ in their likelihood of developing ulcerative colitis.
3. The reversibility of acute renal failure differs according to whether the cause was prolonged vomiting or severe diarrhoea.
4. The chances of developing carcinoma of the cervix differ according to whether there is a history of genital herpes or no history of genital herpes.
5. Women smokers and non-smokers differ in their susceptibility to carcinoma of the bladder.

In each case, the results could go in either of two directions: for example, malignant tumours of the lung could be more likely to metastasise *or* malignant tumours of the rectum could be more likely to metastasise; similarly, smokers *or* non-smokers could be more susceptible to carcinoma of the bladder. It might be a good idea at this point to look at the hypothesis you are interested in testing and decide whether it is one- or two-tailed.

What is the relevance of this? If you turn again to Table C.2 you will see the levels of significance for one- and two-tailed hypotheses at the top of the table:

Level of significance for one-tailed test
0.05      0.025      0.01      0.005

Level of significance for two-tailed test
0.10      0.05      0.02      0.01

As you will remember these figures represent the probability that random error could account for your results, e.g. 0.05 means that there is a 5% chance that your results could be due to random errors or fluctuations. If you look at the levels of significance, you will see that the figures given for two-tailed hypotheses are, in each case, *twice* those given in the corresponding column for one-tailed hypotheses. Therefore, a significance level of 0.05 for a two-tailed hypothesis corresponds with a significance level of 0.025 for a one-tailed hypothesis. Therefore, any given result from the calculations of the statistical test would have a different significance level according to whether your hypothesis was one- or two-tailed. Let us take a quick example. Suppose you had used ten subjects in your study, had analysed the data with the Wilcoxon test and had ended up with a figure of 5 from your calculations. This figure must now be looked up in the probability tables associated with the Wilcoxon test to see if it represents a significant result. If you look down the left-hand column ($N$) to find the number of subjects (10) and then look across the row of four figures, you will see that a result of 5 is associated with a 0.01 level of significance for a one-tailed hypothesis and a 0.02 level of significance for a two-tailed hypothesis. This means that if your hypothesis had been specific in its predictions (i.e. one-tailed), the chances of random error accounting for your results would be 1 in 100; however, had your hypothesis been less precise (i.e. two-tailed), the chances of random error accounting for your results would be 2 in 100. In other words the chances of random error accounting for your results from a two-tailed hypothesis are always *double* those for a one-tailed hypothesis. This is why it is important for you to know whether your hypothesis is one- or two-tailed, because it will affect the significance of your results.

What are the reasons behind this? The answer is quite simple. If you are predicting that there is only *one* possible outcome for your results, then there is only one set of opportunities for random error to explain your results. If you make a two-tailed hypothesis, i.e. there could be either of two possible outcomes for the results, then there are *two* sets of opportunities for random error to explain your results. Hence the probability of random errors being responsible for the outcome of your study is *always* twice as great for a two-tailed hypothesis.

How do you decide whether to make your hypothesis one-tailed or two-tailed in the first place? Prior to carrying out any research you should look at all the relevant work that has been carried out in relation to your project (see Clifford and Gough 1990). The information in the associated research should guide you in formulating your hypothesis. For example, you might be interested in looking at staffing levels in relation to absenteeism in ward staff. So you would probably look at all the literature on workload, stress and job satisfaction in nursing. The chances are you would find reports that suggest that high workload, high stress and low job satisfaction all lead to high absenteeism rates. From this, you might reasonably conclude that low staffing levels would certainly increase workload and stress, while reducing job satisfaction, and so you would predict that low staffing levels would lead to high absenteeism. Thus your hypothesis would be: 'Low staffing levels lead to high absenteeism in nursing staff'.

This is a *one*-tailed hypothesis because you are making a very specific prediction about the outcome of your study (low staffing levels equal high absenteeism), and this prediction is based on the available literature on the subject, which has guided your expectations about the results.

On the other hand, you might want to look at the efficacy of skin creams on the prevention of stretch marks in obese patients. Again, you would look at the available research articles on the topic, but this time you might well find that the indications for your hypothesis are less clear-cut. For example, you would probably find some research (possibly carried out by the manufacturers of the product!) which suggests that certain skin creams are effective in stretch-mark prevention. On the other hand, there may also be considerable evidence that these preparations are useless (or even made the stretch marks worse). Here you have a divided body of evidence which gives you no indication as to the outcome of your own study. With this sort of split in the background literature, you should make a two-tailed hypothesis because the anticipated outcome of your study is less clear-cut.

---

**Summary of key points**

1 All the research designs covered in this book have their own appropriate statistical test with which to analyse the data. If the wrong test is used for a given design then the results will be null and void.

2 To select the correct test, four questions have to be asked:
   (a) Has an experimental or correlational design been used?
   (b) How many conditions (or sets of data) were involved?
   (c) Was a different, same or matched subject design selected (experimental designs only)?
   (d) What level of measurement was used?

3 For most of the research designs, either a parametric or a non-parametric statistical test will be appropriate.

4 Parametric tests are more sensitive instruments of analysis, but before they can be used, four conditions must be fulfilled, the most important of which is that the data must be on an interval/ratio scale.

5 When the data have been subjected to statistical analysis, the results of the calculations are looked up in a set of probability tables to find out whether the data supported the experimental hypothesis.

6 Before the results of the calculations can be looked up, the nature of the hypothesis's predictions must be ascertained, i.e. whether it is a one- or two-tailed hypothesis.

7 One-tailed hypotheses make specific predictions that there will be one outcome only to the results. Two-tailed hypotheses make vaguer predictions that there could be either of two outcomes to the study.

8 This point is relevant to the significance level of the results. Because two-tailed hypotheses predict either of two outcomes, there are two sets of opportunities for random errors to account for the results and so the probability of the results from a two-tailed hypothesis being the result of random error is twice that of for a one-tailed hypothesis.

---

*Exercises (answers on page 314)*

**1** Look at the following descriptions of research and decide which statistical test(s) is (or are) appropriate to analyse the data.

(a) You want to find out whether cleft palates are more common in children of mothers who drank more than three units of alcohol per day for the first twelve weeks of pregnancy. You randomly select thirty women who are tee-total, and a further thirty who drink in excess of your criterion and simply allocate their babies into either the 'cleft palate' category or the 'no cleft palate' category.

(b) You are interested in the large numbers of people from the local paint factory who develop lung diseases. You decide to test the hypothesis that there is a greater incidence of lung infection among this group than among the general population. You select two groups of twenty subjects each, one from the paint factory and one from the general population matched on certain key variables, such as age, sex, previous medical history, area of residence and smoking and compare them on the number of occasions they have experienced lung infections over the past five years.

(c) In order to test the hypothesis that the closer the proximity of beds on a ward, the greater the patient's stress levels, you select a number of patients whose beds vary in distance (as measured in feet and inches) from the adjacent beds and assess their stress levels on a six point scale.

(d) In order to test the efficacy of a new health education campaign aimed at encouraging people to stop smoking, you select one group of twenty-five people and assess how heavily they smoke (on a six point scale) *prior* to being exposed to the campaign and again following the campaign.

**2** Look at the following hypotheses and decide whether they are one- or two-tailed.

(a) The greater the physical contact sick infants receive, the faster their recovery.

(b) Women are more likely to be prescribed tranquillisers by male GPs than by female GPs.

(c) The amount of information given to orthopaedic patients about their treatment alters their fear levels.

(d) Ototoxic drugs taken during pregnancy increase the likelihood of deafness in the child.

(e) Direct flap and tubed pedicle flap skin grafts are differentially successful.

**3** Looking at the above hypotheses again, change those you have designated one-tailed into two-tailed hypotheses and those designated two-tailed into one-tailed hypotheses. (It does not matter in this latter instance what your predictions are.)

## 7.5  Key words and terms

Degrees of freedom
Non-parametric test
Normal distribution
One- and two-tailed hypotheses
Parametric test

# 8 Statistical tests for same and matched subject designs with two conditions only (experimental designs)

The statistical tests covered in this chapter are used to analyse data collected from experiments which have either same subject or matched subject designs:

1. A same-subject design, i.e. *one* group of subjects only is used; this group is tested on two occasions, or conditions. In other words, you would have collected two sets of data from one group of subjects. These two sets of data need to be compared using a statistical test to see if the group's performance differs on the two occasions. You have the design shown in Figure 8.1. This sort of design is a typical 'before and after' type experiment, common in medical and paramedical research.

2. A matched subjects design using two groups only, i.e. *two* groups of *matched* subjects are used; each group is tested on one occasion, or condition, only and the two sets of data thus derived are compared for differences between them, using a statistical test. You have the design shown in Figure 8.2. Do remember that a matched subject design means that pairs of subjects are selected who are 'identical' to each other on certain critical variables which might influence the results. One subject is then allocated to one group and the other subject to the second group. If you still feel unclear about this point, re-read Section 4.2.3.

In this chapter, three statistical tests are described: the McNemar test, the Wilcoxon test and the related *t* test. The first two tests are non-parametric while the related *t* test is parametric (look back to Section 7.2 if you need to refresh your memory on this distinction). Essentially, however, the type of measurement you

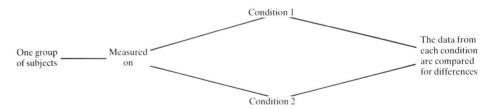

**Fig. 8.1** Same subject design.

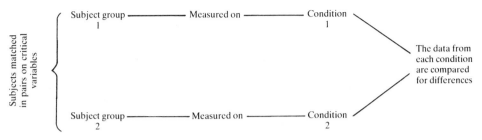

**Fig. 8.2** Matched subject design.

**Table 8.1** Summary table outlining the conditions required by the relevant statistical tests

| Test | Type of design | Conditions for use |
|------|----------------|--------------------|
| McNemar test | Same subject and matched subject designs | Can *only* be used with:<br>(a) nominal data;<br>(b) *two* sets of data. |
| Wilcoxon test | Same subject and matched subject designs | Can *only* be used with:<br>(a) ordinal or interval ratio data;<br>(b) *two* sets of data;<br>and should be used when the conditions required by a parametric test cannot be met (see section 7.2) |
| Related *t* test | Same subject and matched subject designs | Can *only* be used with:<br>(a) interval/ratio data;<br>(b) *two* sets of data;<br>and when the other conditions required by a parametric test can be met |

have will determine which of these three tests you can use, since the McNemar test can only be used with nominal data (i.e. your subjects' responses or scores can only be allocated to *named* categories, e.g. pass/fail, low/high, good/bad, etc.); the Wilcoxon test can only be used with ordinal or interval/ratio data and the related *t* test can only be used with interval/ratio data.

We summarise this information in Table 8.1. Despite these slightly different requirements, all the tests do the same basic job – they compare the two sets of data derived from either design for significant differences between them.

## 8.1 Non-parametric tests

### The McNemar test for the significance of changes

The McNemar test should be used under the following conditions:

1. Where *two* sets of data have been obtained from either a same subject or matched subjects design.
2. The data are of a *nominal* type.

The McNemar test tells you whether there are significance differences between the two sets of data.

When you compute the McNemar test you will end up with a numerical value for $\chi^2$, which you then look up in the probability tables associated with the McNemar test. This will tell you how probable it is that your results are due to random error. If this probability is 5% or less, your results are said to be *significant* since it is unlikely that random error could account for them. The experimental hypothesis would be supported and the null (no relationship) hypothesis could be rejected. If it is greater than 5%, your results *could* be explained by random error, and you would have to accept your null (no relationship) hypothesis. The results in this case would not be significant.

**Example**
Burns patients must have dead skin removed in order to allow new skin to grow. The process of removal is known as 'debriding' and is often very painful. However, you have noticed that if burns patients are allowed to help with the debriding, then the degree of pain they experience appears to be reduced – possibly because they feel more in control of the situation. You decide to test out this observation in an experiment.

*Hypotheses*

> $H_1$: Self-debriding reduces the level of pain experienced by burns patients.
>
> $H_0$: There is no relationship between self-debriding and the degree of pain experienced by burns patients.

The $H_1$ is a *one*-tailed hypothesis because we are predicting a *reduction* in pain with self-debriding.

*Brief outline of study*
Fourteen burns patients are selected and their reported level of pain monitored when debriding is carried out by a nurse. They are classified as having either moderate (M) or severe (S) pain. This is a nominal level of measurement since you have simply allocated the patients to non-numerical, i.e. named, categories. You repeat this assessment of their pain level when they do their own debriding. You have the design shown in Figure 8.3. (If you were carrying out this experiment in reality, you would need to consider the constant errors, such as type, extent and position of burn, as well as any possible order effects if nurse debriding and self-debriding are to be carried out on different days.) You are interested in whether pain levels differ according to who does the debriding.

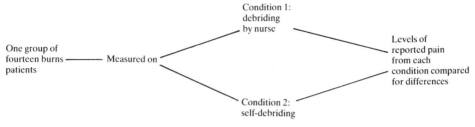

One group of
fourteen burns —— Measured on
patients

Condition 1:
debriding
by nurse

Condition 2:
self-debriding

Levels of
reported pain
from each
condition compared
for differences

**Fig. 8.3**

*Results*

You obtain the results in Table 8.2 (M = moderate; S = severe).

**Table 8.2**

| Subject | Condition 1: pain during debriding by nurse | Condition 2: pain during self-debriding |
|---------|---------------------------------------------|-----------------------------------------|
| 1 | S | M |
| 2 | S | M |
| 3 | S | M |
| 4 | S | S |
| 5 | S | M |
| 6 | S | S |
| 7 | S | S |
| 8 | S | M |
| 9 | S | M |
| 10 | S | M |
| 11 | S | M |
| 12 | S | M |
| 13 | S | S |
| 14 | S | M |

*Calculating the McNemar test*

PROCEDURE

1. Count up the total number of subjects who changed from moderate pain (M) in condition 1 to severe pain (S) in condition 2.

2. Count up the total number of subjects who changed from severe pain (S) in condition 1 to moderate pain (M) in condition 2.

WORKED EXAMPLE

1. Total number of subjects who changed from M to S = 0

2. Total number of subjects who changed from S to M = 10

3. Count up the total number of subjects who had moderate pain (M) in both conditions.
4. Count up the total number of subjects who had severe pain (S) in both conditions.
5. Check your calculations are correct using the following formula: step 1 + step 2 + step 3 + step 4 = total number of subjects in study.

3. Total number of patients who remained in category M on conditions 1 and 2 = 0
4. Total number of patients who stayed in category S in conditions 1 and 2 = 4
5. $0 + 10 + 0 + 4 = 14$

Arrange your calculations in the following table:

|  | Condition 1 (debriding by nurse) | |
|---|---|---|
|  | M | S |
| Condition 2 (self-debriding) S | Cell A | Cell B |
| Condition 2 (self-debriding) M | Cell C | Cell D |

|  | Condition 1 (debriding by nurse) | |
|---|---|---|
|  | M | S |
| Condition 2 (self-debriding) S | 0 | 4 |
| Condition 2 (self-debriding) M | 0 | 10 |

where  cell A contains results of Step 1
       cell B contains results of Step 4
       cell C contains results of Step 3
       cell D contains results of Step 2

---

*Special point 1*

You *must* arrange your table such that the top left-hand cell does *not* have the same label for condition 1 as for condition 2. In other words, the following arrangements are wrong:

|  | Condition 1 | |
|---|---|---|
|  | M | S |
| Condition 2 M |  |  |
| Condition 2 S |  |  |

|  | Condition 1 | |
|---|---|---|
|  | S | M |
| Condition 2 S |  |  |
| Condition 2 M |  |  |

Therefore, the following arrangements *only* are acceptable when organising your table.

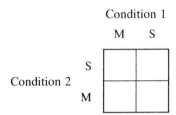

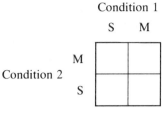

However, you must ensure that you put in the appropriate figures for each cell.

## Special point 2

Try to ensure that $\frac{1}{2}$(Cell A + Cell D) = at least 5. If it does not, collect more data until this condition is satisfied. Here,

$$\frac{0 + 10}{2} = 5$$

∴ the McNemar calculations on these data may continue.

6. Find $\chi^2$ from the following formula:

$$\chi^2 = \frac{([A - D^*] - 1)^2}{A + D}$$

where $A$ is the number in Cell A
$D$ is the number in Cell D

6.

$$\chi^2 = \frac{([0 - 10] - 1)^2}{0 + 10}$$

where $A = 0$
$D = 10$

$$\chi^2 = \frac{9^2}{10}$$

∴ $\chi^2 = 8.1$

−10 becomes +10 after evaluating the square brackets.

\* If you should get a minus figure from the calculations in the square brackets, ignore the minus sign and treat the number as a plus number.

7. Find the degrees of freedom (d.f.). For the McNemar test, d.f. always equals 1.

7. d.f. = 1

*Looking up the results in the probability tables*
The results of our calculations have produced a $\chi^2$ value of 8.1, with d.f. = 1. This value must be looked up in the probability tables (Table C.1) related to the

McNemar test, to see if it represents a significant difference in the patients' pain levels.

Turn to Table C.1. Across the top, you will see the words 'Level of significance for a two-tailed test', with probability levels from 0.10 (10%) to 0.001 (0.1%) underneath. Down the left-hand column you will see degrees of freedom values from 1 to 30.

Look down this column for our d.f. value of 1. To the right of this are five numbers:

$$2.71 \quad 3.84 \quad 5.41 \quad 6.64 \quad 10.83$$

These numbers are called 'critical values' of $\chi^2$ and each is associated with the probability level indicated at the top of its column (e.g. the probability value for 6.64 is 0.01 or 1%), For our $\chi^2$ value of 8.1 to be associated with one of these probability levels, it must be *equal to* or *larger than* the associated critical value.

If the $\chi^2$ value is *equal* to one of the critical values, then the associated probability level would be *exactly* the same as that for the critical value. For example, if you had obtained $\chi^2 = 3.84$, the relevant probability level would be exactly 0.05. This would be expressed as $p = 0.05$. If, however, your $\chi^2$ value is larger than the critical value, then the associated probability level is even less than that indicated. For example if you had obtained $\chi^2 = 4.02$, the relevant probability would be *even less than* 0.05 because 4.02 is larger than 3.84. You would express this as

$$p < 0.05 \ (< \text{ means 'less than')}$$

Our $\chi^2$ value is larger than 6.64 which means that its associated probability level is *even less than* 0.01 or 1% (because 8.1 is larger than 6.64).

*But* if you look again at the top of the $\chi^2$ table, you will see that it is entitled 'Levels of significance for a *two*-tailed test'. We have a *one*-tailed test or hypothesis, because a *reduction* in pain level was predicted. If you look back to Section 7.4, you will see that a two-tailed hypothesis has twice the probability of random error that a one-tailed hypothesis has. In other words, for a one-tailed hypothesis we have to *halve* the probability value associated with the two-tailed hypothesis.

Therefore, for $\chi^2 = 8.1$, the associated probability for a two-tailed test is <0.01, *but* for a one-tailed test it is <0.005, i.e. *half*. Therefore, the *p*-value associated with our $\chi^2$ value for this hypothesis is <0.005.

So if you have a one-tailed hypothesis, simply halve the appropriate two-tailed probability value.

*What do the results mean?*
The probability value associated with our results is less than 0.5%. This means that there is less than 0.5 chance in 100 that our results could be explained by random error. In Chapter 5 it was noted that the usual cut-off point for claiming results to be significant is 5% or less, as long as the subjects are in no danger.

Using this cut-off point in the present study, we can claim that our results are significant. But before we can assume that the experimental hypothesis has been supported, the original data *must* be checked, to see if pain levels were lower for self-debriding, as predicted. The reason for doing this is simple. Sometimes it is possible to obtain results which are statistically significant, but which are in the *opposite* direction from that predicted in the hypothesis. In other words, self-debriding could have *increased* pain, and the results could still have been significant, since all the McNemar does is to identify whether the two sets of results differ significantly from each other. So you must check your data to see if it is in the direction predicted. In the present example, there were *no* changes from M to S, but ten changes from S to M, so the data confirm the hypothesis. Should your results ever be significant, but your data in a direction opposite to that predicted, your hypothesis has *not* been supported. In this example, the null (no relationship) hypothesis can be rejected. Therefore, self-debriding does lead to a reduction in the pain level experienced by burns patients. The experimental hypothesis has been supported.

When you carry out this test on your own data, some queries might arise.

*Q1: What would it mean if the obtained $\chi^2$ value had been smaller than any of the critical values?*

Had your $\chi^2$ value been, say, 1.94 this would mean that it is smaller than the smallest critical value of 2.71. For the results from an experiment to be significant, the obtained $\chi^2$ value must be equal to or larger than the critical values in the table. In cases where the obtained $\chi^2$ value is smaller than the critical values, the results are not significant, and the null (no relationship) hypothesis would have to be accepted. Random error would account for your results.

*Q2: What would I do when looking up my results in the probability tables, had my hypothesis been two-tailed?*

If you had a two-tailed hypothesis you would look up the obtained $\chi^2$ value as described earlier; if it was equal to or larger than one of the critical values in the table, you would take the probability associated with that value, as specified at the top of the table. So, if you obtained a $\chi^2$ value of 5.41 for a two-tailed hypothesis, the associated *p*-value would be 0.02.

---

*Exercises (answers on page 315)*

**1** Look up the following $\chi^2$ values in the relevant probability table and state whether or not they are significant and at what probability level:
(a) $\chi^2 = 2.91$, d.f. $= 1$, two-tailed
(b) $\chi^2 = 3.84$, d.f. $= 1$, one-tailed
(c) $\chi^2 = 5.96$, d.f. $= 1$, one-tailed
(d) $\chi^2 = 4.03$, d.f. $= 1$, two-tailed
(e) $\chi^2 = 2.71$, d.f. $= 1$, one-tailed

**2** Calculate a McNemar test on the following results; state the $\chi^2$, d.f. and *p*-values and clarify what the results mean.

> **H$_1$:** Poorly controlled diabetic patients are more likely to follow a strict diet after being informed *in writing* of the potential problems of poor control.
>
> **H$_0$:** There is no relationship between the method of information – exchange and the likelihood of following a strict diet among poorly controlled diabetics.

*Brief outline of study*

A group of twelve middle-aged, poorly controlled diabetic women are randomly selected from a diabetic out-patient clinic. From their clinical notes, they are assessed on how closely they follow the recommended dietary guidelines. If they follow them closely, they are given a tick and if they do not follow them, they are given a cross. The subjects are then given a leaflet which outlines in words and pictures, the possible long-term complications of poor control. After one month, the subjects are reassessed in the same way, on how closely they follow the recommended diet. The results in Table 8.3 are obtained ($\checkmark$ = diet followed closely, $\times$ = not followed closely).

**Table 8.3**

| Subject | Condition A: before leaflet | Condition B: after leaflet |
|---|---|---|
| 1 | $\times$ | $\checkmark$ |
| 2 | $\checkmark$ | $\times$ |
| 3 | $\times$ | $\checkmark$ |
| 4 | $\times$ | $\checkmark$ |
| 5 | $\times$ | $\checkmark$ |
| 6 | $\times$ | $\checkmark$ |
| 7 | $\checkmark$ | $\times$ |
| 8 | $\times$ | $\checkmark$ |
| 9 | $\times$ | $\checkmark$ |
| 10 | $\times$ | $\checkmark$ |
| 11 | $\times$ | $\times$ |
| 12 | $\checkmark$ | $\times$ |

## 8.1.2 The Wilcoxon signed ranks test

The Wilcoxon test is used under the following conditions:

1. Where *two* sets of data have been obtained either from a same subject design or from a matched subjects design.
2. The data are ordinal or interval/ratio.

Like the McNemar test, the Wilcoxon simply compares the two sets of data obtained from your subjects and tells you whether there are significant differences between them.

When you calculate the Wilcoxon test, you find a numerical value for $T$, which you then look up in the probability tables associated with the Wilcoxon. If the probability associated with the obtained $T$ value is 5% or less, then your results are said to be *significant* as random error is unlikely to account for them. The experimental hypothesis would have been supported, and the null hypothesis could be rejected. If it is greater than 5%, then the results are not significant and the null (no relationship) must be accepted.

**Example**
Let us take an example to illustrate the Wilcoxon test. There have been great developments in colostomy appliances over the past few years. However, digestive tract enzymes may still produce considerable skin irritation around the stoma, where the bag comes into contact with the skin. Imagine that a manufacturer has developed a new adhesive – Stomaid (a fictitious name!) – which is claimed will virtually eliminate this problem. Given that such claims are likely to be heavily biased in the manufacturer's favour, it would be extremely useful to test the claim in an objective way. So, you set up the following hypotheses:

*Hypotheses*

> **H₁:** Stomaid reduces the degree of skin irritation around the stoma in colostomy patients.

> **H₀:** There is no relationship between Stomaid and the degree of skin irritation around the stoma in colostomy patients.

The $H_1$ is a one-tailed hypothesis because it predicts a *reduction* in skin irritation.

*Brief outline of study*
You select twelve colostomy patients, all of whom have some degree of skin irritation around the stoma. You assess this on a five point scale thus:

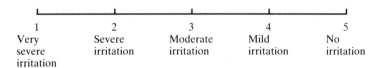

| 1 | 2 | 3 | 4 | 5 |
|---|---|---|---|---|
| Very severe irritation | Severe irritation | Moderate irritation | Mild irritation | No irritation |

This is an *ordinal* level of measurement, since it is ordering the patients' reactions on a dimension of irritation.

You then give the subjects (Ss) Stomaid and a set of instructions on its use. After two weeks you reassess the degree of irritation around the stoma. (What might the constant errors be and how would you eliminate them?) You have the design shown in Figure 8.4. This is a typical before and after type design, using one group of

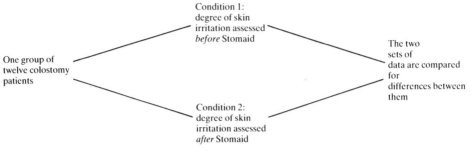

**Fig. 8.4**

subjects. Your interest is in whether the degree of skin irritation is reduced after using Stomaid.

*Results*
You end up with the data given in Table 8.4; fill in your calculations in the appropriate column. (Bold type figures are the raw data from the study, while the lighter type figures are the calculations from the Wilcoxon test.)

**Table 8.4**

| Subject | Data from experiment | | d | Calculations from Wilcoxon test | | |
|---|---|---|---|---|---|---|
| | *Condition A\*: degree of irritation before Stomaid* | *Condition B: degree of irritation after Stomaid* | | *Rank of d* | *Rank of +d values* | *Rank of −d values* |
| 1 | **1** | **3** | −2 | 5.5 | | −5.5 |
| 2 | **2** | **4** | −2 | 5.5 | | −5.5 |
| 3 | **1** | **4** | −3 | 8.5 | | −8.5 |
| 4 | **1** | **5** | −4 | 11.0 | | −11.0 |
| 5 | **3** | **2** | +1 | 2.5 | +2.5 | |
| 6 | **2** | **3** | −1 | 2.5 | | −2.5 |
| 7 | **1** | **4** | −3 | 8.5 | | −8.5 |
| 8 | **2** | **3** | −1 | 2.5 | | −2.5 |
| 9 | **2** | **5** | −3 | 8.5 | | −8.5 |
| 10 | **3** | **4** | −1 | 2.5 | | −2.5 |
| 11 | **1** | **1** | 0 | | | |
| 12 | **1** | **4** | −3 | 8.5 | | −8.5 |
| | ΣA = **20** | ΣB = **42** | | | Σ + d = 2.5 | Σ − d = 63.5 |
| | $\bar{x}$A = **1.67** | $\bar{x}$B = **3.5** | | | | |

\* It does not matter which condition is called A and which B, although in before/after designs like this one, it makes logical sense to call 'A' the 'before' condition.

Σ means 'total' and $\bar{x}$ means 'average'.

*Calculating the Wilcoxon test*

| PROCEDURE | WORKED EXAMPLE |
|---|---|
| 1. Add up the scores for condition A and enter the result as $\Sigma A$. | 1. $\Sigma A = 20$ |
| 2. Find the mean score for condition A and enter the result as $\bar{x}A$. | 2. $\bar{x}A = 1.67$ |
| 3. Add up the scores for condition B and enter the result as $\Sigma B$. | 3. $\Sigma B = 42$ |
| 4. Find the mean score for condition B and enter the result as $\bar{x}B$. | 4. $\bar{x}B = 3.5$ |
| 5. For each S, take their condition B score away from their condition A score and enter it in the column '$d$'. Put in the appropriate $+$ and $-$ signs. | 5. See '$d$' column above. |
| 6. Eliminate from the study any S whose $d$ score is 0. | 6. Eliminate subject 11 from all further calculations. |
| 7. Ignoring the plus and minus signs, rank order the remaining $d$ scores, giving a rank of 1 to the smallest, a rank of 2 to the next smallest and so on. Use the tied rank procedure (see Section A.2) if two or more $d$ values are the same. This procedure involves applying an average rank to those scores which are identical. | 7. See 'Rank of $d$' column. |
| 8. Put all the ranks associated with positive $d$ values in the 'Rank of $+d$ values' column. | 8. See 'Rank of $+d$ values' column. |
| 9. Put all the ranks associated with negative $d$ values in the column 'Rank of $-d$ values'. | 9. See 'Rank of $-d$ values' column. |
| 10. Add up the total of the ranks in the column 'Rank of $+d$ values' to give $\Sigma + d$. | 10. $\Sigma + d = +2.5$ |
| 11. Add up the total of the ranks in the column 'Rank of $-d$ values' to give $\Sigma - d$. | 11. $\Sigma - d = -63.5$ |
| 12. *Ignoring* the plus and minus signs, select the smaller of the two rank totals to be the value of $T$. | 12. $T = 2.5$ |

find the value of $N$, i.e. the total          13. $N = 11$
number of Ss (or pairs of matched
Ss) who took part in the study,
minus any who were eliminated at
step 6 (i.e. $d$ values of 0).

*Looking up the results in the probability tables*
The calculations from this Wilcoxon test have produced $T = 2.5$, with $N = 11$. This
$T$ value must be looked up in the relevant probability tables to see if it represents a
significant difference between the two sets of scores.

Turn to Table C.2. Down the left-hand column, you will see values of $N$, from 5
to 50. Look down this column until you see our $N$ of 11. To the right of this are the
following numbers:

$$14 \quad 11 \quad 7 \quad 5$$

These numbers are called 'critical values of $T$' and each is associated with the
probability level (or level of significance) indicated at the top of its column, such
that 11 is associated with a $p$ value of 0.025 (2%) for a one-tailed hypothesis, and
0.05 (5%) for a two-tailed hypothesis.

For our $T$ value to be significant at one of these levels, it must be either equal to
or less than one of the four critical values to the right of $N = 11$. If the $T$ value is
*equal* to one of these critical values, then it has an associated probability which is
*exactly* the same as that of the critical value. For example, $T = 11$ (for $N = 11$ and a
one-tailed hypothesis) has a probability level of 0.025 exactly. This is expressed as
$p = 0.025$ (or 2½%).

However, had the $T$ value been smaller than one of the critical values then the
probability value would be *even less* than the one specified. For instance, for $T = 9$,
$N = 11$ and a one-tailed hypothesis, the probability value is *even less* than 0.025,
because $T = 9$ is less than the critical value of 11. This is expressed as $p < 0.025$
($<$ means 'less than').

Our $T$ of 2.5 is less than the smallest critical value of 5. As we had a one-tailed
hypothesis, this means that $T$ also has an associated probability of even less than
0.005 (½%). This is expressed thus: $p < 0.005$.

*What do the results mean?*
The calculations of the Wilcoxon test on our data have produced a $p$ value of $<$½%
(or $<0.005$). This means that there is less than half a percent chance that the results
could be due to random error. Using the standard cut-off point of 5% (see Section
5.6), this means that there are significant differences between the two sets of data in
our experiment. The direction of these differences must be checked, however,
before we can conclude that our hypothesis has been supported. This is essential
because, on occasions, significant results can be obtained which are directly
opposite to those predicted, and so, of course, would *not* support the experimental
hypothesis. (This outcome is quite possible, because all of the Wilcoxon does is find
out whether there are significant differences between the two sets of data.)

If we look at the mean scores for each condition, we can see that $\bar{x}A = 1.67$ and $\bar{x}B = 3.5$, which means that there was less skin irritation after using Stomaid. This supports the experimental hypothesis, and so we can conclude that Stomaid reduces skin irritation in colostomy patients. The null (no relationship) hypothesis can be rejected.

Some questions follow which might arise when you perform a Wilcoxon on your own set of data.

> Q1. *What does it mean if the obtained* T *value is larger than any of the relevant critical values?*

To be significant at one of the levels stated at the top of the table, your *T* value has to be equal to, or smaller than, any of the relevant critical values. If it is larger, then your *T* value does not represent significant results. In other words, there is a high probability that random error could account for your results. Your experimental hypothesis would have to be rejected and the null (no relationship) hypothesis accepted.

> Q2. *Supposing I end up with a rank total of 0. Do I count this as the smaller value and use it as* T*?*

Yes – even if one of the rank totals is 0, it is still the smaller value and so would be used as the *T* value which you would look up in your probability tables.

---

*Exercises (answers on page 315)*

3 Look up the following *T* values in the probability tables and state whether or not they are significant and at what probability level:
   (a) $T = 37$, $N = 18$, two-tailed
   (b) $T = 68$, $N = 25$, two-tailed
   (c) $T = 12$, $N = 10$, one-tailed
   (d) $T = 281$, $N = 40$, two-tailed
   (e) $T = 29$, $N = 19$, one-tailed
4 Calculate a Wilcoxon test on the following results; state the *T*, *N* and *p* values. Explain what the results mean.

> $H_1$: Asthmatic patients suffer less severe asthma attacks when artificial E additives are eliminated from their diets.
>
> $H_0$: There is no relationship between the severity of asthma attacks and diet.

*Brief outline of study*
Fifteen asthma patients are randomly selected and the severity of their asthma attacks whilst on a normal diet are rated along a six point scale:

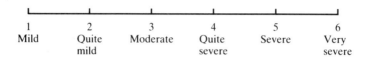

| 1 | 2 | 3 | 4 | 5 | 6 |
|---|---|---|---|---|---|
| Mild | Quite mild | Moderate | Quite severe | Severe | Very severe |

After one month on an additive-free diet, the severity of their attacks is again rated and the data in Table 8.5 are obtained.

**Table 8.5**

| Subject | Condition A: before additive-free diet | Condition B: after additive-free diet |
|---|---|---|
| 1 | 4 | 2 |
| 2 | 5 | 2 |
| 3 | 3 | 3 |
| 4 | 6 | 4 |
| 5 | 6 | 6 |
| 6 | 5 | 6 |
| 7 | 4 | 2 |
| 8 | 5 | 3 |
| 9 | 4 | 3 |
| 10 | 6 | 3 |
| 11 | 6 | 4 |
| 12 | 5 | 1 |
| 13 | 4 | 2 |
| 14 | 3 | 6 |
| 15 | 5 | 1 |

## 8.2 Parametric test

### 8.2.1 The related t test

The related $t$ test should be used when under the following conditions:

1. Two sets of data *only* have been collected from a same subject or matched subject design.
2. The data are of an interval/ratio nature.
3. The remaining three conditions required by a parametric test can be (reasonably) fulfilled.

The related $t$ test compares the two sets of data derived from either a same or matched subject design for significant differences between them. When calculating the related $t$ test, you end up with a numerical value for $t$ which is then looked up in the appropriate probability tables. The probability value obtained by doing this will tell you whether or not your results are significant. In other words, if this $p$ value is 5% or less, the results are said to be significant since it is unlikely that random error can explain them. The experimental hypothesis would be supported, and the null hypothesis rejected. If it is greater than 5%, the difference between the two sets of data is not significant. In this case, you would have to accept the null (no relationship) hypothesis and reject the experimental hypothesis. Random error would account for these results.

**Example**

Mentally handicapped people may show a progressive intellectual deterioration if they are institutionalised, unless care is taken to encourage and stimulate their activities. You decide to look at the effects of a short concentrated period of one-to-one interaction on the IQ of a group of mentally handicapped adults.

*Hypotheses*

> $H_1$: Short periods of concentrated one-to-one interaction increase the IQ level of mentally handicapped adults.
>
> $H_0$: There is no relationship between concentrated one-to-one interaction and IQ level among mentally handicapped adults.

The experimental hypothesis is one-tailed because an *increase* in IQ is anticipated.

*Brief outline of study*

Ten mentally handicapped adults are randomly selected and their IQ assessed. For one hour every alternate day, a nurse interacts intensively with each patient individually (playing, talking, singing, etc.). After three months, each patient's IQ is reassessed. You have the design shown in Figure 8.5. This design is a same subject before and after type design and the level of data is interval/ratio. Therefore a related $t$ test is appropriate (as long as the remaining three conditions required by a parametric test are not wildly violated).

(You should note the above description is a *very* brief resumé of the study. If it were to be carried out properly, a host of constant errors would need to be considered, e.g. age, type, sex of patient and nature of interaction.) You are interested in finding out whether the IQ levels are raised after the period of concentrated interaction.

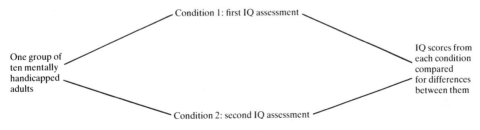

**Fig. 8.5**

*Results*

The results in Table 8.6 are obtained. (The bold type figures are the raw data from the experiment, while the lighter type figures are the calculations from the related *t* test. It does not matter which condition is A and which is B though, in a before/after design like this one, it makes sense to call the 'before' condition 'A'.)

**Table 8.6**

| Subject | Results from experiment | | Calculations from the related *t* test | |
| --- | --- | --- | --- | --- |
| | *Condition A:* *IQ scores before concentrated interaction* | *Condition B:* *IQ scores after concentrated interaction* | d *(A − B)* | d² |
| 1 | **54** | **58** | −4 | 16 |
| 2 | **49** | **59** | −10 | 100 |
| 3 | **60** | **65** | −5 | 25 |
| 4 | **62** | **60** | +2 | 4 |
| 5 | **58** | **62** | −4 | 16 |
| 6 | **45** | **50** | −5 | 25 |
| 7 | **55** | **64** | −9 | 81 |
| 8 | **63** | **67** | −4 | 16 |
| 9 | **47** | **59** | −12 | 144 |
| 10 | **50** | **58** | −8 | 64 |
| | **ΣA = 543** | **ΣB = 602** | Σd(A − B) = −59 | Σd² = 491 |
| | **x̄A = 54.3** | **x̄B = 60.2** | | |

Σ means 'total' and $\bar{x}$ means 'average'.

*Calculating the related* t *test*

Do not forget that parametric tests are more difficult to calculate than non-parametric ones. Keep calm when you look at the formula – as long as you follow the necessary steps you are unlikely to go wrong.

| PROCEDURE | WORKED EXAMPLE |
| --- | --- |
| 1. Add up the scores for condition A to give ΣA. | 1. ΣA = 543 |
| 2. Calculate the mean score for condition A to give $\bar{x}_A$. | 2. $\bar{x}_A$ = 54.3 |
| 3. Add up the scores for condition B to give ΣB. | 3. ΣB = 602 |
| 4. Calculate the mean score for condition B to give $\bar{x}_B$. | 4. $\bar{x}_B$ = 60.2 |
| 5. Take the condition B score from the condition A score for each subject to give $d(A - B)$. Put in the plus and minus signs as appropriate. | 5. See column $d(A - B)$ above. |

6. Add up the $d$ values in column $d(A - B)$ to give $\Sigma d$. Remember to take account of the plus and minus signs.

6. $\Sigma d = -59$

7. Square the $\Sigma d(A - B)$ value to give $(\Sigma d)^2$.

7. $(\Sigma d)^2 = 3481$

8. Square each $d$ value to give $d^2$.

8. See column $d^2$ above.

9. Add up the $d$ values to give $\Sigma d^2$.

9. $\Sigma d^2 = 491$

---

**Special point 1**

It is essential to distinguish between $(\Sigma d)^2$ (step 7) and $\Sigma d^2$ (step 9). $(\Sigma d)^2$ is the total of the $d(A - B)$ values, *squared*, while $\Sigma d^2$ is the total of all the individually squared $d$ values.

$\therefore (\Sigma d)^2 = (-59)^2$
$\quad \Sigma d^2 = 491$

---

10. Find the value of $t$ from the formula

$$t = \frac{\Sigma d}{\sqrt{\left(\dfrac{N\Sigma d^2 - (\Sigma d)^2}{N - 1}\right)}}$$

Where
$\Sigma d$ = the total of the $d$ values
$(\Sigma d)^2$ = the total of the $d$ values, squared
$\Sigma d^2$ = the total of the individually squared $d$ values
$N$ = the number of subjects or pairs of matched subjects
$\sqrt{\ }$ = the square root of the number resulting from all calculations under this sign

10.

$$t = \frac{-59}{\sqrt{\left(\dfrac{10 \times 491 - 3481}{10 - 1}\right)}}$$

Where
$\Sigma d = -59$
$(\Sigma d)^2 = 3481$

$\Sigma d^2 = 491$

$N = 10$

$$t = \frac{-59}{\sqrt{\left(\dfrac{4910 - 3481}{9}\right)}}$$

$$= \frac{-59}{\sqrt{158.78}}$$

$$= \frac{-59}{12.60}$$

$$= -4.68$$

> ### Special point 2
>
> If you end up with a negative *t* value just ignore the minus sign: $t = -4.68$ becomes $t = 4.68$.

| | |
|---|---|
| 11. Calculate the d.f. from $N - 1$ where $N$ is the number of Ss or pairs of matched Ss. | 11. d.f. $= 9$ |

*Looking up the results in the probability table*

The results of the related *t* test on these data have produced a *t*-value of 4.68. This value must be looked up in Table C.3, which is the probability value associated with the related *t* test. This will tell us the probability that our results are due to random error.

To look up $t = 4.68$, d.f. $= 9$, turn to Table C.3. Across the top you will see levels of significance for one- and two-tailed tests (or hypotheses), while down the left-hand side are d.f. values from 1 to infinity ($\infty$). Find our d.f. of 9 and to the right of this figure you will see the following numbers:

$$1.383 \quad 1.833 \quad 2.262 \quad 2.821 \quad 3.250 \quad 4.781$$

These are called 'critical values of *t*' and each critical value is associated with the *p* value indicated at the top of its column. So the critical value 2.821 is associated with a probability of 0.01 for a one-tailed test and 0.02 for a two-tailed test.

For our $t = 4.68$ to be associated with one of the probability levels indicated at the top of the table, it must be *equal to* or *larger than* the associated critical value. If the obtained *t* value is *exactly the same* as one of the critical values then it has exactly the same probability value. For example, had the obtained *t* value been 2.262 (one-tailed hypothesis) then the *p* value would have been 0.025 exactly. This would be expressed as $p = 0.025$. However, if the obtained *t* value had been 2.543, then it would have been larger than the critical value 2.262, and so the associated probability would have been even less than 0.025. This would be expressed as $p < 0.025$ ($<$ means 'less than').

Our *t* value of 4.68 is larger than the critical value of 3.250, and so it has a *p* value of *even less* than 0.005, for a one-tailed hypothesis.

*What do the results mean?*

The *p* value associated with $t = 4.68$, for a d.f. of 9 and a one-tailed hypothesis is less than 0.005. This means that there is less than ½% chance that the results obtained could be due to random error. Using the standard cut-off point of 5% or less (see Section 5.6), this suggests that the difference between the two sets of IQ scores is significant. However before we can say that the results support our experimental hypothesis, we must check that the difference in IQ scores is in the direction predicted. (It is possible to obtain significant results in the opposite direction to the one predicted and this would not support our hypothesis.) It was predicted that IQ scores would be *higher* after the period of intensive interaction and if we look at the mean scores for the before and after conditions, we can see that they do, in fact, support the prediction (mean IQ before = 54.3; mean IQ after = 60.2). We can conclude that concentrated one-to-one interaction leads to an increase in the IQ scores of mentally handicapped adults. The experimental hypothesis has been supported and the null (no relationship) hypothesis can be rejected.

Some queries might arise when you perform the related *t* test on your own data.

> Q1. *What happens if the obtained* t *is smaller than any of the appropriate critical values?*

This simply means that your results are not significant. Suppose you had obtained a *t* value of 0.987, for d.f. = 11. This is smaller than even the smallest critical value to the right of d.f. = 11. Given that the obtained *t* value must be *equal to* or *larger* than these critical values in order for the results to be significant, then we would have to conclude that our results were not significant. In this case the null (no relationship) hypothesis would have been supported, since random error would be accounting for your results.

---

*Exercises (answers on page 315)*

**5** Look up the following *t* values in the relevant probability tables and state whether or not they are significant and at what probability level.
 (a) *t* = 2.043, d.f. = 14, one-tailed
 (b) *t* = 1.478, d.f. = 19, one-tailed
 (c) *t* = 2.797, d.f. = 24, two-tailed
 (d) *t* = 4.179, d.f. = 11, one-tailed
 (e) *t* = 1.943, d.f. = 15, two-tailed
**6** Calculate a related *t* test on the results given in Table 8.7; state the *t*, d.f. and *p* values. Explain what the results mean.

> $H_1$: Relaxation techniques are effective in altering blood pressure in hypertensive patients.
>
> $H_0$: There is no relationship between relaxation techniques and blood pressure in hypertensive patients.

*Brief outline of study*
A group of twelve middle-aged hypertensive women were randomly selected for study. Their blood pressure was measured at the start of the study; they were then given daily relaxation exercises to carry out for six weeks. At the

end of that time, blood pressure was monitored again. Diastolic pressures only are to be compared. The results are as given in Table 8.7.

**Table 8.7**

| Subject | Diastolic pressure before relaxation | Diastolic pressure after relaxation |
|---------|--------------------------------------|-------------------------------------|
| 1       | 90                                   | 80                                  |
| 2       | 95                                   | 85                                  |
| 3       | 100                                  | 100                                 |
| 4       | 90                                   | 90                                  |
| 5       | 100                                  | 95                                  |
| 6       | 100                                  | 90                                  |
| 7       | 105                                  | 100                                 |
| 8       | 95                                   | 95                                  |
| 9       | 100                                  | 95                                  |
| 10      | 95                                   | 100                                 |
| 11      | 100                                  | 95                                  |
| 12      | 105                                  | 105                                 |

# 9 Statistical tests for same and matched subject designs with three or more conditions (experimental designs)

The statistical tests described in this chapter are all used to analyse data collected from studies that use either of the following designs:

1. A same subject design (i.e. *one* group of subjects only). One group of subjects is tested on *three or more* occasions (or conditions) and the results from each testing are compared for differences. In other words, there are three or more sets of data which have been derived from one group of subjects. The design is shown in Figure 9.1.
2. A matched subject design using *three or more* groups of *matched* subjects. Each group of subjects is tested on one occasion or condition only and the data derived from each group are compared for differences. Again, there are three or more sets of data, but each set is derived from one of three or more groups of matched subjects.

The design is shown in Figure 9.2. This sort of matched subject design is exceedingly hard to achieve, since it requires the experimenter to select 'triplets' of subjects (or if there are more than three groups, 'quads', 'quins' and so on), with the members of each set of triplets being 'identical' to each other on the key factors which could bias the results. Each member of the triplets is allocated to a different subject group. Because of the difficulties involved in using this design, a same subject design might be more suitable. Chapter 4 outlines these issues in more detail.

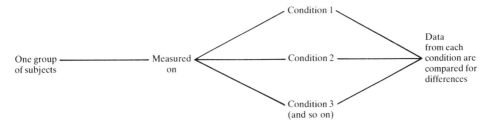

**Fig. 9.1** A same subject design with three or more conditions.

**Fig. 9.2**   A matched subject design using three or more matched groups.

This chapter deals with three non-parametric tests – the Cochran $Q$ test, the Friedman test and Page's $L$ trend test – and one parametric test – the one-way analysis of variance (anova) for related designs. In addition, a further test is described which can only be used *in conjunction* with the anova (the Scheffé multiple range test).

The conditions required by each test (and in particular, the level of data) are

**Table 9.1**  Summary table outlining the conditions required by the relevant statistical tests

| Test | Type of design | Conditions for use |
|------|----------------|--------------------|
| Cochran $Q$ test | Same subject and matched subject designs | Can *only* be used with:<br>(a)  nominal data;<br>(b)  three or more sets of data. |
| Friedman test | Same subject and matched subject designs | Can *only* be used with:<br>(a)  ordinal or interval/ratio data;<br>(b)  three or more sets of data;<br>and should be used when the conditions required by a parametric test cannot be met. |
| Page's $L$ trend test | Same subject and matched subject designs | Can *only* be used with:<br>(a)  ordinal or interval/ratio data;<br>(b)  three or more sets of data;<br>and when a *trend* in the results is predicted. |
| One-way analysis of variance (anova) for related designs | Same subject and matched subject designs | Can *only* be used with:<br>(a)  interval/ratio data;<br>(b)  three or more sets of data;<br>and when the remaining conditions required by a parametric test can be met (see section 7.2) |
| Scheffé multiple range test for use with related anovas | Same subject and matched subject designs | Can *only* be used:<br>(a)  when the anova for related designs has been calculated;<br>(b)  the results of the anova are significant. |

slightly different. These conditions are given in Table 9.1. While these conditions might look a bit overwhelming, they are really quite simple to understand, and will be described in more detail in the relevant sections that follow.

## 9.1 Non-parametric tests

### 9.1.1 The Cochran Q test

The Cochran $Q$ test should be used when *three or more* sets of *nominal* data have been obtained from either *one* group of subjects or from three or more groups of *matched* subjects. This test will tell you whether or not there are significant differences between the sets of data. It cannot tell you whether the results from one condition are better or worse than those from another. Because no direction to the results can be established using this test, any hypothesis associated with it must be *two-tailed*.

When calculating the Cochran $Q$ test, you find a numerical value for $Q$. This number is then looked up in the probability tables associated with the Cochran $Q$ test which will tell you the probability of your results being due to random error. If this probability (or $p$ value) is 5% or less, then it is unlikely that random error can account for your results. Your results are said to be *significant*, and the experimental hypothesis would have been supported. The null hypothesis can be rejected. If it is more than 5%, then random error could explain your results. Your experimental hypothesis would not have been supported and the null (no relationship) hypothesis would have to be accepted instead.

The Cochran $Q$ test should be used with a reasonable number of subjects or matched subjects. No specific official recommendation regarding numbers has been made, but a minimum of about fifteen is probably sufficient.

**Example**
Multiple sclerosis patients often have to catheterise themselves in order to ensure complete drainage of the bladder. However, this can lead to urinary tract infections, especially in female sufferers. You decide to assess the relative efficacy of rigid v. flexible catheters in preventing infection, since you think that the more flexible the catheter, the less the likelihood of damage to the urethra, with a consequent reduction in infection.

*Hypotheses*

$H_1$: The degree of flexibility of catheters affects the incidence of urinary tract infection in female multiple sclerosis sufferers.

$H_0$: There is no relationship between the use of flexible catheters and urinary tract infection in female multiple sclerosis sufferers.

The $H_1$ is a two-tailed hypothesis since it predicts only that there will be differences between the sets of data, and does not specify the nature of these differences.

*Brief outline of study*

Twenty-one female multiple sclerosis sufferers between the ages of 30 and 40 are randomly selected for study. All have been referred to the continence advisor because they have reached a stage in their condition when they will require regular catheterisation. The subjects are all requested to use a rigid, semi-rigid and flexible

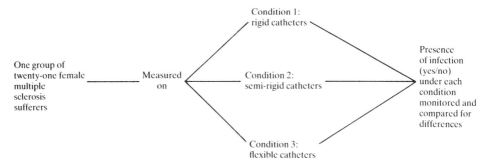

**Fig. 9.3**

**Table 9.2**

| Subject | Condition 1: rigid | Condition 2: semi-rigid | Condition 3: flexible | $T_S$ | $T_S^2$ |
|---------|------|------|------|------|------|
| 1 | 1 | 0 | 0 | 1 | 1 |
| 2 | 1 | 1 | 0 | 2 | 4 |
| 3 | 0 | 0 | 0 | 0 | 0 |
| 4 | 1 | 0 | 1 | 2 | 4 |
| 5 | 0 | 0 | 0 | 0 | 0 |
| 6 | 1 | 1 | 1 | 3 | 9 |
| 7 | 1 | 0 | 0 | 1 | 1 |
| 8 | 1 | 1 | 1 | 3 | 9 |
| 9 | 1 | 0 | 1 | 2 | 4 |
| 10 | 0 | 1 | 0 | 1 | 1 |
| 11 | 1 | 1 | 0 | 2 | 4 |
| 12 | 1 | 0 | 0 | 1 | 1 |
| 13 | 0 | 1 | 0 | 1 | 1 |
| 14 | 0 | 0 | 0 | 0 | 0 |
| 15 | 1 | 1 | 1 | 3 | 9 |
| 16 | 0 | 0 | 0 | 0 | 0 |
| 17 | 1 | 1 | 0 | 2 | 4 |
| 18 | 1 | 0 | 0 | 1 | 1 |
| 19 | 1 | 0 | 0 | 1 | 1 |
| 20 | 1 | 1 | 0 | 2 | 4 |
| 21 | 1 | 0 | 0 | 1 | 1 |
| | $\Sigma_1 = 15$ | $\Sigma_2 = 9$ | $\Sigma_3 = 5$ | $\Sigma T_S = 29$ | $\Sigma T_S^2 = 59$ |

$\Sigma$ means 'total'

catheter for one week each, but the sequence of use is counterbalanced to minimise order effects. At the end of each one-week period, each subject is assessed as to whether they have had any urinary tract infection. The responses are allocated to a yes or no category. As this is a nominal level of measurement and the twenty-one subjects are assessed on all of the three conditions (rigid, semi-rigid and flexible) the essential conditions required by the Cochran $Q$ test have been fulfilled. The design is shown in Figure 9.3. (There are a number of constant errors which would need to be controlled for if you were carrying out this experiment properly, e.g. individual skill with the catheter and presence of a helper). The focus of your interest is whether infection rates vary with the type of catheter used.

*Results*
The results in Table 9.2 are obtained (1 = yes, presence of infection; 0 = no, no infection). The bold type figures are raw data from the study, while the lighter type figures are from the calculations. It does not matter which condition is called 1, 2 or 3.

*Calculating the Cochran Q test*

PROCEDURE

1. Count up the total number of 'Yesses' in condition 1 to give $\Sigma_1$.
2. Count up the total number of 'Yesses' in condition 2 to give $\Sigma_2$.
3. Count up the total number of 'Yesses' in condition 3 to give $\Sigma_3$.
4. Add up the total number of 'Yesses' for each S, e.g. $S_1 = 1$, $S_2 = 2$. Enter the scores in $T_S$ column.
5. Square each $T_S$ value, e.g. for $S_1$, $1^2 = 1$. Enter the results in $T_s^2$ column.
6. Add up the scores in $T_S$ column, to give $\Sigma T_s$.
7. Add up the scores in the $T_s^2$ column, to give $\Sigma T_s^2$.
8. Calculate $Q$ from the formula

$$Q = \frac{(C - 1)[C(\Sigma_1^2 + \Sigma_2^2 + \Sigma_3^2) - (\Sigma T_s)^2]}{C\Sigma T_S - \Sigma T_s^2}$$

Where
   $C$ = the number of conditions
   $\Sigma_1^2$ = the total for condition 1, squared

WORKED EXAMPLE

1. $\Sigma_1 = 15$

2. $\Sigma_2 = 9$

3. $\Sigma_3 = 5$

4. See $T_S$ column in Table 9.2

5. See $T_s^2$ column in Table 9.2

6. $\Sigma T_s = 29$

7. $\Sigma T_s^2 = 59$

8.

$$Q = \frac{(3 - 1)[3(15^2 + 9^2 + 5^2) - 29^2]}{3 \times 29 - 59}$$

Where
   $C = 3$
   $\Sigma_1^2 = 15^2$
        $= 225$

$\Sigma_2^2$ = the total for condition 2, squared

$\Sigma_3^2$ = the total for condition 3, squared

$\Sigma T_S$ = the total of $T_s$ column

$(\Sigma T_S)^2$ = the total of $T_s$ column, squared

$\Sigma T_S$ = the total of $T_S^2$ column

$\Sigma_2^2 = 9^2$
$= 81$
$\Sigma_3^2 = 5^2$
$= 25$
$\Sigma T_S = 29$
$(\Sigma T_S)^2 = 29^2$
$= 841$
$\Sigma T_S^2 = 59$

$$Q = \frac{2[3(225 + 81 + 25) - 841]}{87 - 59}$$

$$= \frac{2[993 - 841]}{28}$$

$$\therefore Q = 10.86$$

---

**Special point**

Do make sure you take account of the difference between $(\Sigma T_S)^2$ and $\Sigma T_S^2$. $(\Sigma T_S)^2$ is the total of the $T_S$ column, squared, while $\Sigma T_S^2$ is the total of the $T_S^2$ column.

---

9. Calculate the degrees of freedom (d.f.) using $C - 1$ where $C$ is the number of conditions.

9. d.f. = 2

*Looking up the results in the probability tables*
The results of the calculations yielded a $Q$ of 10.86, with d.f. = 2. To see whether this value represents significant differences between the data from the three conditions, it must be looked up in the relevant probability tables. Turn to Table C.1. Across the top is 'level of significance for a two-tailed test' while down the left-hand column are the degrees of freedom (d.f.) values. If you look down this column until you find our d.f. of 2, you will see five numbers to the right:

4.60    5.99    7.82    9.21    13.82

These are called 'critical values' and each is associated with the significance level (or $p$ value) at the head of its column. So the critical value 9.21 is associated with a probability of 0.01 (or 1%) for a two-tailed test. For our $Q$ value of 10.86 to be significant at one of these probability levels, it must be *equal to* or *larger* than the related critical value. If the $Q$ value is equal to the critical value, then the associated probability is exactly the same as that for the critical value. In other words, if you obtain a $Q$ value of 7.82 exactly, then the associated $p$ value is 0.02 exactly. However, had you obtained a $Q$ value of 8.41, this is larger than the critical value 7.82, and so the associated probability is even less than 0.02. This is expressed as $p < 0.02$ (< means less than).

Our $Q$ of 10.86 is *larger* than the critical value 9.21 and so the associated probability is *less than* 0.01. (Remember the $H_1$ had to be two-tailed).

*What do the results mean?*
The probability of our results being due to random error is less than 0.01, which, using the usual cut-off point of 5% or less, means the results are significant. There is less than a 1% chance that the outcome of the experiment could be due to random error. This means that the experimental hypothesis has been supported – the flexibility of catheters influences the urinary tract infection rate in female multiple sclerosis sufferers.

Some questions might arise when you calculate the Cochran $Q$ test on your own data:

> Q1: What would it mean had my Q value been smaller than the relevant critical values in the table?

This simply means that your results are not significant and could be explained by random error. In this case your null (no relationship) hypothesis would have been supported, and the experimental hypothesis would have to be rejected.

---

*Exercises (answers on page 315)*

**1** Look up the following $Q$ values in the relevant probability tables and state whether or not they are significant and at what level.
(a) $Q = 6.91$, d.f. = 3, two-tailed
(b) $Q = 7.91$, d.f. = 2, two-tailed
(c) $Q = 5.99$, d.f. = 2, two-tailed
(d) $Q = 8.43$, d.f. = 4, two-tailed
(e) $Q = 13.96$, d.f. = 3, two-tailed
**2** Calculate a Cochran $Q$ test on the following results; state the $Q$, d.f. and $p$ values and clarify what your results mean.

> $H_1$: There is a relationship between quality of student nurse performance and type of clinical placement (surgical, medical and psychiatric).
>
> $H_0$: There is no relationship between type of ward placement and student nurse performance.

*Brief outline of study*
In order to assess student nurse performance on the psychiatric, surgical and medical modules of training, a group of twenty nurses is randomly selected and their pass/fail practical assessments on each of the above, following ward placements is noted. The data are given in Table 9.3: (1 = pass; 0 = fail)

**Table 9.3**

| Subject | Condition 1: psychiatric | Condition 2: surgical | Condition 3: medical | Condition 4: geriatric |
|---------|--------------------------|-----------------------|----------------------|------------------------|
| 1 | 0 | 1 | 1 | 0 |
| 2 | 0 | 1 | 1 | 0 |
| 3 | 1 | 0 | 1 | 0 |
| 4 | 0 | 1 | 0 | 1 |
| 5 | 1 | 0 | 1 | 0 |
| 6 | 0 | 1 | 1 | 1 |
| 7 | 0 | 1 | 1 | 1 |
| 8 | 0 | 1 | 0 | 1 |
| 9 | 1 | 0 | 1 | 0 |
| 10 | 0 | 1 | 0 | 1 |
| 11 | 1 | 1 | 0 | 1 |
| 12 | 1 | 1 | 1 | 1 |
| 13 | 0 | 1 | 0 | 0 |
| 14 | 0 | 0 | 0 | 1 |
| 15 | 0 | 1 | 1 | 1 |
| 16 | 0 | 1 | 1 | 0 |
| 17 | 1 | 0 | 0 | 0 |
| 18 | 0 | 1 | 1 | 1 |
| 19 | 0 | 1 | 1 | 0 |
| 20 | 0 | 1 | 1 | 1 |

## 9.1.2 The Friedman test

The Friedman test is used under the following conditions:

1. With same subject or matched subject designs.
2. When there are three or more sets of data or conditions.
3. The data are ordinal or interval/ratio.

The Friedman test compares three or more sets of data to see whether there are differences between them. When calculating this test, a numerical value for $\chi_r^2$ is found, which is then looked up in the probability tables associated with the Friedman test. This will give the probability (or $p$ value) of the results being due to random error. If this $p$ value is 5% or less, then the results are said to be significant, since it is unlikely that random error can account for them. The experimental hypothesis would have been supported, and the null hypothesis can be rejected. If it is more than 5%, then random error could explain the results. In this case the results are not significant, and the null (no relationship) hypothesis would have been supported.

The Friedman test will only tell you whether or not there are significant differences between the conditions or sets of data. It cannot tell you whether the results in one of the conditions are significantly better than in the others. Because of this, any hypothesis associated with the test must be two-tailed.

**Example**
Osteoarthritis is a very common and painful condition and treatment is not wholly satisfactory. There is, however, some evidence to suggest that the use of cognitive strategies to direct the patient's attention away from the pain may modify the pain experience. You decide to compare the efficacy of some of these strategies with the use of standard analgesics in a group of osteoarthritis patients.

*Hypotheses*

>   **H₁:** The use of cognitive strategies v. analgesia differentially affects the experience of pain in osteoarthritis patients.

>   **H₀:** There is no relationship between the use of analgesia/cognitive strategies and the experience of pain in osteoarthritis patients.

The $H_1$ is a two-tailed hypothesis (which is essential for the Friedman test) as it predicts no direction to the results.

*Brief outline of study*
A group of nine post-menopausal women with primary osteoarthritis of the neck and shoulders is randomly selected for study. All have been on standard analgesia for over a year. Before the start of the study each patient's pain level is assessed on a five point scale:

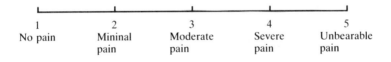

| 1 | 2 | 3 | 4 | 5 |
|---|---|---|---|---|
| No pain | Mininal pain | Moderate pain | Severe pain | Unbearable pain |

These results provide a baseline pain level while on analgesics only, with which the cognitive strategies can be compared. The group is then randomly split into two sub-groups of five and four. One sub-group is asked firstly to try to concentrate their attention fully on distractions or events going on around them, for example, on the radio, or on any activity in which they are engaged. After one week of this, the patients' pain levels are again assessed. The group is then asked to construct pleasant fantasies rather than focussing attention on the environment, as a means of modifying the pain. After a week, their pain levels are again monitored. For the second sub-group of four patients, the order of these two cognitive strategies is reversed. All the subjects continue to take analgesics during the cognitive strategies. As *one* group of subjects has been assessed under all *three* conditions (analgesia only, i.e. baseline pain level, attention to the environment and analgesia, fantasies and analgesia) and the data are on an *ordinal* scale, the conditions required by the Friedman test have been fulfilled.

The design is shown in Figure 9.4. (All the constant errors would have to be

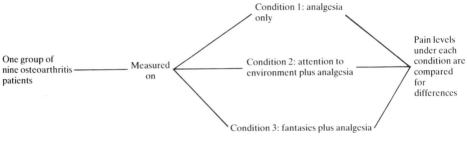

**Fig. 9.4**

eliminated if this study were to be carried out in reality.) What you are interested in finding out is whether the subjects' reported pain levels differ between the conditions.

*Results*
The results in Table 9.4 are obtained (it does not matter which condition is A, B or C).

**Table 9.4**

| Subject | Condition A: analgesia | Rank | Condition B: attention to environment plus analgesia | Rank | Condition C: fantasies plus analgesia | Rank |
|---------|---------|------|---------|------|---------|------|
| 1 | 3 | 2.5 | 3 | 2.5 | 2 | 1 |
| 2 | 4 | 2.5 | 3 | 1 | 4 | 2.5 |
| 3 | 4 | 2.5 | 4 | 2.5 | 3 | 1 |
| 4 | 3 | 2 | 2 | 1 | 5 | 3 |
| 5 | 3 | 2.5 | 3 | 2.5 | 2 | 1 |
| 6 | 5 | 3 | 4 | 1.5 | 4 | 1.5 |
| 7 | 4 | 2.5 | 3 | 1 | 4 | 2.5 |
| 8 | 4 | 3 | 2 | 1 | 3 | 2 |
| 9 | 3 | 1 | 5 | 3 | 4 | 2 |
| | $\Sigma A = 33$ | $\Sigma R_A = 21.5$ | $\Sigma B = 29$ | $\Sigma R_B = 16.0$ | $\Sigma C = 31$ | $\Sigma R_C = 16.5$ |
| | $\bar{x}A = 3.67$ | | $\bar{x}B = 3.22$ | | $\bar{x}C = 3.44$ | |

Notes: ($\Sigma$ means total and $\bar{x}$ means average.)
Figures in bold type are the raw data and those in lighter type are the calculations.

*Calculating the Friedman test*

| PROCEDURE | WORKED EXAMPLE |
|-----------|----------------|
| 1. Add up the scores in condition A to give $\Sigma A$. | 1. $\Sigma A = 33$ |

2. Add up the scores in condition B to give $\Sigma B$.

3. Add up the scores in condition C to give $\Sigma C$.

4. Calculate the means for each condition to give $\bar{x}A$, $\bar{x}B$, $\bar{x}C$.

5. Rank the scores from all conditions for each *subject* (i.e. rank the scores across each row) giving a rank of 1 to the smallest score, 2 to the next smallest, etc. Where two or more scores are the same, use the tied ranks procedure as outlined in Section A.2, e.g. for S1, the score of 2 in condition 3 gets a rank of 1, and the scores of 3 each get a rank of 2.5.

2. $\Sigma B = 29$

3. $\Sigma C = 31$

4. $\bar{x}A = 3.67$
   $\bar{x}B = 3.22$
   $\bar{x}C = 3.44$

5. See the 'Rank' columns above.

---

*Special point*

Do make sure you rank the scores for each subject, i.e. *across* the page, rather than for each condition.

---

6. Add up the ranks for each condition to give $\Sigma R_A$, $\Sigma R_B$, $\Sigma R_C$.

7. Find the value of $\chi_r^2$ from the formula:

$$\chi_r^2 = \left[\left(\frac{12}{NC(C+1)}\right)(\Sigma R_A^2 + \Sigma R_B^2 + \Sigma R_C^2)\right] - 3N(C+1)$$

where
  $N$ = the number of subjects (or sets of matched Ss)
  $C$ = the number of conditions
  $\Sigma R_A^2$ = the rank total for condition A, squared
  $\Sigma R_B^2$ = the rank total for condition B, squared

6. $\Sigma R_A = 21.5$
   $\Sigma R_B = 16.0$
   $\Sigma R_C = 16.5$

7.

$$\chi_r^2 = \left[\left(\frac{12}{9 \times 3(3+1)}\right)(21.5^2 + 16^2 + 16.5^2)\right] - 3 \times 9(3+1)$$

where
  $N = 9$

  $C = 3$
  $\Sigma R_A^2 = 21.5^2$
      $= 462.25$
  $\Sigma R_B^2 = 16.0^2$
      $= 256$

$\Sigma R_C^2$ = the rank total for condition C, squared

$\Sigma R_C^2 = 16.5^2$
$= 272.25$

$\chi_r^2 = [0.11 \times 990.5] - 108$
$= 108.96 - 108$

$\therefore \chi_r^2 = 0.96$

*Looking up the results in the probability tables*
The results of the Friedman test have produced a $\chi_r^2$ value of 0.96. In order to find out whether this value represents significant differences in reported pain levels between the three pain-control conditions, it has to be looked up in the probability values associated with the Friedman test. Turn to Tables C.4(a) and C.4(b). The first table is used for designs with three conditions and two to nine subjects per condition, while the second table is for designs with four conditions and two to four subjects per condition. If you have more than four conditions and/or more than the number of subjects specified at the top of these tables, you must use Table C.1, which is also associated with the $\chi^2$ test. When using this table, you will need the d.f. value in order to look up your $\chi_r^2$ figure. The d.f. value for Friedman test is $C - 1$ (i.e. the number of conditions minus 1). The procedure for looking up the $\chi_r^2$ value in this situation is outlined in Q2 below.

However, in this study, we have used three conditions, with nine subjects per condition, which means that Table C.4(a) is appropriate. Look across the top until you find $N = 9$. Beneath this are two columns, one giving critical values of $\chi_r^2$ and the other giving the relevant probabilities alongside. Thus, a $\chi_r^2$ value of 6.222 is associated with a probability of 0.048.

For our $\chi_r^2$ value of 0.96 to be significant at one of these probability levels, it has to be *equal to* or *larger* than one of the $\chi_r^2$ figures in the column for $N = 9$. If our $\chi_r^2$ value is equal to one of these figures, then it has exactly the same probability as that figure. So had we obtained a $\chi_r^2$ value of 10.889 for $N = 9$, then the associated probability is 0.0029 exactly. On the other hand, had the obtained $\chi_r^2$ value been 10.947, the associated probability would have been *even less* than 0.0029, because 10.947 is *larger* than the critical value 10.889. This is expressed as $p < 0.0029$.

Our $\chi_r^2$ value is larger than the critical value of 0.889, which means that the associated probability is less than 0.865.

*What do the results mean?*
The probability of our results being due to random error is less than 0.865, or 86%. As the critical cut-off point for claiming results as significant is 5% or less, this means that our results are not significant, and that they can be explained by random error. We can conclude that there is no difference in the reported pain levels under analgesia and the two cognitive strategies in a group of primary osteoarthritis patients. The null (no relationship) hypothesis must be accepted, and the experimental hypothesis rejected.

If you have been calculating the Friedman test on your own data, then some queries may have arisen:

*Q1: I used four conditions in my study. How do I look up the $\chi_r^2$ value?*

The same procedure is used in four-condition studies as for three-condition studies when looking up the $\chi_r^2$ value, although Table C.4(b) is used instead. If you turn to Table C.4(b), you will see that it makes provision for four-conditions and two to four subjects only. If your design meets these specifications then you simply repeat the procedure outlined in the example when looking up the obtained $\chi_r^2$ value.

*Q2: I used more than the number of subjects and conditions catered for by Tables C.4(a) and (b). How do I look up my $\chi_r^2$ value?*

Because you have more than the number of subjects and conditions appropriate for Tables C.4(a) and (b), you must use Table C.1 instead. When you use this table, you also need a d.f. value, calculated by the formula $C - 1$ (where $C$ is the number of conditions in the experiment). Imagine you used five conditions and twelve subjects and obtained a $\chi_r^2$ value of 12.2. The d.f. in this instance is 4. Turn to Table C.1. Down the left-hand side are d.f. values and along the top the probability values. Look down the d.f. column in the table until you find d.f. = 4. To the right of this are five figures, called 'critical values':

$$7.78 \quad 9.49 \quad 11.67 \quad 13.28 \quad 18.46$$

Each of these critical values is associated with the probability indicated at the top of its column (e.g. 9.49 has a $p$ value of 0.05). For your $\chi_r^2$ value to be associated with one of these probabilities it has to be *equal to* or *larger* than the relevant critical value. Our $\chi_r^2$ value is larger than the critical value of 11.67, and so its associated probability is even less than 0.02 or 2%. This means that these results would be significant, using the usual 5% or less cut-off point.

---

*Exercises (answers on page 316)*

**3** Look up the following $\chi_r^2$ values, and state whether or not they are significant and at what level.
   (a) $\chi_r^2 = 6.7$, $C = 3$, $N = 4$
   (b) $\chi_r^2 = 8.41$, $C = 8$, $N = 4$
   (c) $\chi_r^2 = 6.82$, $C = 8$, $N = 3$
   (d) $\chi_r^2 = 8.01$, $C = 5$, $N = 5$
   (e) $\chi_r^2 = 7.8$, $C = 4$, $N = 4$
**4** Calculate a Friedman test on the following data; state the $\chi_r^2$ and $p$ values. Clarify what your results mean.

   **H₁:** Stroke victims make differential progress according to whether they live alone, in sheltered accommodation or with their families.

**H₀:** There is no relationship between the progress of stroke victims and the nature of their home support structure.

*Brief outline of study*
Select three *matched* groups each of six stroke victims, ensuring that they are similar on factors such as nature and severity of stroke, age and sex. Of these groups, one is discharged from hospital into sheltered housing, the second into their own homes, without any family support, and the third group into their family's homes. After one month the progress of these groups is assessed on a six-point scale (1 = no progress, 6 = total recovery). The results are given in Table 9.5.

**Table 9.5**

| Subject | Condition A: sheltered housing | Condition B: alone | Condition C: family |
|---------|--------------------------------|--------------------|---------------------|
| 1       | 3                              | 1                  | 2                   |
| 2       | 2                              | 2                  | 2                   |
| 3       | 4                              | 2                  | 3                   |
| 4       | 3                              | 3                  | 1                   |
| 5       | 2                              | 1                  | 2                   |
| 6       | 4                              | 1                  | 3                   |

## 9.1.3 Page's L trend test

Page's *L* trend test is used under the following conditions:

1. With a same or matched subject design.
2. When there are three or more sets of data.
3. When the data are ordinal or interval/ratio.
4. When a *trend* in the results is predicted.

The last point is important. While the Friedman test is used to identify whether there are *differences* between the data from each condition, without specifying which conditions are better or worse than the others, Page's *L* trend test is used when the researcher makes a definite prediction about a trend in the results – in other words, the hypothesis specifies which of the conditions will be better or worse than the others. As a result, hypotheses associated with Page's *L* trend test must be one-tailed. An example may clarify this. In Exercise 4 (for the Friedman test) the hypothesis under investigation was:

**H₁:** Stroke victims make differential progress according to whether they live alone, in sheltered accommodation or with their families.

Here, there was no specific prediction about the results – all that was predicted was that there would be a *difference* in progress according to the type of home situation the patient had. However, suppose the hypothesis had been more precise in its predictions:

> **H₁:** Stroke victims make most progress in sheltered accommodation, followed by living with families. Living alone produces least progress.

This hypothesis is predicting a very definite trend in the results with sheltered accommodation better than living with families, which is in turn better than living alone. Where such a trend is predicted, and the other three conditions have been fulfilled, Page's $L$ trend test should be used to analyse the data.

When calculating Page's $L$ trend test, a numerical value for $L$ is found, which is then looked up in the probability tables associated with Page's $L$ trend test. This will give the probability of the results being due to random error. If the probability is 5% or less, then it is unlikely that random error can account for the results, which are then said to be *significant*. The experimental hypothesis would have been supported, and the null hypothesis can be rejected. If it is more than 5%, random error could explain the results; in this case, the null (no relationship) hypothesis would have been supported.

Remember that because Page's $L$ trend test is only used when a specific trend is predicted, the hypothesis must be one-tailed. In addition, you should restrict yourself to a maximum of six conditions and twelve subjects per condition for this test, as the probability tables do not allow for larger numbers.

**Example**
Much orthopaedic treatment involves immobilising the affected part of the body, often for long periods of time. This can mean lengthy hospital stays, which may be difficult for the patient to cope with, particularly since he or she may otherwise be in good health. Suppose the problems of the orthopaedic patient are your focus of interest. In particular, you are concerned with identifying those aspects of the condition which cause the patient most distress, so that priorities for nursing care can be established.

*Hypotheses*

> **H₁:** Lack of privacy causes most distress to long-stay orthopaedic patients, followed by boredom and finally fear of treatment procedures.
>
> **H₀:** There is no relationship between aspects of long-term hospitalisation and distress in orthopaedic patients.

The H₁ is one-tailed because it is predicting that lack of privacy causes greatest distress, followed by boredom and then fear.

*Brief outline of study*

A group of eight patients receiving skull traction is randomly selected for study. All have been admitted for the same problem and all have been in hospital for a comparable period of time. Each patient is asked to consider three aspects of their situation: boredom, fear of treatment procedures and lack of privacy and to rate each on a six point scale according to how distressing they find them:

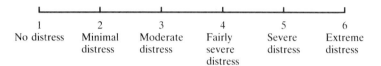

| 1 | 2 | 3 | 4 | 5 | 6 |
|---|---|---|---|---|---|
| No distress | Minimal distress | Moderate distress | Fairly severe distress | Severe distress | Extreme distress |

In this study, one group of subjects is tested under each of three conditions (boredom, fear, lack of privacy); the data are ordinal and a trend in the results has been predicted. Therefore all four conditions required by Page's *L* trend test have been fulfilled. The design is shown in Figure 9.5. (Remember that there are a number of constant errors that would have to be controlled for, such as time of day when questioned and immediate previous events.) What you are interested in here is whether aspects of the condition produce different degrees of distress.

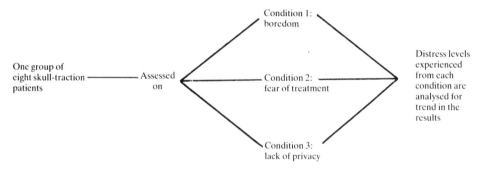

**Fig. 9.5**

*Results*

The following results are obtained and should be set out in a table such that the condition whose scores are anticipated to be the *smallest* is placed on the left, and those with the largest on the right. The remaining conditions should be arranged in between accordingly (Table 9.6).

*Calculating Page's L trend test*

PROCEDURE

1. Add up all the scores for each condition to give $\Sigma C_1$, $\Sigma C_2$ and $\Sigma C_3$.

WORKED EXAMPLE

1. $\Sigma C_1 = 14$
   $\Sigma C_2 = 32$
   $\Sigma C_3 = 37$

2. Find the mean scores for each condition to give $\bar{x}_1$, $\bar{x}_2$ and $\bar{x}_3$.

2. $\bar{x}_1 = 1.75$
$\bar{x}_2 = 4$
$\bar{x}_3 = 4.63$

3. Rank each subject's set of scores, *across the row*, giving a rank of 1 to the smallest, 2 to the next smallest and so on. Where two or more scores are the same use the tied rank procedure outlined in Section A.2.

3. See 'Rank' columns in Table 9.6

---

*Special point 1*

Remember to rank each individual subject's set of scores going across the rows. Do *not* rank the scores for each condition.

---

4. Add up the ranks for each condition to give $\Sigma R_1$, $\Sigma R_2$ and $\Sigma R_3$.

4. $\Sigma R_1 = 8$
$\Sigma R_2 = 19$
$\Sigma R_3 = 21$

5. Find the value of $L$ from the following formula:

$$L = \Sigma[(\Sigma R_1 \times c) + (\Sigma R_2 \times c) + (\Sigma R_3 \times c)]$$

5. $L = \Sigma[(8 \times 1) + (19 \times 2) + (21 \times 3)]$

**Table 9.6**

| Subject | Condition 1: fear | Rank | Condition 2: boredom | Rank | Condition 3: lack of privacy | Rank |
|---|---|---|---|---|---|---|
| 1 | 2 | 1 | 4 | 2 | 5 | 3 |
| 2 | 3 | 1 | 5 | 3 | 4 | 2 |
| 3 | 1 | 1 | 3 | 2 | 6 | 3 |
| 4 | 2 | 1 | 4 | 3 | 3 | 2 |
| 5 | 1 | 1 | 4 | 2 | 6 | 3 |
| 6 | 3 | 1 | 5 | 2.5 | 5 | 2.5 |
| 7 | 1 | 1 | 3 | 2 | 4 | 3 |
| 8 | 1 | 1 | 4 | 2.5 | 4 | 2.5 |
| | $\Sigma C_1 = 14$ | $R_1 = 8$ | $\Sigma C_2 = 32$ | $\Sigma R_2 = 19.0$ | $\Sigma C_3 = 37$ | $\Sigma R_3 = 21.0$ |
| | $\bar{x}_1 = 1.75$ | | $\bar{x}_2 = 4$ | | $\bar{x}_3 = 4.63$ | |

Notes: ($\Sigma$ means total and $\bar{x}$ means average.)
Figures in bold type are the raw data and those in lighter type are the calculations.

where

$\Sigma R_1$ = the rank total for condition 1

$\Sigma R_2$ = the rank total for condition 2

$\Sigma R_3$ = the rank total for condition 3

$c$ = the number allotted to the condition, with 1 as the left-hand condition, 2 as the next and so on.

$$L = \Sigma[9 + 38 + 63]$$
$$L = 110$$

6. Identify the number of conditions ($C$) and the number of Ss or sets of matched Ss ($N$)

6. $C = 3$
   $N = 8$

*Looking up the results in the probability tables*

An $L$ value of 110 has been obtained from the calculations of Page's $L$ trend test. To find out whether this represents a significant trend in the results, it must be looked up in Table C.5 which is the probability table associated with the Page's $L$ trend test. Across the top, you will see the number of conditions ($C$) from three to six, while down the side are the $N$ values (i.e. the number of subjects or matched subjects). To the right of the table are the probability values (note the $<$, or less than, sign here at the top of the column). If we take our $N$ of 8 and $C$ of 3, you will see three figures at the intersection point: 109, 106 and 104. These numbers are called 'critical values of $L$', and each is associated with the $p$ value to its right. In other words:

109 has a $p$ value of $< 0.001$
106 has a $p$ value of $< 0.01$
104 has a $p$ value of $< 0.05$

For our $L$ of 110 to be significant at one of these probability levels, it has to be *equal to* or *larger than* one of the given critical values. The obtained value of 110 is larger than 109 and so the associated probability is $<0.001$ or 0.01%. Remember that Page's $L$ trend test requires a one-tailed hypothesis, so this probability table refers to one-tailed hypotheses only.

*What do the results mean?*

The probability of these results being due to random error is less than 0.1% which means that if we use the usual cut-off point of 5% (or less), our results are significant. The null (no relationship) hypothesis can be rejected, as the experimental hypothesis has been supported. The outcome of our calculations has confirmed that lack of privacy causes most distress to long-stay orthopaedic patients, followed by boredom and fear of treatment.

If you have calculated this test on your own data, there may be some issues you wish to raise.

*Q1: I obtained an* L *value which is smaller than any of the three critical values at the intersection point of my* C *and* N. *What does this mean?*

If you get an *L* value which is smaller than any of the three relevant critical values, your results are not significant. In this case, the null (no relationship) hypothesis would have to be accepted as there is no trend in the results.

---

## Exercises (answers on page 316)

5 Look up the following *L* values and state whether or not they are significant and at what level. (Note the $<$ sign by *p* at the top of the table.)
   (a) $L = 251, C = 5, N = 5$
   (b) $L = 260, C = 4, N = 10$
   (c) $L = 166, C = 3, N = 12$
   (d) $L = 490, C = 6, N = 6$
   (e) $L = 105, C = 3, N = 8$
6 Calculate a Page's *L* trend test on the following data; state the *L, C, N* and *p* values. Clarify what the results mean.

   **H₁:** Student nurses are most likely to take sick leave during their psychiatric ward experience, followed by geriatric and paediatric ward experience respectively.

   **H₀:** There is no relationship between sick leave taken by student nurses and type of ward experience.

*Brief outline of study*
Randomly select a group of twelve second-year nurses who have already completed their psychiatric, geriatric and paediatric ward placements. Count up the number of days' sick leave each student took while completing the placements. The results are given in Table 9.7.

Note how the conditions have been set out with the condition anticipated to have the smallest scores on the left.

**Table 9.7**

| Subject | Condition 1: paediatric | Condition 2: geriatric | Condition 3: psychiatric |
|---------|-------------------------|------------------------|--------------------------|
| 1 | 3 | 2 | 5 |
| 2 | 0 | 2 | 3 |
| 3 | 2.5 | 3 | 3 |
| 4 | 5 | 0 | 7 |
| 5 | 7 | 0 | 8 |

| | | | |
|---|---|---|---|
| 6 | 0 | 4 | 5 |
| 7 | 1 | 3 | 3 |
| 8 | 0 | 2.5 | 2 |
| 9 | 0 | 3 | 5 |
| 10 | 2 | 5 | 5 |
| 11 | 1 | 2 | 6 |
| 12 | 3 | 4 | 7 |

## 9.2 Parametric tests

### 9.2.1 One-way analysis of variance (anova) for related designs

The one-way anova for related designs is used under the following conditions:

1. A same subject or matched subject design has been used.
2. There are three or more conditions (or sets of data) to be compared.
3. The data are of an interval/ratio level.
4. The remaining three conditions required by a parametric test can be (more or less) fulfilled (see Section 7.2).

The one-way anova is the parametric equivalent of the Friedman test, and compares the sets of data to see if there are any differences between them. When calculating this test, two numerical values for $F$ are found which are then looked up in the probability tables associated with the anova. This will tell you whether the probability associated with either $F$ value is 5% or less, which means that it is unlikely that random error can explain the results. The results are then said to be *significant*. The experimental hypothesis would have been supported. The null hypothesis can be rejected. If it is more than 5%, it means that the results are *not* significant because they can be explained by random error. In this case the null (no relationship) hypothesis would have been supported.

    The one-way anova for related designs will only tell you whether or not there are differences between the sets of data (or conditions); it cannot say whether subjects perform significantly better to worse under one condition than another. If you want to know this, then the Scheffé multiple range test (see next section) should be performed *after* the anova has been calculated. Because the anova only specifies whether or not there are differences between the conditions and not the direction of these differences, hypotheses associated with it must be two-tailed.

    The anova is a rather complex test to calculate, because a number of stages are involved. The process can be clarified by a brief explanation of how the anova works. In order to do this, reference will be made to an example.

**Example**
There is some evidence that patients are treated differently according to their physical appearance, with attractive patients receiving more positive treatment (see

Darbyshire 1986). While the underlying reasons for this need not concern us at the moment, the finding clearly has important implications for nursing care. Let us imagine you are a sister on a male medical ward, on which there are a variety of patients. During the course of your duties you have noticed that the time spent by the nurses with the erythrodermic psoriasis patients appears to be less than that spent with either the chronic bronchitics or the stroke patients, and you suspect that this may be attributable to the rather unattractive nature of the psoriatic condition. You decide to carry out some research to see whether your observation is correct.

*Hypotheses*

**H₁:** There is a relationship between the amount of time spent by nurses in non-nursing related interaction and the nature of the patient's illness (erythrodermic psoriasis, chronic bronchitis or stroke).

**H₀:** There is no relationship between the amount of time spent by nurses in non-nursing related interaction and the nature of the patient's condition.

The $H_1$ is two-tailed (of necessity) as it simply predicts a relationship between the two variables, without specifying the nature of this relationship.

*Brief outline of study*
Six staff nurses on a male medical ward are selected for study. The total amount of time each one spends in a non-nursing interaction with each patient suffering from erythrodermic psoriasis is monitored over three randomly selected days. The average time spent in each interaction with the psoriasis patients is then calculated for each nurse. This procedure is repeated for the chronic bronchitis and stroke patients. In this way, each nurse has a score for the average amount of time spent in a single non-nursing interaction with each category of patient.

The design is shown in Figure 9.6. Because one group of nurses is measured

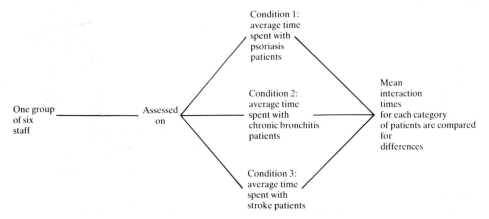

**Fig. 9.6**

under all three conditions and the measurement (in minutes) is interval/ratio, the necessary requirements for a one-way anova for related designs have been fulfilled. Note, however, that if this study were to be carried out properly, a number of constant errors would have to be controlled for.

Your concern here is whether or not nurses spend different amounts of time in non-nursing interaction according to the type of patient they are dealing with.

*Results*
The mean times in minutes given in Table 9.8 were obtained. (The figures in bold type are the data obtained from the study, while those in light type are part of the calculations from the anova. It does not matter which is condition 1, 2 or 3.)

Because you are anticipating that the interaction will differ according to the category of patient, you should expect to find differences between the condition $(T_C)$ totals. This is called the *between conditions comparison*, and, if your hypothesis is correct, should be a major source of variation in the results. However, the anova also calculates another possible source of variation in the results – the differences *between the subjects* in terms of their overall interaction times. In other words, each subject's total interaction time under all three conditions is compared and this is known as the *between subjects comparison*. Each subject's total interaction time for all three conditions is shown as the $T_S$ totals to the right of the table. Finally, as with all statistical tests, there is also the possibility of random error influencing the results.

So, the one-way anova for related designs is concerned with three types of variation in the data:

1. The variation between conditions (i.e. differences in interaction time between the psoriasis, bronchitis and stroke patients).
2. The variation between subjects (i.e. differences between the subjects in terms of their *overall* interaction times).
3. Variation due to random error.

What we would hope to find is a significant difference between the conditions since

**Table 9.8**

| Subject | Condition 1: psoriasis | Condition 2: bronchitis | Condition 3: stroke | Total for each S ($T_S$) |
|---|---|---|---|---|
| 1 | **3** | **6** | **6** | 15 |
| 2 | **5** | **7** | **5** | 17 |
| 3 | **5** | **8** | **8** | 21 |
| 4 | **3** | **11** | **8** | 22 |
| 5 | **4** | **8** | **7** | 19 |
| 6 | **6** | **9** | **7** | 22 |
| Total for each condition ($T_C$) | $T_C1 = 26$ | $T_C2 = 49$ | $T_C3 = 41$ | $\Sigma x = 116$ |

$\Sigma$ means 'total'

**Table 9.9** Sample format of an anova table

| Source of variation | Sums of squares (SS) | Degrees of freedom (d.f.) | Mean squares (MS) | F ratios (F) |
|---|---|---|---|---|
| Variation between conditions (i.e. types of patient) | $SS_{bet}$ | $d.f._{bet}$ | $MS_{bet}$ | $F_{bet}$ |
| Variation between subjects' overall performance (i.e. between the nurses in their total interaction time) | $SS_{subj}$ | $d.f._{subj}$ | $MS_{subj}$ | $F_{subj}$ |
| Variation due to random error | $SS_{error}$ | $d.f._{error}$ | $MS_{error}$ | |
| Total | $SS_{tot}$ | $d.f._{tot}$ | | |

this was what the $H_1$ predicted, but no significant differences between the subjects because this would suggest that individual differences among the subjects might account for the results.

If we just think about this a bit more, the presence of two very chatty nurses in the sample could distort the results, such that any differences between the conditions might not be the result of patient type but of the garrulous nature of some of the subjects. Clearly, then, we would hope for no significant differences between the subjects.

Where there are significant differences between the subjects, we cannot be sure that the experimental hypothesis is correct, since the results could be explained by the individual differences in the subject group. In such cases, it might be advisable to repeat the study with a different subject sample, selected more carefully to eliminate constant errors of this type. And of course, we do not want random error to account for the results as this would negate the experimental hypothesis.

In the introduction to this section, it was said that two numerical values for $F$ are calculated. One of these relates to the between conditions comparison, while the other relates to the between subjects comparison. In order to compute these, three values for each source of variation are required: sums of squares (SS), degrees of freedom (d.f.) and mean squares (MS). Once these values have been calculated they are entered into a table such as Table 9.9. The steps required to calculate these values are described below.

*Calculating the one-way anova for related designs*

PROCEDURE

1. Calculate the totals for each condition to give $T_C1$, $T_C2$ and $T_C3$.

WORKED EXAMPLE

1. $T_C1 = 26$
   $T_C2 = 49$
   $T_C3 = 41$

2. Caculate the totals for each subject to give $T_s1$, $T_s2$, $T_s3$, $T_s4$, $T_s5$, and $T_s6$.

2. $T_s1 = 15$
$T_s2 = 17$
$T_s3 = 21$
$T_s4 = 22$
$T_s5 = 19$
$T_s6 = 22$

3. Calculate the grand total ($\Sigma x$) by adding up all the $T_C$ values.

3. $\Sigma x = 116$

4. To calculate the SS for each source of variation, the following values are needed:

4.

$\Sigma T_C^2$ = the total of each squared $T_C$ value

$\Sigma T_C^2 = 26^2 + 49^2 + 41^2$
$= 4758$

$\Sigma T_s^2$ = the total of each squared $T_s$ value

$\Sigma T_s^2 = 15^2 + 17^2 + 21^2 + 22^2$
$+ 19^2 + 22^2$
$= 2284$

$n$ = the number of subjects or sets of matched subjects

$n = 6$

$C$ = the number of conditions

$C = 3$

$N$ = the total number of scores (i.e. $n \times C$)

$N = 18$

$\Sigma x$ = the grand total

$\Sigma x = 116$

$(\Sigma x)^2$ = the grand total squared

$(\Sigma x)^2 = 116^2$
$= 13\,456$

$\dfrac{(\Sigma x)^2}{N}$ = a constant to be subtracted from all SS values

$\dfrac{(\Sigma x)^2}{N} = \dfrac{13\,456}{18}$
$= 747.56$

$x$ = each individual score

$\Sigma x^2$ = the total of each squared individual score

$\Sigma x^2 = 3^2 + 6^2 + 6^2 + 5^2 + 7^2 +$
$5^2 + 5^2 + 8^2 + 8^2 + 3^2 +$
$11^2 + 8^2 + 4^2 + 8^2 + 7^2 +$
$6^2 + 9^2 + 7^2$
$= 822$

---

*Special point*

Note the difference between $(\Sigma x)^2$ which is the grand total squared, and $\Sigma x^2$ which is the total of each squared score.

---

5. Calculate the SS between conditions ($SS_{bet}$) from the formula

5.

$$\frac{\Sigma T_C^2}{n} - \frac{(\Sigma x)^2}{N}$$

$$\frac{4758}{6} - \frac{13\,456}{18}$$

$$= 793 - 747.56$$

$$\therefore SS_{bet} = 45.44$$

6. Calculate the SS between subjects ($SS_{subj}$) from

$$\frac{\Sigma T_S^2}{c} - \frac{(\Sigma x)^2}{N}$$

6.

$$\frac{2284}{3} - \frac{13\,456}{18}$$

$$= 761.33 - 747.56$$

$$\therefore SS_{subj} = 13.77$$

7. Calculate the $SS_{tot}$ from

$$\Sigma x^2 - \frac{(\Sigma x)^2}{N}$$

7.

$$822 - \frac{13\,456}{18}$$

$$\therefore SS_{tot} = 74.44$$

8. Calculate the $SS_{error}$ from

$$SS_{tot} - SS_{bet} - SS_{subj}$$

8.

$$74.44 - 45.44 - 13.77$$

$$\therefore SS_{error} = 15.23$$

9. Calculate the degrees of freedom (d.f.) values for each source of variation, using the following formulae:

9.

$$d.f._{bet} = C - 1$$

where $C$ is the number of conditions

$$d.f._{bet} = 3 - 1$$
$$= 2$$

$$d.f._{subj} = n - 1$$

where $n$ is the number of subjects or sets of matched subjects

$$d.f._{subj} = 6 - 1$$
$$= 5$$

$$d.f._{tot} = N - 1$$

where $N$ is the total number of scores

$$d.f._{tot} = 18 - 1$$
$$= 17$$

$$d.f._{error} = d.f._{tot} - d.f._{bet} - d.f._{subj}$$

$$d.f._{error} = 17 - 2 - 5$$
$$= 10$$

10. Calculate the mean squares (MS) values for each source of variation using the following formulae:

$$MS_{bet} = \frac{SS_{bet}}{d.f._{\cdot bet}} \qquad\qquad MS_{bet} = \frac{45.44}{2}$$

$$= 22.72$$

$$MS_{subj} = \frac{SS_{subj}}{d.f._{\cdot subj}} \qquad\qquad MS_{subj} = \frac{13.77}{5}$$

$$= 2.75$$

$$MS_{error} = \frac{SS_{error}}{d.f._{\cdot error}} \qquad\qquad MS_{error} = \frac{15.32}{10}$$

$$= 1.52$$

11. Calculate the $F$ values using the    11.
following formulae:

$$F_{bet} = \frac{MS_{bet}}{MS_{error}} \qquad\qquad F_{bet} = \frac{22.72}{1.52}$$

$$= 14.95$$

$$F_{subj} = \frac{MS_{subj}}{MS_{error}} \qquad\qquad F_{subj} = \frac{2.75}{1.52}$$

$$= 1.81$$

Enter these values in the table as shown in Table 9.10.

*Looking up the results in the probability tables*
Two values of $F$ have been obtained, one for the between conditions comparison ($F = 14.95$) and the other for the between subjects comparison ($F = 1.81$). To see if either of these represents significant differences in the data, they must be looked up in the appropriate probability tables (Tables C.6(a)–(d)). Each table is associated

**Table 9.10** Anova table using figures derived from the one-way analysis of variance for related designs

| Source of variation in scores | Sums of squares (SS) | Degrees of freedom (d.f.) | Mean squares (MS) | F ratios (F) |
|---|---|---|---|---|
| Variation in scores between conditions | 45.44 | 2 | 22.72 | 14.95 |
| Variation in scores between subjects | 13.77 | 5 | 2.75 | 1.81 |
| Variation in scores due to random error | 15.23 | 10 | 1.52 | |
| Total | 74.44 | 17 | | |

with different probability levels and each has values of $v_1$ across the top and $v_2$ down the left-hand side, which are d.f. values. To find out whether the differences between *conditions* (or type of patient) are significant, the $F$ of 14.95 must be looked up in these tables. When looking up this $F_{bet}$ value, the d.f.$_{bet}$ and d.f.$_{error}$ figures are required. Here these are 2 and 10 respectively. Turn to Table C.6(a). Across the top are $v_1$ figures which here refer to the d.f.$_{bet}$ number, while the $v_2$ figures down the left-hand side refer to the d.f.$_{error}$ value. Locate our d.f.$_{bet}$ of 2 across the top, and our d.f.$_{error}$ of 10 down the side. At their intersection point is the number 4.10 which is called the critical value of $F$. If the obtained $F_{bet}$ value is *equal to* or *larger than* the critical value at the intersection point, then the differences between conditions are significant at the level indicated at the top of the table. Our $F_{bet}$ of 14.95 is *larger* than the critical value 4.10, which means that the differences in interaction times for each category of patient are significant at $p <$ 0.05. However, it is possible that the probability of random error accounting for these results is even less than the 0.05 level. In other words, we might be able to get a 'better' $p$ value for our results. To find out if this is the case, repeat the procedure with Table C.6(b) (probabilities of <0.025). Here the critical value at the intersection point is 5.46. Our obtained $F_{bet}$ value is larger than this, so the results are significant at $p < 0.025$. Repeat the process with Tables C.6(c) and (d). The critical values at the intersection points are 7.56 and 14.91 respectively, which means that since the obtained $F_{bet}$ of 14.95 is larger than 14.91, the probability level associated with our results is <0.001, for a two-tailed hypothesis.

To find out whether the $F_{subj}$ value represents significant differences in the *overall* interaction times of the subjects, this value must also be looked up in Tables C.6(a)–(d). To do this, the d.f.$_{subj}$ value (which is the $v_1$ figure at the top of the table) and the d.f.$_{error}$ value (which is the $v_2$ value down the side of the table) are needed. On Table C.6(a), locate d.f.$_{subj}$ = 5 across the top and d.f.$_{error}$ = 10 down the side. At their intersection point is the critical value 3.33. To be significant at the level indicated at the top of the table, the obtained $F_{subj}$ value must be *equal to* or *larger than* the critical value at the intersection point. Our $F_{subj}$ of 1.81 is smaller than 3.33 and so is not significant at $p < 0.05$. In other words there is no significant difference between subjects' overall interaction times.

*What do the results mean?*
The probability of the between conditions results being due to random error is less than 0.001 or 0.1%. Using the standard cut-off point of 5% (or less) we can conclude that these results are significant, and that the experimental hypothesis has been supported. In other words the amount of time spent in non-nursing interaction varies according to the type of patient involved.

However, the $F_{subj}$ value was not significant ($p$ is greater than 0.05). This means that the nurses in the study did not differ significantly from one another in terms of the *overall* amount of non-nursing interaction they engaged in. This result in very important, since a significant $F_{subj}$ result would have meant that definite conclusions about the validity of the experimental hypothesis could not confidently have been drawn (see page 129).

We can conclude, therefore, that staff nurses do spend differing amounts of time in non-nursing related interaction according to the type of patient involved. Given that these differences cannot be ascribed to any individual differences among the sample of nurses, we must conclude that the nature of the patient's condition is responsible for the results. The experimental hypothesis has therefore been supported and the null hypothesis can be rejected.

If you re-read the background to this study, you will see that the patient's physical attractiveness was assumed to be the key factor in whether a nurse engaged in much interaction. On this basis, then, we expected that the erythrodermic psoriasis patients would experience less contact time with the nurses. However, we cannot establish whether this is so simply from the anova results, since all an anova does is to highlight whether there are general, overall differences between the conditions; it does not specify which condition did better or worse than the others. If this knowledge would be useful to you (as it would here) then a further test must be performed on the data – the Scheffé multiple comparison of means. This test allows you to assess whether any one condition differs significantly from the others. It is described in the next section.

If you are using the one-way anova to analyse your own data, there may be some points you wish to raise:

Q1: My $F_{bet}$ value was smaller than the critical value of F on the $p <$ 0.05 table. What does this mean?

If your obtained $F_{bet}$ figure is smaller than the critical value of $F$ at the intersection point on the $p < 0.05$, this means that there is greater than a 5% chance that random error could account for your results. The differences between the conditions in your study would *not* be significant, and the null (no relationship) hypothesis would have to be accepted.

Q2: My $F_{bet}$ value is larger than the critical value in Table C.6(b), but smaller than that in Table C.6(c). What does this mean?

This simply means that your results are significant at the level indicated at the top of Table C.6(b), but not at the level of Table C.6(c). In other words, the differences between your conditions are significant at <0.025 level.

Q3: Both my $F_{bet}$ and $F_{subj}$ values are significant. What does this mean?

This means that there are wide variations between your subjects which could account for the significant between conditions results. Because of this, you could not conclude definitively that the experimental hypothesis had been supported. You would have to conclude that the study was not well executed and that all the results may be suspect.

Q4: I have $v_1$ and $v_2$ values which are not in the table. How do I look up the obtained F?

When there is no exact number for your d.f. values, you should use the next *lowest* number. For d.f.s which are much larger than those shown in the table, use the row for infinity, marked by the ∞ symbol.

*Exercises (answers on page 316)*

7 Look up the following *F* values and state whether or not they are significant and at what level. (Note the < sign at the top of Tables A.6(a)–(d).
   (a) $F_{bet} = 3.51$, $F_{subj} = 1.93$, d.f.$_{bet} = 3$, d.f.$_{subj} = 7$, d.f.$_{error} = 21$
   (b) $F_{bet} = 5.94$, $F_{subj} = 5.63$, d.f.$_{bet} = 5$, d.f.$_{subj} = 9$, d.f.$_{error} = 46$
   (c) $F_{bet} = 15.31$, $F_{subj} = 2.24$, d.f.$_{bet} = 2$, d.f.$_{subj} = 11$, d.f.$_{error} = 23$
   (d) $F_{bet} = 2.91$, $F_{subj} = 3.15$, d.f.$_{bet} = 4$, d.f.$_{subj} = 5$, d.f.$_{error} = 15$
   (e) $F_{bet} = 4.32$, $F_{subj} = 2.56$, d.f.$_{bet} = 2$, d.f.$_{subj} = 14$, d.f.$_{error} = 29$
8 Calculate a one-way anova for related designs on the following data. State the MS, SS, d.f. and *F* ratios, placing them in a table like Table 9.10. State the *p* values and clarify what the results mean.

   **H₁:** Student nurses perform differently on the examination in nursing studies according to their stage of training.

   **H₀:** There is no relationship between student nurses' performances on the examination in nursing studies and stage of training.

*Brief outline of study*
Randomly select ten student nurses completing their final year of training and collect their nursing studies marks for each of the end-of-year exams (marks out of 20). In this way each S has three nursing studies marks. The results are given in Table 9.11.

**Table 9.11**

| Subject | Condition 1: year 1 marks | Condition 2: year 2 marks | Condition 3: year 3 marks |
|---|---|---|---|
| 1 | 10 | 12 | 13 |
| 2 | 11 | 12 | 11 |
| 3 | 9 | 11 | 12 |
| 4 | 11 | 12 | 14 |
| 5 | 11 | 14 | 14 |
| 6 | 10 | 11 | 14 |
| 7 | 10 | 12 | 14 |
| 8 | 9 | 10 | 13 |
| 9 | 12 | 11 | 13 |
| 10 | 13 | 12 | 14 |

## 9.2.2 The Scheffé multiple range test for use with one-way anovas for related designs

The Scheffé multiple range test is used under the following conditions:

1. *Following* the calculations of an anova only.
2. Only if those calculations yield a significant $F_{bet}$ value.

While the anova tells you whether there are general overall differences between the conditions in your study, the Scheffé test goes one stage further – it allows the researcher to find out which conditions did significantly 'better' or 'worse' than the others. In the example from the anova, our calculations produced a significant $F_{bet}$ value which means that staff nurses do spend different amounts of time in interaction with psoriasis, bronchitis and stroke patients. However, from the anova alone we do not know whether significantly less time was spent with the psoriasis patients than with the others. Given that this assumption was underlying the study in the first place (see page 126), this sort of information would be useful. The Scheffé test can provide this information by comparing the mean scores from each condition. In this way, we can compare the following:

1. The mean interaction score for the psoriasis patients with the mean interaction score for the bronchitis patients.
2. The mean interaction score for the psoriasis patients with the mean interaction score for the stroke patients.
3. The mean interaction score for the bronchitis patients with the mean interaction score for the stroke patients.

From these comparisons, we can find out whether pairs of conditions differ significantly from each other.

Do remember, however, that the Scheffé can only be used *after* an anova has been calculated and if the $F_{bet}$ value is significant. Note also that the Scheffé test for related and same subject designs has the same formula as the Scheffé for unrelated designs, given in Section 11.2.2. (I have included the formula for the Scheffé in each chapter, because it makes it easier for the beginning researcher.)

When calculating the Scheffé, numerical values for $F$ and $F'$ are found. If the $F$ value is equal to or larger than the obtained $F'$ value then that comparison is significant. The procedure is outlined in more detail below.

**Example**
We will calculate the Scheffé test on the data and anova results given in the worked example in the last section. There, significant differences were found in interaction times for each of three categories of patients.

*Calculating the Scheffé multiple comparisons of means test for related subject designs*

PROCEDURE

WORKED EXAMPLE

1. Calculate the mean score for each condition to give $\bar{x}_1$, $\bar{x}_2$ and $\bar{x}_3$.

1. $\bar{x}_1 = 4.33$
   $\bar{x}_2 = 8.17$
   $\bar{x}_3 = 6.83$

2. To compare condition 1 with condition 2 use the following formula:

2. To compare the interaction times with psoriasis and bronchitis patients

$$F = \frac{(\bar{x}_1 - \bar{x}_2)^2}{(MS_{error}/n_1) + (MS_{error}/n_2)}$$

$$F = \frac{(4.33 - 8.17)^2}{(1.52/6) + (1.52/6)}$$

where

where

$\bar{x}_1$ = the mean score for condition 1

$\bar{x}_1 = 4.33$

$\bar{x}_2$ = the mean score for condition 2

$\bar{x}_2 = 8.17$

$MS_{error}$ = the mean squares value for the between subjects variation (from anova table)

$MS_{error} = 1.52$

$n_1$ = the number of Ss in condition 1

$n_1 = 6$

$n_2$ = the number of Ss in condition 2

$n_2 = 6$

$$F = \frac{(-3.84)^2}{0.25 + 0.25}$$

$$= \frac{14.75}{0.5}$$

$$\therefore F = 29.5$$

3. To compare condition 1 with condition 3, use the following formula

3. To compare the interaction times with psoriasis and stroke patients

$$F = \frac{(\bar{x}_1 - \bar{x}_3)^2}{(MS_{error}/n_1) + (MS_{error}/n_3)}$$

$$F = \frac{(4.33 - 6.83)^2}{(1.52/6) + (1.52/6)}$$

where

where

$\bar{x}_1$ = the mean score for condition 1

$\bar{x}_1 = 4.33$

$\bar{x}_3$ = the mean score for condition 3

$\bar{x}_3 = 6.83$

$MS_{error}$ = the mean squares value for the between subjects variation (from the anova table)

$n_1$ = the number of Ss in condition 1

$n_3$ = the number of Ss in condition 3

$MS_{error}$ = 1.52

$n_1 = 6$

$n_3 = 6$

$$F = \frac{(-2.5)^2}{0.25 + 0.25}$$

$$= \frac{6.25}{0.5}$$

$$\therefore F = 12.5$$

4. To compare condition 2 with condition 3, use the following formula:

$$F = \frac{(\bar{x}_2 - \bar{x}_3)^2}{(MS_{error}/n_2) + (MS_{error}/n_3)}$$

where

$\bar{x}_2$ = the mean score for condition 2

$\bar{x}_3$ = the mean score for condition 3

$MS_{error}$ = the mean squares value for the between subjects variation (from the anova table)

$n_2$ = the number of Ss in condition 2

$n_3$ = the number of Ss in condition 3

4. To compare the interaction times with bronchitis and stroke patients:

$$F = \frac{(8.17 - 6.83)^2}{(1.52/6) + (1.52/6)}$$

where

$\bar{x}_2 = 8.17$

$\bar{x}_3 = 6.83$

$MS_{error} = 1.52$

$n_2 = 6$

$n_3 = 6$

$$F = \frac{(1.34)^2}{0.25 + 0.25}$$

$$= \frac{1.8}{0.5}$$

$$\therefore F = 3.6$$

---

*Special point*

If you have more than three conditions that you wish to compare, simply substitute the appropriate $\bar{x}$ and $n$ values in the formula.

5. To calculate $F'$, take the following steps: select the d.f.$_{bet}$ and d.f.$_{subj}$ values from the anova table to be $v_1$ and $v_2$ respectively.

6. Turn to Table C.6(a) (<0.05 values) and find the intersection figure for the $v_1$ and $v_2$ values.

7. Calculate $F'$ using:

$$F' = (C - 1)F^0$$

where $C$ is the number of conditions

$F^0$ is the value at the intersection point

5. $v_1 = $ d.f.$_{bet} = 2$
   $v_2 = $ d.f.$_{subj} = 5$

6. Intersection figure = 5.79

7. $F' = (3 - 1)5.79 = 2 \times 5.79$
   $\therefore F' = 11.58$

*Finding out whether the results are significant*
To find out whether the $F$ values obtained from the three comparisons are significant, each $F$ has to be compared with the $F'$ calculated above. If $F$ is *equal to* or *larger than* $F'$ then the comparison is significant at <0.05 (because the <0.05 table was used to calculate $F$).

The obtained $F$ values were:

condition 1 v. condition 2 = 29.5
condition 1 v. condition 3 = 12.5
condition 2 v. condition 3 =  3.6

With $F'$ equalling 11.58, it can be seen that only conditions 1 and 2, and 2 and 3, differ significantly from each other.

*What do the results mean?*
Looking back to the mean scores for each condition, we can conclude that staff nurses spend significantly less time talking to psoriasis patients than to stroke or bronchitis patients. The chance of random error accounting for these results is less than 5%. However, there is no significant difference in interaction time between bronchitis and stroke patients.

There may be some questions you want to ask about the Scheffé test.

*Q1: Given that highly significant results were obtained from the anova, how is it that only two comparisons of means were significant?*

There are two explanations for this. The first one relates to the nature of the Scheffé test itself. The test is a very stringent one which means that there is a greater likelihood of *non-significant* results. Secondly, the results of the anova are the product of the interaction between all the differences in scores. These general, overall differences contribute to the significance of the results. The differences obtained between the psoriasis and stroke and bronchitis conditions are a second specific source of differences and would also contribute to the significant anova results.

*Q2: Why is the <0.05 table used to calculate F'?*

The reason for this again relates to the stringency of the Scheffé test. If smaller probabilities were used it would be highly unlikely that you would ever get significant differences in the comparisons of means. However, should you ever find that your obtained *F* values are very much bigger than *F'* when table <0.05 has been used, you can recalculate *F'* using the relevant critical value from the smaller probability tables (i.e. Tables C.6(b), (c) or (d)).

*Q3: What is the implication of significant anova results, but insignificant Scheffé results?*

Where all the comparisons of means produce insignificant results, this simply means that the outcome of the anova was due to the total contribution of all the differences in scores without any individual condition being responsible.

---

*Exercise (answers on page 317)*

**9** Calculate a Scheffé on the following data, stating all the *F* values, as well as the *F'*, $F^0$ and *p* values. Clarify what the results mean. (The hypothesis is derived from Exercise 8 for the anova.)

**$H_1$:** Student nurses perform differently on the nursing studies examination depending on the stage of training.

**$H_0$:** There is no relationship between performance on the nursing studies examination and stage of training.

*Brief outline of study*
See Exercise 8 for the anova (page 135).

*Results*
Note that these too are derived from the sample data given in Exercise 8.

$$\text{d.f.}_{\text{bet}} = 2$$
$$\text{d.f.}_{\text{subj}} = 9$$
$$\text{MS}_{\text{error}} = 0.89$$

$$\bar{x}_1 = 10.6$$
$$\bar{x}_2 = 11.7$$
$$\bar{x}_3 = 13.2$$

# 10    Statistical tests for different subject designs with two conditions only (experimental designs)

This chapter covers three statistical tests which can be used to analyse the results from experimental designs which use *two different* or separate groups of subjects. Each group is measured on one occasion (or condition) only and the resulting two sets of data are compared for differences. In other words, you are interested in comparing the performance of two totally separate groups of subjects. The design is shown in Figure 10.1.

This sort of design is essential if two inherently different groups are being compared, for instance males/females, Asians/West Indians, Crohn's disease patients/diverticulitis patients. Remember, too, that because two different groups of subjects are involved, you do not need equal numbers in each group. If you are still unclear about when and why this design is used re-read Section 4.2.1.

The three tests described in this chapter all do essentially the same task – they compare the two sets of data to see if there are differences between them. Two of the tests – the chi-squared ($\chi^2$) test and the Mann-Whitney $U$ test are non-parametric, which means that they are less sensitive to any differences between the data, but are easier to calculate and require only minimal conditions to be fulfilled before they can be used. In contrast, the parametric unrelated $t$ test is quite complex, more sensitive and requires four conditions to be met before it can be used, the most important of which is that the data must be of an interval/ratio type (see Section 7.2 for a fuller discussion of these points.)

A basic consideration, however, when deciding which of these three tests to use

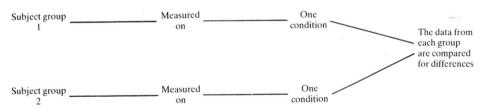

**Fig. 10.1**   A different subject design using two groups of subjects only.

**Table 10.1** Summary table outlining the conditions required by the relevant statistical tests

| Test | Type of design | Conditions for use |
|------|----------------|--------------------|
| Chi-squared ($\chi^2$) test | Different subject groups | Can *only* be used with:<br>(a) nominal data;<br>(b) two sets of data;<br>(c) at least twenty subjects Ss per group. |
| Mann-Whitney $U$ test | Different subject groups | Can *only* be used with:<br>(a) ordinal or interval/ratio data;<br>(b) two sets of data.<br>Should be used when conditions required by a parametric test cannot be met. |
| Unrelated $t$ test | Different subject groups | Can *only* be used with:<br>(a) interval/ratio data;<br>(b) two sets of data.<br>Can only be used when the remaining three conditions required by a parametric test can be fulfilled. |

concerns the level of the data that has been derived from your experiment since the following hold:

1. The $\chi^2$ test can *only* be used with nominal data (i.e. your subjects' test scores are allocated to named categories, such as pain/no pain; high/low; smoker/non-smoker).
2. The Mann-Whitney $U$ test can *only* be used with ordinal or interval/ratio data.
3. The unrelated $t$ test can *only* be used with interval/ratio data.

You should also note that the $\chi^2$ test should only be used when there are at least twenty subjects in each group.

All these requirements are summarised in Table 10.1. Although each test has slightly different specifications, they all do the same job – they compare the performances of two different subject groups for differences between them.

## 10.1 Non-parametric tests

### 10.1.1 The chi-squared test

The chi-squared test (pronounced 'kie') should be used under the following conditions:

1. The experimental design uses two *different* (separate) subject groups. Each of the two groups provides one set of data and their results are compared for differences.

2. The data derived from the experiment are of a nominal type.
3. There at least twenty subjects in each of the two groups.

The $\chi^2$ test tells you whether or not the scores from your two subject groups differ significantly from one another.

When you calculate the $\chi^2$ test, you find a numerical value for $\chi^2$ which is then looked up in the relevant probability tables. This will tell you the probability of your results being due to random error. If this probability is 5% or less, then it is unlikely that random error could account for the results, and your results are said to be *significant*. The experimental hypothesis would have been supported and the null hypothesis can be rejected. If it is more than 5%, then random error could be responsible for your results. In this case the results are *not significant*, and the null (no relationship) hypothesis would have to be accepted.

**Example**

Imagine you are working as a practice nurse and have decided to set up a 'well-man' clinic. Amongst the checks that will be on offer are cholesterol readings. Given the established link between high cholesterol and heart disease, it would be desirable if cholesterol levels could be reduced in at-risk patients. You decide to see what effect a high fibre diet would have on the cholesterol levels of these patients.

*Hypotheses*

$H_1$: High fibre diets affect levels of cholesterol in males.
$H_0$: There is no relationship between high fibre diets and cholesterol in males.

The $H_1$ is a two-tailed hypothesis because no direction to the results is predicted.

*Brief outline of study*

Forty male patients with high cholesterol levels are randomly selected from the well-man clinic. All are of similar ages, have comparable cholesterol levels and are between 5% and 10% overweight. Of these forty, twenty are randomly selected and are given a fact-sheet and advice on how to increase the fibre levels in their diet, with a daily target of 30 g. The remaining twenty patients are simply advised to be more careful about what they eat. Both groups return to the clinic after two months to have their cholesterol levels assessed again.

These results are simply classified as 'cholesterol reduced' or 'cholesterol not reduced'. This is a nominal level of measurement because the subjects are only being allocated to a named category. Because we have used two separate groups of twenty subjects and are allocating their responses to nominal categories, the necessary conditions for the $\chi^2$ test have been fulfilled.

You have the design shown in Figure 10.2. (Remember that there would be a number of additional constant errors that would have to be considered, e.g. exercise, whether the diets were followed, if this study were to be carried out properly.) You are interested in finding out whether there is any difference in cholesterol levels between the two groups at the end of your study.

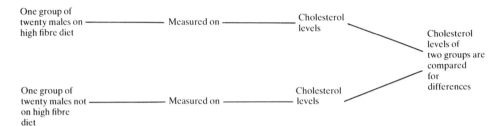

**Fig. 10.2**

*Results*
The results are given in Table 10.2.

**Table 10.2**

|  | Cholesterol reduced | Cholesterol not reduced |
|---|---|---|
| Group 1: twenty males on high fibre diet | 13 | 7 |
| Group 2: twenty males not on high fibre diet | 6 | 14 |

Although equal numbers of subjects have been used in each group here, this is not essential for the $\chi^2$ test.

*Calculating the $\chi^2$ test*

PROCEDURE

1. Set your results out in a table, with the subject groups down the side and the nominal categories across the top. Label the cells A–D *exactly* as shown

|  | Nominal category | |
|---|---|---|
| Subject groups | A | B |
|  | C | D |

WORKED EXAMPLE

1.

|  | Cholesterol | |
|---|---|---|
|  | Reduced | Not reduced |
| High fibre diet | A          13 | B          7 |
| Not high fibre diet | C          6 | D          14 |

Subject groups

---

*Special point 1*

It does not matter which nominal category is on the left, nor which subject group is at the top.

*Special point 2*

You *must* put your nominal categories across the *top* and the subject groups down the *side*, and you *must* label the cells A–D in the order shown.

---

2. Add up
   (a) cell A + cell B
   (b) cell C + cell D
   (c) cell A + cell C
   (d) cell B + cell D
3. Find the grand total $N$ by adding up the numbers in each cell, i.e. cell A + cell B + cell C + cell D.
4. Find $\chi^2$ from the formula:

$$\chi^2 = \frac{N[(AD - BC^*) - (N/2)]^2}{(A + B)(C + D)(A + C)(B + D)}$$

where $N$ is the grand total
   $AD$ = cell A × cell D
   $BC$ = cell B × cell C
   $A + B$ = cell A + cell B
   $C + D$ = cell C + cell D
   $A + C$ = cell A + cell C
   $B + D$ = cell B + cell D

\* If you ever get a minus figure from the calculation $(AD - BC)$, ignore the minus sign and treat the figure as a plus.

5. Find the degrees of freedom from the formula:
   d.f. = $(C - 1)(R - 1)$
   where $C$ = the number of columns (or vertical lines of figures in the table)
   $R$ = the number of rows (or horizontal lines of figures in the table)

2.
   (a) $13 + 7 = 20$
   (b) $6 + 14 = 20$
   (c) $13 + 6 = 19$
   (d) $7 + 14 = 21$
3. $N = 13 + 7 + 6 + 14$
   $= 40$
4.

$$\chi^2 = \frac{40[(182 - 42) - (40/2)]^2}{20 \times 20 \times 19 \times 21}$$

$N = 40$
$AD = 182$
$BC = 42$
$A + B = 20$
$C + D = 20$
$A + C = 19$
$B + D = 21$

$$\chi^2 = \frac{40[140 - 20]^2}{159\,600}$$

$$= \frac{40 \times 14\,400}{159\,600}$$

$\therefore \chi^2 = 3.61$

5. d.f. = $(2 - 1)(2 - 1)$
   $= 1$

*Looking up the results in the probability tables*
The results of the calculations have produced a $\chi^2$ value of 3.61. This must be looked up in the probability tables associated with the $\chi^2$ test to see if this figure represents a significant difference in the cholesterol levels of the two groups.

If you turn to Table C.1, you will see that across the top are the levels of significance for a two-tailed test, while down the left-hand side are degrees of freedom (d.f.) values. Firstly, find our d.f. value of 1. To the right of this are five numbers (called critical values of $\chi^2$)

$$2.71 \quad 3.84 \quad 5.41 \quad 6.64 \quad 10.83$$

Each critical value is associated with the probability level specified at the top of its column, e.g. the *p* level for the critical value 10.83 is 0.001. For an obtained $\chi^2$ value to represent significant differences between the two groups, then it must be *equal* to or *larger* than one of the critical values to the right of d.f. = 1. If the $\chi^2$ value is exactly the same as one of these critical values, then its *p* value is exactly the same as that value's. For example, if the $\chi^2$ value was 5.41 exactly then it would have an associated *p* value of 0.02 exactly. This would be expressed as $p = 0.02$.

If, however, the $\chi^2$ value is larger than the relevant critical value, then its *p* value is even less than that associated with the critical value. For instance, a $\chi^2$ value of 5.42 is larger than the critical value 5.41 and so its *p* value is even less than 0.02. This would be expressed as $p < 0.02$.

Our $\chi^2$ value of 3.61 is larger than the critical value 2.71, and so the associated *p* value is less than 0.10, i.e. $p < 0.10$. (Our hypothesis was two-tailed.)

*What do the results mean?*
Our $\chi^2$ value has a probability value of less than 0.10. This means that there is less than a 10% chance that the results from the study could be accounted for by random error. However, using the usual cut-off point of 5% or less (see Section 5.6), this means our results are not significant. In other words, the difference in the cholesterol levels of the two groups is not a significant one and could be explained by random error. Therefore, the experimental hypothesis has to be rejected and the null (no relationship) hypothesis must be accepted. We must conclude that a high fibre diet does not significantly affect high cholesterol levels in males.

The data in this example were not significant. However, if you have a one-tailed hypothesis and you find that your own results *are* significant when you look them up in the table, you must check your raw data before concluding that your experimental hypothesis has been supported. The reason for this is simply that it is possible to obtain results which are significant, but which are directly *opposite* to those predicted in a one-tailed hypothesis. In such cases, the hypothesis would not have been supported. So, it is essential to check the figures in your table to ensure that their direction corresponds to the one predicted.

If you have carried out the $\chi^2$ test on your own data there may be some issues you wish to raise.

*Q1: My hypothesis is one-tailed, but the probability levels in the table are only for two-tailed hypotheses. What should I do?*

If you look back at Section 7.4 you will see that the results from a two-tailed hypothesis are twice as likely to be caused by random error as those from a one-tailed hypothesis. Translated into probability levels, this means that the $p$ value for a two-tailed hypothesis is always twice the $p$ value of a one-tailed hypothesis. Therefore, if you have a one-tailed hypothesis, you simply look up your $\chi^2$ value as normal and *halve* the associated probability value. For instance, you might have a $\chi^2$ value of 2.71. This has a $p$ value of 0.10 for a two-tailed test (see Table C.1) and consequently of 0.05 for a one-tailed test. So for a one-tailed test, simply halve the two-tailed $p$ value.

---

*Exercises (answers on page 317)*

1  Look up the following $\chi^2$ values in the appropriate table and state whether or not they are significant and at what probability level.
(a) $\chi^2 = 4.79$, d.f. $= 1$, one-tailed
(b) $\chi^2 = 3.05$, d.f. $= 1$, one-tailed
(c) $\chi^2 = 5.41$, d.f. $= 1$, two-tailed
(d) $\chi^2 = 1.91$, d.f. $= 1$, one-tailed
(e) $\chi^2 = 11.24$, d.f. $= 1$, two-tailed
2  Calculate a $\chi^2$ test on the following results; state the $\chi^2$, d.f. and $p$ values and clarify what the results mean.

**H₁:** 'Women who have a lumpectomy for carcinoma are less likely to suffer from severe depression in the six months following surgery, when compared with women who have a radical mastectomy.'

**H₀:** 'There is no relationship between the type of surgery for carcinoma of the breast and subsequent depression.'

*Brief outline of study*
One group of twenty-five lumpectomy patients and a further group of twenty radical mastectomy patients are randomly selected for study. During the six

**Table 10.3**

|                                      | *Depression* | *No depression* |
|--------------------------------------|:------------:|:---------------:|
| Group 1: lumpectomy patients         | 8            | 17              |
| Group 2: radical mastectomy patients | 15           | 5               |

months following surgery they are monitored for severe depression (defined as whether or not medical intervention has been required). If the patient experienced severe depression they are allocated to the 'Depression' category, and if they did not, they are allocated to the 'No depression' category. The results in Table 10.3 are obtained. Calculate a $\chi^2$ test on these results.

### 10.1.2 The Mann-Whitney U test

The Mann-Whitney $U$ test is used under the following conditions:

1. Two different (or separate) groups of subjects are being compared. Each group is tested on one occasion (or condition) only and the two sets of data that result are compared for differences between them.
2. The data from the experiments are ordinal or interval/ratio.

The Mann-Whitney $U$ test tells you whether the scores obtained from the two groups of subjects differ significantly from each other.

When calculating the Mann-Whitney $U$ test, a numerical value for $U$ is found. This value is then looked up in the probability tables associated with this test to see whether it represents a significant difference in the data from the two groups. If the probability ($p$) level associated with the $U$ value is 5% or less, then the difference between the two sets of data is *significant* and the results cannot be explained by random error. The experimental hypothesis can be accepted and the null hypothesis rejected. If it is greater than 5%, then the difference between the two sets of data is *not significant* and could be explained by random error. In such cases, the null (no relationship) hypothesis would have been supported.

**Example**
Imagine you are working on a specialist plastic surgery ward and you have noticed that tubed pedicle flaps are more likely to become oedematous in the forty-eight hours after surgery than are direct flap grafts. Obviously, excess oedema may result in the loss of the flap and so it would be valuable to compare the two techniques of this issue. You decide to make a more systematic investigation of this.

*Hypotheses*

> **H₁:** Tubed pedicle flaps are more likely to become oedematous than are direct flap grafts.
>
> **H₀:** There is no relationship between the type of skin graft and the degree of oedema.

The $H_1$ is one-tailed because it is predicted that tubed pedicle grafts are *more* likely to become oedematous.

*Brief outline of study*

Over a period of three months you randomly select twelve patients who have received tubed pedicle grafts to the forearm and ten who have had direct flap grafts to the forearm. (Because the Mann-Whitney test is used for experimental designs with two different subject groups, it is not essential to have equal numbers of subjects in each group.) All patients were of comparable age, and had similar size grafts. (Other constant errors would need to be taken into account if this study were to be carried out properly, e.g. presence of diabetes, surgical skill, age and condition of patient.) During the forty-eight hours following surgery, the grafts are assessed on a six point scale for the degree of oedema that occurs:

| 1 | 2 | 3 | 4 | 5 | 6 |
|---|---|---|---|---|---|
| Not at all oedematous | Very slightly oedematous | Fairly oedematous | Moderately oedematous | Very oedematous | Extremely oedematous |

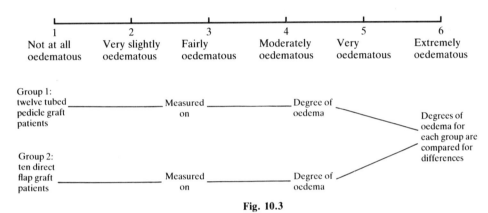

Fig. 10.3

**Table 10.4**

| Subject | Condition A: tubed pedicle graft patients | Rank A | Subject | Condition B: direct graft patients | Rank B |
|---|---|---|---|---|---|
| 1 | 6 | 22 | 1 | 2 | 3.5 |
| 2 | 3 | 9 | 2 | 4 | 14.5 |
| 3 | 4 | 14.5 | 3 | 1 | 1 |
| 4 | 5 | 19 | 4 | 2 | 3.5 |
| 5 | 5 | 19 | 5 | 3 | 9 |
| 6 | 3 | 9 | 6 | 3 | 9 |
| 7 | 2 | 3.5 | 7 | 2 | 3.5 |
| 8 | 5 | 19 | 8 | 4 | 14.5 |
| 9 | 3 | 9 | 9 | 3 | 9 |
| 10 | 4 | 14.5 | 10 | 3 | 9 |
| 11 | 5 | 19 | | | |
| 12 | 5 | 19 | | | |
| $\Sigma A = 50$ | | $\Sigma R_A = 176.5$ | $\Sigma B = 27$ | | $\Sigma R_B = 76.5$ |
| $\bar{x}_A = 4.17$ | | | $\bar{x}_B = 2.7$ | | |

$\Sigma$ means 'total' and $\bar{x}$ means 'average'

This is an ordinal scale of measurement, since it orders the results along the dimension of oedema and since a score of six does *not* necessarily mean twice the degree of oedema as a score of three. As two separate groups of subjects and an ordinal scale of measurement have been used, the prerequisite conditions of the Mann-Whitney have been fulfilled. The experimental design is shown in Figure 10.3 (see previous page). You are interested in establishing whether there are any differences in the degree of oedema between the two groups.

*Results*
The results in Table 10.4 (see previous page) are obtained (bold type represents the results from the study; light type are the calculations from the Mann-Whitney test. It does not matter which is condition A and which condition B.)

*Calculating the Mann-Whitney* U *test*

| PROCEDURE | WORKED EXAMPLE |
|---|---|
| 1. Add up all the scores in condition A to give $\Sigma A$. | 1. $\Sigma A = 50$ |
| 2. Calculate the mean score for condition A to give $\bar{x}_A$. | 2. $\bar{x}_A = 4.17$ |
| 3. Add up all the scores in condition B to give $\Sigma B$. | 3. $\Sigma B = 27$ |
| 4. Calculate the mean score for condition B to give $\bar{x}_B$. | 4. $\bar{x}_B = 2.7$ |
| 5. Taking the whole set of twenty-two scores (i.e. the scores for *both* groups together), rank order these scores, giving a rank of 1 to the lowest score, a rank of 2 to the next lowest and so on. Where two or more scores are the same, use the tied rank procedure described in Section A.2. | 5. See the 'Rank' columns in Table 10.4. |
| 6. Put the ranks in the columns entitled Rank A and Rank B. Add up the ranks in the Rank A column to give $\Sigma R_A$. | 6. $\Sigma R_A = 176.5$ |
| 7. Add up the ranks in the Rank B column to give $\Sigma R_B$. | 7. $\Sigma R_B = 76.5$ |
| 8. Select the *larger* of the two rank totals to use in the Mann-Whitney formula. | 8. $\Sigma R_A = 176.5$ is the larger rank total. |
| 9. Calculate *U* from the formula: | 9. |

$$U = n_A n_B + \frac{n_x(n_x + 1)}{2} - \Sigma R_x$$

Where

$n_A$ = the number of Ss in condition A

$n_B$ = the number of Ss in condition B

$n_x$ = the number of Ss in the condition with the larger rank total

$\Sigma R_x$ = the larger rank total

10. As we used *different* numbers of subjects in each group, these calculations must be repeated, substituting the *smaller* rank total for $\Sigma R_x$ and the number of Ss in the condition with the smaller rank total for $n_x$. From these calculations we shall find the value of $U'$.

11. Select the smaller of the two values $U$ and $U'$ to be looked up in the probability tables.

$$U = 12 \times 10 + \frac{12(12 + 1)}{2} - 176.5$$

Where

$n_A = 12$

$n_B = 10$

$n_x = 12$

$\Sigma R_x = 176.5$

$$U = 120 + \frac{12 \times 13}{2} - 176.5$$

$$= 120 + 78 - 176.5$$

$$\therefore U = 21.5$$

10. With $\Sigma R_x = 76.5$
$$n_x = 10$$

$$U' = 12 \times 10 + \frac{10(10 + 1)}{2} - 76.5$$

$$= 120 + \frac{10 \times 11}{2} - 76.5$$

$$= 120 + 55 - 76.5$$

$$\therefore U' = 98.5$$

11. Select $U = 21.5$ to be looked up in the probability tables.

---

*Special point 1*

Remember to take the *whole* set of scores for *both* groups when you carry out the ranking in stage 5. Do *not* rank each group's scores separately.

*Special point 2*

If you used equal numbers of Ss in each group you need only calculate the first formula, which uses the larger rank total ($\Sigma R_x$) and the number of Ss in the condition with the larger rank total as $n_x$. However, if you use unequal numbers of Ss in each group then you must also calculate the value of $U'$, substituting for $\Sigma R_x$, the smaller rank total and for $n_x$ the number of Ss in the condition with the smaller rank total.

> *Special point 3*
>
> Remember to select the *smaller* of the two values U and U' to look up in the
> probability tables.

*Looking up the results in the probability tables*
The results of the Mann-Whitney test have produced a $U$ value of 21.5, which must
now be looked up in the relevant probability tables, to see whether this figure
represents a significant difference in the degree of oedema in the two types of skin
graft. Turn to Tables C.7(a)–(d). Each of these tables is associated with a different
probability level. Table C.7(a) represents the most significant $p$ values and Table
C.7(d) the least significant ones.

Start with Table C.7(a). Across the top you will see values for $n_1$ and down
the side values for $n_2$; $n_1$ is the number of subjects in condition 1 (or A) and $n_2$
is the number of subjects in condition 2 (or B). Our $n_1$ is 12 and $n_2$ is 10. Look
across the $n_1$ values in the table for 12 and down the $n_2$ values for 10. Then look
across to the figure at their intersection point in the table: 21. For our $U = 21.5$ to
be significant at the $p$ level indicated by the table, it has to be *equal to* or *less* than
the figure at the intersection point. If the obtained $U$ value is *exactly the same* as the
figure at the intersection point, then the associated probability level is exactly the
same as the one indicated at the top of the table. This is expressed as $p = 0.005$ (or
0.01 if the hypothesis is two-tailed). However, if the obtained $U$ value is less than
the figure at the intersection point, then the associated probability is even less than
the one indicated at the top of the table. This is expressed as $p < 0.005$ (or $p < 0.01$
if the hypothesis is two-tailed).

Our $U$ value of 21.5 is larger than the intersection figure of 21 and so it is not
significant on Table C.7(a).

Move on, then, to Table C.7(b) and repeat the process. Here the number at the
intersection point is 24, which, since our $U$ value is smaller, and we had a one-tailed
hypothesis, means that our results are significant at *even less* than 0.01. This is
expressed as $p < 0.01$. Had our $U$ value been larger than the figure at the intersection point, we would have repeated the process for Tables C.7(c) and (d). If the
$U$ value was larger than the appropriate figure in Table C.7(d) we would have to
conclude that our results were not significant.

*What do the results mean?*
Our $U$ value has an associated probability of less than 0.01 for a one-tailed
hypothesis, which means that there is less than a 1% chance that our results could
be explained by random error. Taking the standard cut-off point of 5% or less for
claiming results as significant (see Section 5.6), this means that the results from our
study *are* significant and could not be accounted for by random error. In other
words, there is a significant difference in the degree of oedema according to the
type of skin graft performed. However, our hypothesis predicted that tubed pedicle
flaps would produce *more* oedema than direct flap grafts, so we must check the
original data to see whether the difference in the actual results was the same as the

one predicted. (Checking the raw data is an essential step before you can state unequivocally that your hypothesis has been supported, since it is possible to get significant differences in the results which are directly *opposite* to those predicted.) Here the mean score for tubed pedicle grafts is 4.17 and for direct flap grafts is 2.7. This means that tubed pedicle grafts *do* produce significantly more oedema than direct flap grafts. As this is what was predicted, we can state the experimental hypothesis has been supported, consequently the null (no relationship) hypothesis can be rejected. We can conclude that tubed pedicle grafts produce significantly more oedema than direct flap grafts in patients having forearm skin grafting.

---

### Exercises (answers on page 317)

**3** Look up the following $U$ values in the probability tables and state whether or not they are significant, and at what probability level.
  (a) $U = 43.5$, $n_1 = 15$, $n_2 = 14$, one-tailed
  (b) $U = 60$, $n_1 = 16$, $n_2 = 12$, two-tailed
  (c) $U = 41.5$, $n_1 = 12$, $n_2 = 12$, one-tailed
  (d) $U = 97$, $n_1 = 20$, $n_2 = 18$, two-tailed
  (e) $U = 49$, $n_1 = 14$, $n_2 = 14$, two-tailed
**4** Calculate a Mann-Whitney $U$ test on the following results, stating the $U$, $U'$, $n_1$, $n_2$ and $p$ values. State what the results mean.

  **H$_1$:** People who live in areas with high levels of aluminium in the water supply are more likely to suffer a severe degree of Alzheimer's disease than are people living in areas with low levels of aluminium in the water.

  **H$_0$:** There is no relationship between the level of aluminium in the water supply and severity of Alzheimer's disease.

*Brief outline of study*
Randomly select fifteen elderly people, all with Alzheimer's disease, from a geriatric ward in an area with high levels of aluminium in the water. Randomly select a further fourteen subjects, again all with Alzheimer's disease, from a geriatric ward in an area with low levels of aluminium in the water supply. Assess each subject along a five point scale of severity of Alzheimer's:

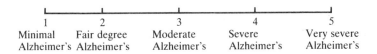

| 1 | 2 | 3 | 4 | 5 |
|---|---|---|---|---|
| Minimal | Fair degree | Moderate | Severe | Very severe |
| Alzheimer's | Alzheimer's | Alzheimer's | Alzheimer's | Alzheimer's |

You obtain the results given in Table 10.5.

**Table 10.5**

| Subject | Condition A: high levels of aluminium | Subject | Condition B: low levels of aluminium |
|---------|---------------------------------------|---------|--------------------------------------|
| 1  | 4 | 1  | 3 |
| 2  | 3 | 2  | 3 |
| 3  | 5 | 3  | 5 |
| 4  | 3 | 4  | 1 |
| 5  | 2 | 5  | 4 |
| 6  | 4 | 6  | 5 |
| 7  | 2 | 7  | 1 |
| 8  | 3 | 8  | 2 |
| 9  | 5 | 9  | 2 |
| 10 | 4 | 10 | 1 |
| 11 | 4 | 11 | 3 |
| 12 | 3 | 12 | 5 |
| 13 | 2 | 13 | 3 |
| 14 | 3 | 14 | 3 |
| 15 | 3 |    |   |

## 10.2  Parametric test

### 10.2.1  The unrelated t test

The unrelated *t* test should be used under the following conditions:

1. The experimental design uses *two* different (separate) groups of subjects. Each group provides *one* set of data, which are then compared to find out whether the performance of the two groups differs significantly.
2. The data derived from the experiment are of an interval/ratio type.
3. The remaining three conditions required by a parametric test can be (more or less) fulfilled.

The unrelated *t* test tells you whether the scores from your two groups are significantly different or whether the results are simply due to random error.

When you calculate the unrelated *t* test, you find a numerical value for *t* which is then looked up in the probability tables associated with the unrelated *t* test. This will give you the probability that your results are due to random error. If this probability is 5% or less, then your results are *significant* and it is unlikely that random error can account for them. The experimental hypothesis would have been supported, and the null hypothesis can be rejected. If it is more than 5%, then random error could explain your results and so they are *not significant*. The null (no relationship) hypothesis would have been supported.

The unrelated *t* test is quite difficult to calculate and produces large numbers. Do not attempt it without a calculator. (Note that it is unnecessary with this test to use equal numbers of subjects in each group.)

**Example**
Imagine you are a Senior Nurse Tutor in a large school of nursing. When each new intake of trainee nurses arrives, you randomly allocate half of them to tutor A for physiology and the remaining half to tutor B. Over the last five intakes, however, you have noticed that students with tutor A do consistently less well than students with tutor B, despite comparable marks in other subjects. You suspect tutor A of poor teaching, but before doing anything, you want to establish some facts. You decide to compare the physiology exam marks for each tutor, to see if, in fact, tutor A's students do perform significantly less well.

*Hypotheses*

> $H_1$: Physiology exam performance of students with tutor A is signi-
> ficantly lower than that of students with tutor B.
>
> $H_0$: There is no relationship between exam performance in physiology
> and tuition.

The $H_1$ is a one-tailed hypothesis because we are predicting *lower* exam marks with tutor A.

*Brief outline of study*
From the past two years, one intake of twenty-three student nurses is randomly selected. The exam marks in the first year physiology exam are collected, and sorted according to whether the student was taught by tutor A or tutor B. (Remember that the initial allocation to tutor A or B was made upon entry to the course, so no further random selection is necessary.) Since two separate groups of subjects have been used, the data are of an interval/ratio type (because percentages are used) and the remaining three conditions required by parametric tests have not been violated, the necessary conditions for the unrelated *t* test have been met.

You are interested in whether there are differences between the two groups on the physiology exam. You have the design shown in Figure 10.4.

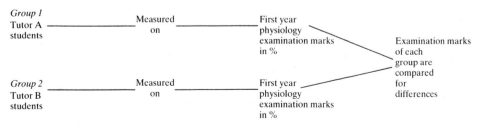

Fig. 10.4

*Results*

The results in Table 10.6 are obtained (figures in bold type are the raw data from the study, while those in light type are the calculations from the unrelated *t* test. It does not matter which condition is A and which B.)

**Table 10.6**

| Subject | Condition A: students of tutor A | $A^2$ | Subject | Condition B: students of tutor B | $B^2$ |
|---------|----------------------------------|-------|---------|----------------------------------|-------|
| 1 | **45** | 2025 | 1 | **53** | 2809 |
| 2 | **50** | 2500 | 2 | **61** | 3721 |
| 3 | **61** | 3721 | 3 | **65** | 4225 |
| 4 | **59** | 3481 | 4 | **64** | 4096 |
| 5 | **35** | 1225 | 5 | **69** | 4761 |
| 6 | **49** | 2401 | 6 | **40** | 1600 |
| 7 | **58** | 3364 | 7 | **49** | 2401 |
| 8 | **64** | 4096 | 8 | **62** | 3844 |
| 9 | **68** | 4624 | 9 | **55** | 3025 |
| 10 | **50** | 2500 | 10 | **39** | 1521 |
| 11 | **53** | 2809 | 11 | **71** | 5041 |
| 12 | **56** | 3136 | | | |
| $\Sigma A = \mathbf{648}$ | | $\Sigma A^2 = 35\,882$ | $\Sigma B = \mathbf{628}$ | | $\Sigma B^2 = 37\,044$ |
| $\bar{x}_A = \mathbf{54}$ | | | $\bar{x}_B = \mathbf{57.09}$ | | |

$\Sigma$ means 'total' and $\bar{x}$ means average

*Calculating the unrelated* t *test*

| PROCEDURE | WORKED EXAMPLE |
|-----------|----------------|
| 1. Add up all the scores in condition A to give $\Sigma A$. | 1. $\Sigma A = 648$ |
| 2. Find the mean score for condition A to give $\bar{x}_A$. | 2. $\bar{x}_A = 54$ |
| 3. Add up all the scores in condition B to give $\Sigma B$. | 3. $\Sigma B = 628$ |
| 4. Find the mean score condition B to give $\bar{x}_B$. | 4. $\bar{x}_B = 57.09$ |
| 5. Square *each* score in condition A and enter the result in the '$A^2$' column, e.g. $45^2 = 2025$. | 5. See '$A^2$' column |
| 6. Square *each* score in condition B and enter the result in the '$B^2$' column, e.g. $53^2 = 2809$. | 6. See '$B^2$' column |
| 7. Add up all the squared scores in $A^2$ column to give $\Sigma A^2$. | 7. $\Sigma A^2 = 35\,882$ |
| 8. Add up all the squared scores in $B^2$ column to give $\Sigma B^2$. | 8. $\Sigma B^2 = 37\,044$ |

9. Square $\Sigma A$ to give $(\Sigma A)^2$.

10. Square $\Sigma B$ to give $(\Sigma B)^2$.

9. $(\Sigma A)^2 = 648^2$
$= 419\,904$

10. $(\Sigma B)^2 = 628^2$
$= 394\,384$

---

**Special point 1**

It is essential to distinguish between $\Sigma A^2$ (step 7) and $(\Sigma A)^2$ (step 9). $\Sigma A^2$ is the total of all the individually squared scores. $(\Sigma A)^2$ is the total of all the scores, which is then squared.

---

11. Calculate $t$ from the formula:

$$t = \frac{\bar{x}_A - \bar{x}_B}{\sqrt{\left[\frac{\left(\Sigma A^2 - \frac{(\Sigma A)^2}{n_1}\right) + \left(\Sigma B^2 - \frac{(\Sigma B)^2}{n_2}\right)}{(n_1 - 1) + (n_2 - 1)}\left(\frac{1}{n_1} + \frac{1}{n_2}\right)\right]}}$$

11.

$$t = \frac{54 - 57.09}{\sqrt{\left[\frac{\left(35\,882 - \frac{419\,904}{12}\right) + \left(37\,044 - \frac{394\,384}{11}\right)}{11 + 10}\left(\frac{1}{12} + \frac{1}{11}\right)\right]}}$$

Where

$\bar{x}_A$ = the mean score for condition A

$\bar{x}_B$ = the mean score for condition B

$(\Sigma A)^2$ = the total of the scores from condition A, squared

$(\Sigma B)^2$ = the total of the scores from condition B, squared

$\Sigma A^2$ = the total of all the individually squared scores from condition A

$\Sigma B^2$ = the total of all the individually squared scores from condition B

$n_A$ = the number of Ss in condition A

$n_A$ = the number of Ss in condition B

$\sqrt{}$ = the square root of the result of the calculations under this sign

$\bar{x}_A = 54$

$\bar{x}_B = 57.09$

$(\Sigma A)^2 = 419\,904$

$(\Sigma B)^2 = 394\,384$

$\Sigma A^2 = 35\,882$

$\Sigma B^2 = 37\,044$

$n_A = 12$

$n_B = 11$

$$t = \cfrac{-3.09}{\sqrt{\left[\cfrac{890 + 1190.91}{21} \times 0.17\right]}}$$

$$= \cfrac{-3.09}{\sqrt{[99.09 \times 0.17]}}$$

$$= \cfrac{-3.09}{4.11}$$

$$\therefore t = -0.75$$

---

*Special point 2*

If you end up with a minus value for $t$, ignore the minus sign and treat the figure as positive, $\therefore t = -0.75$ becomes $t = 0.75$

---

12. Calculate the degrees of freedom (d.f.) from the following formula:
    d.f. $= (n_A - 1) + (n_B - 1)$

12. d.f. $= (12 - 1) + (11 - 1)$
    $= 21$

*Looking up the results in the probability tables*
The results of our calculations have produced a $t$ value of 0.75. To see whether this figure represents a significant difference in the exam marks of each group, it must be looked up in the relevant probability tables. This will provide a probability ($p$) value for $t = 0.75$, which essentially will tell us the likelihood of the results being due to random error.

To look up $t = 0.75$, we also need the d.f. value of 21. Turn to Table C.3. Across the top are the levels of significance (or probabilities) for both one- and two-tailed hypotheses, while down the side are d.f. values from 1 to infinity ($\infty$). Look down the left-hand column for our d.f. of 21. To the right of this are six numbers, called 'critical values of $t$':

$$1.323 \quad 1.721 \quad 2.080 \quad 2.518 \quad 2.831 \quad 3.819$$

Each critical value is associated with the probability value stated at the top of its column. So 1.323 is associated with a $p$ value of 0.10 for a one-tailed hypothesis and 0.20 for a two-tailed hypothesis.

For our $t$ of 0.75 to be significant at one of the levels indicated at the top of the table, it has to be *equal to* or *larger than* the associated critical value. If the obtained $t$ value is *exactly the same* as the critical value, then it has exactly the same probability, e.g. $t = 2.080$ has a $p$ value of exactly 0.025 (one-tailed) (or 0.05, two-tailed). This is expressed as $p = 0.025$ (or $p = 0.05$).

However, if the obtained $t$ value is larger than the critical value, the associated probability is even less than that of the critical value. So, an obtained $t$ of 2.115 is larger than the critical value 2.080, and so the associated $p$ value is even less than 0.025 one-tailed (or 0.05, two-tailed). This is expressed as $p < 0.025$ (or $<0.05$).

Our $t$ value of 0.75 is smaller than even the smallest critical value for d.f. = 21, and so is not significant at the 0.10 level for a one-tailed hypothesis.

*What do the results mean?*

For a $t$ of 0.75, d.f. or 21 and a one-tailed hypothesis, our results were not significant at even the 10% probability level. This means that there is greater than a 10% chance that random error could account for the results. Using the usual cut-off point of 5% or less for claiming results to be significant we have to accept that there is no difference in performance between the two groups. The null (no relationship) hypothesis has been supported. We can conclude that students of tutor A did not perform less well than students of tutor B on the 1st year physiology exam.

Some questions which might arise when you calculate the $t$ test on your own data:

> Q1: *Supposing I obtained a* t *value which was equal to or larger than one of the critical values. What does this mean about my results?*

If you obtained a critical value which is equal to or larger than one of the critical values associated with a $p$ level of 5% or less, then it would mean your results were significant. Imagine that the study we have just described yielded a $t$ value of 2.94 (d.f. = 21 and a one-tailed hypothesis). This is even larger than the critical value 2.831 and so has a $p$ value of even less than 0.005 (expressed as $p < 0.005$). This means that there is less than a ½% chance that the results are due to random error, and (using the usual cut-off point of 5% or less) means that there is a significant difference in the exam performance of the two groups. *But* our hypothesis predicted that the students of tutor A would do less well, and so you would have to check the mean scores for each group to see whether that, in fact, was the case before we could claim that our experimental hypothesis had been supported. This is essential since, as the unrelated $t$ test only tells you whether there are significant differences between two sets of scores, it is quite possible to get significant results which are the opposite to those predicted (i.e. tutor A's students doing *better* than tutor B's). If your mean scores confirm that the differences are as predicted, you can conclude that the experimental hypothesis has been supported.

*Exercises (answers on page 317)*

**5** Look up the following $t$ values in the relevant probability tables and state whether or not they are significant and at what probability level.
(a) $t = 2.39$, d.f. = 15, two-tailed
(b) $t = 2.65$, d.f. = 13, two-tailed

(c) $t = 1.53$, d.f. $= 19$, one-tailed
(d) $t = 2.91$, d.f. $= 18$, two-tailed
(e) $t = 4.41$, d.f. $= 24$, one-tailed

**6** Calculate an unrelated $t$ test on the following results; state the $t$, d.f. and $p$ values and explain what the results actually mean.

> **H$_1$:** Patients who have pre-hysterectomy counselling recover more quickly than patients who do not.
>
> **H$_0$:** There is no relationship between whether or not a patient has counselling and the speed of recovery from a hysterectomy.

*Brief outline of study*
From twenty-four women who are due to have a hysterectomy, randomly select twelve to receive two hours' pre-operative counselling about the operation, its effects, the recovery period, etc. (A number of constant errors would need to be controlled for, such as reasons for the operation, complications, age of patient) The remaining twelve women receive no counselling. The number of days each patient takes from the operation to discharge is recorded.

The results are given in Table 10.7.

**Table 10.7**

| Subject | Condition A: counselling | Condition B: no counselling |
|---------|--------------------------|-----------------------------|
| 1 | 5 | 7 |
| 2 | 7 | 7 |
| 3 | 6 | 8 |
| 4 | 8 | 6 |
| 5 | 5 | 7 |
| 6 | 6 | 6 |
| 7 | 7 | 8 |
| 8 | 7 | 9 |
| 9 | 8 | 8 |
| 10 | 6 | 7 |
| 11 | 5 | 7 |
| 12 | 5 | 8 |

# 11 Statistical tests for different subject designs with three or more conditions (experimental designs)

This chapter deals with statistical tests which are used to analyse data derived from experimental designs which use three or more separate or *different* groups of subjects. Each group of subjects is measured on one occasion (or condition) and the results from each group are compared for differences between them. The design is shown in Figure 11.1.

This sort of design is typically used when comparisons are being made between subject groups which differ inherently from each other in some specified way, e.g. children v. adolescents v. adults or Asians v. West Indians v. Caucasians or RGNs v. RMNs v. RSCNs.

The tests described in this chapter all do a similar job – they compare the data from each condition, to establish whether there are significant differences between the groups' performances. However, three of the tests are non-parametric (the extended chi-squared or $\chi^2$, the Kruskal-Wallis and the Jonckheere trend test), while the one-way analysis of variance (anova) for unrelated designs is parametric (see Section 7.2 for a discussion of the differences between parametric and non-parametric tests). The Scheffé multiple range test for use with unrelated designs is included. It should be stressed that this test should *only* be used in conjunction with the anova.

While the tests all perform a similar function, they differ in terms of the conditions required for use. These conditions are outlined in Table 11.1 and will be clarified as each test is described in greater detail.

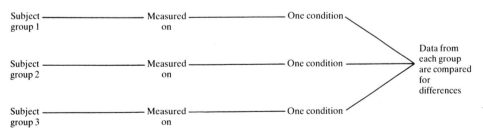

**Fig. 11.1**  A different subject design using three (or more) groups of subjects.

**Table 11.1** Summary table outlining the conditions required by the relevant statistical tests

| Test | Type of design | Conditions for use |
|---|---|---|
| Extended chi-squared ($\chi^2$) test | Different subject designs | Can *only* be used with:<br>(a) three or more subject groups;<br>(b) nominal data, *or alternatively*;<br>(c) two different groups of subjects and three or more nominal categories. |
| Kruskal-Wallis test | Different subject designs | Can *only* be used with:<br>(a) three or more subject groups;<br>(b) ordinal or interval/ratio data. Should be used when the conditions for a parametric test cannot be fulfilled. |
| Jonckheere trend test | Different subject designs | Can *only* be used with:<br>(a) three or more subject groups;<br>(b) ordinal or interval/ratio data; and when the experimental hypothesis predicts a trend in the results. |
| One-way analysis of variance (anova) for unrelated designs | Different subject designs | Can *only* be used with:<br>(a) three or more subject groups;<br>(b) interval/ratio data; and when the remaining conditions required by a parametric test can be fulfilled (see section 7.2). |
| Scheffé multiple range test for use with unrelated anovas | Different subject designs | Can *only* be used:<br>(a) when the one-way anova for unrelated designs has been computed;<br>(b) when the results of the anova are significant. |

## 11.1 Non-parametric tests

### 11.1.1 The extended chi-squared test

The extended $\chi^2$ test should be used under the following conditions:

1. With nominal data.
2. With three or more separate groups of subjects and two or more nominal categories.
3. Alternatively with two separate groups of subjects plus three or more nominal categories.

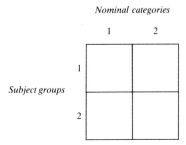

**Fig. 11.2**   A $2 \times 2 \chi^2$ table.

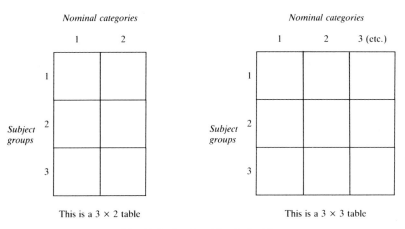

**Fig. 11.3**   $3 \times 2$ and $3 \times 3 \chi^2$ tables.

This might look a bit confusing initially but is really quite simple. If you look back to the ordinary $\chi^2$ (Section 10.1.1), you can see that it is used with two separate groups of subjects whose responses are allocated to two nominal categories only. This yields a $2 \times 2$ table; Figure 11.2. It is called a $2 \times 2$ table because there are two subject groups and two nominal categories.

However the extended $\chi^2$ can be used with three groups of subjects, whose responses can be allocated to two or more nominal categories which would yield the tables in Figure 11.3.

It can also be used with two groups of subjects, whose responses can be allocated to three or more nominal categories. This would produce a $2 \times 3$ (or more) table (Figure 11.4). (Note that this design only has two groups of subjects and therefore appears to contradict the chapter heading. This anomaly is explained on the next page.)

So, if your experimental design uses three or more categories of nominal data plus two groups of subjects or three or more groups of subjects and two or more

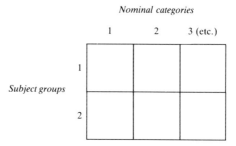

Fig. 11.4   A $2 \times 3 \chi^2$ table.

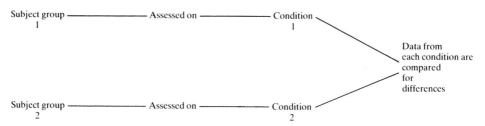

Fig. 11.5   An experimental design which uses two different groups of subjects and three or more nominal categories.

nominal categories of response then the extended $\chi^2$ test should be used to analyse the results.

You might have noticed from this that if you had only two groups of subjects and three or more nominal categories, your design would have looked like Figure 11.5. This design clearly only has two conditions and yet the extended $\chi^2$ test comes into a chapter covering designs with *three or more* conditions. This is because this particular test can be used for designs with only two conditions as long as *three or more* nominal categories are also used. I do hope this does not increase the confusion!

When calculating the extended $\chi^2$ test, a numerical value for $\chi^2$ is calculated which is then looked up in the tables associated with the extended $\chi^2$ test to determine the probability of random error accounting for the results. If the probability is 5% or less, then random errors are unlikely to be able to explain your results. The results are said to be *significant* and the experimental hypothesis *supported*. The null (no relationship) hypothesis can be rejected. If it is more than 5%, then random error could account for the results. In this case the results are *not* significant and the null (no relationship) hypothesis has been supported. The experimental hypothesis must be rejected.

The extended $\chi^2$ test only tells you whether or not there are overall differences between the groups of subjects and not the direction of these differences. Because of this, any hypothesis related to the extended $\chi^2$ test must be two-tailed.

One further proviso should be made with regard to this test. The calculations involve computing those data which would be *expected* to occur if the results were completely random in their distribution. These figures are called the expected frequencies and are compared with the data actually obtained from the study (which are called *observed frequencies*). The larger the disparity between the observed and expected frequencies the more significant the results. However, although it does not matter how big the observed frequency values are, the expected frequency values should always be more than five. The easiest way to ensure this is by having a minimum of twenty subjects in each group. Note that because this is a test for different subject designs, equal numbers of subjects do not have to be used in each group.

**Example**
There has been a significant rise in the incidence of malignant melanoma over recent years, partially attributable to the popularity of sun-tans and package holidays and to the weakening of the ozone layer. Clearly some skin types are more prone to this type of cancer, with fair-skinned people being most vulnerable. Imagine you are working in an oncology unit and that your interest lies with *recurrence* of malignant melanoma. In particular, you think that there seems to be a greater recurrence of the disease among ginger-haired people, as opposed to the fair- or dark-haired, irrespective of how advanced the condition is or of any subsequent protection they might take against the sun.

*Hypotheses*

> $H_1$: There is a relationship between hair colour and recurrence of malignant melanoma.
>
> $H_0$: There is no relationship between hair colour and recurrence of malignant melanoma.

The $H_1$ is two-tailed (of necessity), as it predicts only that there are differences in recurrence rate according to hair colour and not what these differences are.

*Brief outline of study*
Twenty ginger-haired women, twenty fair-haired women and twenty dark haired women all suffering from malignant melanoma are selected for study. In each case the melanoma is at a comparable stage. All the women are advised to use high-protection sun screens following the excising of the tumour, and to keep out of direct and very hot sunlight. They are followed up over a three year period for a recurrence of the malignant melanoma. The results are classified under the nominal categories 'recurrence' or 'no recurrence'.

Because we have three different groups of subjects (ginger-, fair- and dark-haired) and two nominal categories (recurrence or no recurrence) the conditions required by the extended $\chi^2$ test have been fulfilled. The design is shown in Figure 11.6. As there are three subject groups and two nominal categories, a $3 \times 2$ table for the data can be produced.

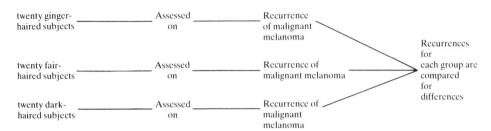

**Fig. 11.6**

A number of constant errors would need to be eliminated if this study were to be carried out, for example, the age of the women, the number of times they had experienced sunburn during childhood, whether or not they conformed to the advice given about subsequent protection, etc.

You are interested in whether the number of recurrences differs for each group.

*Results*
The results in Figure 11.7 are obtained, which should be set out in a $3 \times 2$ table.

*Nominal categories*

|  |  | Recurrence | No recurrence | Totals $(T_s)$ |
|---|---|---|---|---|
| *Subject groups* | Ginger | Cell 1  13  $E = 9$ | Cell 2  7  $E = 11$ | $T_{S1} = 20$ |
|  | Fair | Cell 3  9  $E = 9$ | Cell 4  11  $E = 11$ | $T_{S2} = 20$ |
|  | Dark | Cell 5  5  $E = 9$ | Cell 6  15  $E = 11$ | $T_{S3} = 30$ |
| Totals $(Tc)$ |  | $T_{C1} = 27$ | $T_{C2} = 33$ | $N = 60$ |

**Fig. 11.7**

*Special point*

Remember to arrange your table so that the subject groups are down the side and the nominal categories across the top.

*Calculating the extended $\chi^2$ test*

| PROCEDURE | WORKED EXAMPLE |
|---|---|
| 1. Label your cells 1–6 from top left to bottom right in the manner indicated in Figure 11.7. | 1. See Figure 11.7. |
| 2. Add up the total number of 're-currences' for all groups to give $T_{c1}$. | 2. $T_{c1} = 27$ |
| 3. Add up the total number of 'no recurrences' for all groups to give $T_{c2}$. | 3. $T_{c2} = 33$ |
| 4. Add up the total number of ginger-haired subjects to give $T_S1$. | 4. $T_S1 = 20$ |
| 5. Add up the total number of fair-haired subjects to give $T_S2$. | 5. $T_S2 = 20$ |
| 6. Add up the total number of dark-haired subjects to give $T_S3$. | 6. $T_S3 = 20$ |
| 7. Add up all the $T_S$ totals to give $N$ (i.e. total number of subjects). | 7. $N = 60$ |

8. Calculate the expected frequency value for each cell ($E$) using the following formula:

$$E = \frac{T_S \times T_c}{N}$$

where

$T_S$ is the relevant row total for the cell

$T_c$ is the relevant column total for the cell

$N$ is the grand total

8. Cell 1, $E = \dfrac{20 \times 27}{60} = 9$

Cell 2, $E = \dfrac{20 \times 33}{60} = 11$

Cell 3, $E = \dfrac{20 \times 27}{60} = 9$

Cell 4, $E = \dfrac{20 \times 33}{60} = 11$

Cell 5, $E = \dfrac{20 \times 27}{60} = 9$

Cell 6, $E = \dfrac{20 \times 33}{60} = 11$

9. Enter each cell's expected frequency in the bottom right-hand corner of the cell.

9. See Figure 11.7

10. Calculate $\chi^2$ from the following formula:

$$\chi^2 = \Sigma \frac{(O - E)^2}{E}$$

10. $\chi^2 = \dfrac{(13 - 9)^2}{9} + \dfrac{(7 - 11)^2}{11} +$

$\dfrac{(9 - 9)^2}{9} + \dfrac{(11 - 11)^2}{11} +$

$\dfrac{(5 - 9)^2}{9} + \dfrac{(15 - 11)^2}{11}$

where $O$ is the observed or actual
    data

$E$ is the expected fre-
    quency

$\Sigma$ means 'sum of'

$$= 1.78 + 1.46 + 0 + 0 +$$
$$1.78 + 1.46$$
$$\therefore \chi^2 = 6.48$$

11. Calculate the degrees of freedom (d.f.) using the formula:

$$(r - 1) \times (c - 1)$$

where $r$ is the number of rows or horizontal lines of figures in the table

$c$ is the number of columns or vertical lines of figures in the table

11. d.f. $= (3 - 1) \times (2 - 1)$
    $= 2$

*Looking up the results in the probability tables*
The calculations yielded a $\chi^2$ value of 6.48, with d.f. = 2. To see whether the $\chi^2$ figure represents significant differences between the groups in terms of recurrence of malignant melanoma, this figure has to be looked up in the probability tables associated with the extended $\chi^2$ test. turn to Table C.1, where across the top of the table you will see various levels of significance for a two-tailed test (from 0.10 to 0.001) and down the left-hand side, d.f. values from 1 to 30. If you look down the d.f. column until you find our d.f. of 2, you will see that there are five numbers to the right of this, called 'critical values' of $\chi^2$:

$$4.60 \quad 5.99 \quad 7.82 \quad 9.21 \quad 13.82$$

Each critical value is associated with the probability or level of significance indicated at the top of its column, such that 5.99 has a probability of 0.05. For our $\chi^2$ value of 6.48 to be significant at one of these levels, it has to be *equal to* or *larger than* one of these critical values. If it is *equal to* a critical value then it has *exactly* the same probability as that value. So, had we obtained a $\chi^2$ value of 9.21, the associated probability would have been 0.01 exactly. However, if the obtained $\chi^2$ value is larger than the critical value, the associated probability is even less than that indicated. So had the obtained $\chi^2$ been 10.43, the probability would have been less than 0.01. This is expressed as $p < 0.01$. Our $\chi^2$ value of 6.48 is larger than the critical value 5.99, which means the associated probability is less than 0.05 for a two-tailed hypothesis.

*What do the results mean?*
The probability of the results from our study being due to random error is less than 0.05 or 5 in 100. Taking the usual 5% (or less) cut-off point for claiming results as

significant we can conclude that these results *are* significant and our experimental hypothesis has been supported. In other words, there is a significant relationship between hair colour and recurrence of malignant melanoma. The null (no relationship) hypothesis can be rejected.

By now you might be wanting to raise some issues, particularly if you have been calculating the extended $\chi^2$ on your own data:

> Q1: If I obtain a $\chi^2$ figure which is smaller than 5.99 (the critical value associated with the 0.05 probability level), what does this mean?

If you are using the standard cut-off significance level of 5% for claiming significant results, a $\chi^2$ value smaller than 5.99 would simply mean that your results are not significant. In this case, the null (no relationship) hypothesis would have to be accepted.

---

*Exercises (answers on page 318)*

1 Look up the following $\chi^2$ values in the probability tables and state whether or not they are significant and at what level.
   (a) $\chi^2 = 4.93$,      d.f. = 2,      two-tailed
   (b) $\chi^2 = 10.01$,     d.f. = 3,      two-tailed
   (c) $\chi^2 = 14.68$,     d.f. = 4,      two-tailed
   (d) $\chi^2 = 7.82$,      d.f. = 3,      two-tailed
   (e) $\chi^2 = 17.21$,     d.f. = 3,      two-tailed
2 Calculate an extended $\chi^2$ test on the following data, stating the $\chi^2$, d.f. and *p* values. Explain what the results mean.

> $H_1$: There is a relationship between the degree of maternal in-volvement in caring for a hospitalised child and the mother's experience of stress.
>
> $H_0$: There is no relationship between the degree of maternal in-volvement in caring for a hospitalised child and the mother's experience of stress.

*Brief outline of study*
Over a period of several months, sixty-nine mothers are randomly selected for study. All have a child aged between 6 and 8 years undergoing tonsil-lectomy. Twenty of the mothers simply visit the child during normal visiting hours, twenty-three spend all the afternoon carrying out simple nursing-type activities for the child, and the remaining twenty-six spend all day with the

|  | | Nominal categories | | |
|---|---|---|---|---|
| | | Highly stressed | Moderately stressed | No stress |
| Subject groups | Visiting hours only | 2 | 15 | 3 |
| | Afternoon care only | 2 | 4 | 17 |
| | All day care | 4 | 12 | 10 |

Fig. 11.8

child in similar activities. On the third day after the operation each mother is asked to classify her stress level either as 'highly stressed', 'moderately stressed' or 'no stress'.

The results are as given in Figure 11.8.

## 11.1.2 The Kruskal-Wallis test

The Kruskal-Wallis test is used under the following conditions:

1. When three or more separate groups of subjects are involved in the study.
2. When the data are ordinal or interval/ratio
3. When the conditions required by a parametric test cannot be fulfilled.

The Kruskal-Wallis test compares the data from each group to see if there are significant differences between the groups. Because this test only tells you whether there are overall differences in the results and not the direction of the differences, any hypothesis associated with the Kruskal-Wallis must be two-tailed. The calculations involve finding a numerical value for $H$ which is then looked up in the appropriate probability tables to determine the likelihood of the results being due to random error. If the resulting probability or $p$ value is 5% or less then it is unlikely that the results can be explained by random error and so are said to be *significant*. In this case the experimental hypothesis would be supported and the null (no relationship) hypothesis can be rejected. If it is more than 5% then the results could be explained by random error, and so are said to be *not* significant. The experimental hypothesis would have to be rejected since the null (no relationship) hypothesis has been supported.

Because this test deals with different subject designs, equal numbers of subjects do *not* have to be selected for each group. However, the calculations are easier if the groups are the same size.

**Example**

A number of pilot nurse education schemes were introduced to some schools of nursing in an attempt to broaden the basic RGN training. These schemes were varied in content but all aimed to develop the nurse's initiative and critical awareness, as well as a number of other qualities. Suppose you were interested in evaluating these schemes to see whether, in fact, they were achieving their aims.

*Hypotheses*

> $H_1$: There is a relationship between the type of nurse training and the degree of initiative shown by the trainees.
>
> $H_0$: There is no relationship between type of nurse training and the degree of initiative shown by trainees.

Of necessity the $H_1$ is two-tailed, as it simply predicts a relationship between the two variables, without specifying the nature of this relationship.

*Brief outline of study*

Eight student nurses at the end of the first year of the RGN course are randomly selected from four pilot training schemes. A further eight are selected from the first year of four traditional training schemes and a third group of eight are selected from the first year of four undergraduate nursing courses, thereby giving three groups each of eight student nurses. Each nurse is rated for initiative along a five point scale thus:

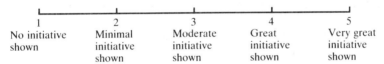

| 1 | 2 | 3 | 4 | 5 |
|---|---|---|---|---|
| No initiative shown | Minimal initiative shown | Moderate initiative shown | Great initiative shown | Very great initiative shown |

In this way, there are three separate groups of student nurses each of whom is assessed on an ordinal scale of initiative. The conditions required by the Kruskal-Wallis have been met. The design is shown in Figure 11.9. Note that there are a

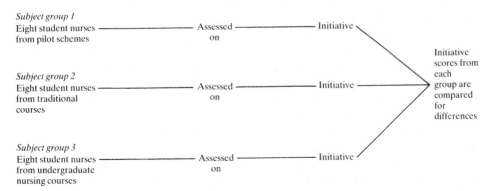

**Fig. 11.9**

number of constant errors which would need to be eliminated if this study were to be carried out in reality (e.g. ability, age, experience of the subjects). Your focus of interest is whether initiative levels differ according to the type of training received.

*Results*

The results are given in Table 11.2. (Bold type are data from study, light type are calculations from the Kruskal-Wallis test). It does not matter which condition is 1, 2 or 3.)

*Calculating the Kruskal-Wallis test*

| PROCEDURE | WORKED EXAMPLE |
|---|---|
| 1. Calculate the total score for each condition to give $T_{C1}$, $T_{C2}$ and $T_{C3}$. | 1. $T_{C1} = 30$<br>$T_{C2} = 21$<br>$T_{C3} = 26$ |
| 2. Calculate the mean score for each condition to give $\bar{x}_1$, $\bar{x}_2$, and $\bar{x}_3$. | 2. $\bar{x}_1 = 3.75$<br>$\bar{x}_2 = 2.63$<br>$\bar{x}_3 = 3.25$ |
| 3. Taking all twenty-four scores together, rank order them, giving a rank of 1 to the lowest, 2 to the next lowest and so on. Where two or more scores are the same, apply the tied rank procedure outline in Section A.2. | 3. See 'Rank' columns. |

---

*Special point*

Do *not* rank the scores from each group separately. Combine the scores from *all* groups and treat them as though they were a single set of scores.

---

| | |
|---|---|
| 4. Add up the total ranks for each condition to give $\Sigma R_1$, $\Sigma R_2$ and $\Sigma R_3$. | 4. $\Sigma R_1 = 127.5$<br>$\Sigma R_2 = 71.5$<br>$\Sigma R_3 = 101.0$ |
| 5. Find the value of $H$ using the following formula: | 5. $H = \left[ \frac{12}{24(24+1)} \left( \frac{127.5^2}{8} + \frac{71.5^2}{8} + \frac{101^2}{8} \right) \right] - 3(24+1)$ |

$$H = \left[ \frac{12}{N(N+1)} \left( \Sigma \frac{R^2}{n_c} \right) \right] - 3(N+1)$$

**Table 11.2**

| Subject | Condition 1: pilot scheme | Rank 1 | Subject | Condition 2: traditional course | Rank 2 | Subject | Condition 3: undergraduate course | Rank 3 |
|---|---|---|---|---|---|---|---|---|
| 1 | 4 | 18.5 | 1 | 3 | 11 | 1 | 3 | 11 |
| 2 | 3 | 11 | 2 | 2 | 4 | 2 | 4 | 18.5 |
| 3 | 5 | 23 | 3 | 3 | 11 | 3 | 5 | 23 |
| 4 | 4 | 18.5 | 4 | 1 | 1 | 4 | 4 | 18.5 |
| 5 | 4 | 18.5 | 5 | 3 | 11 | 5 | 3 | 11 |
| 6 | 3 | 11 | 6 | 4 | 18.5 | 6 | 2 | 4 |
| 7 | 5 | 23 | 7 | 2 | 4 | 7 | 2 | 4 |
| 8 | 2 | 4 | 8 | 3 | 11 | 8 | 3 | 11 |
| Total ($T_C$) | $T_{C1} = 30$ | $\Sigma R_1 = 127.5$ | | $T_{C2} = 21$ | $\Sigma R_2 = 71.5$ | | $T_{C3} = 26$ | $\Sigma R_3 = 101$ |
| Mean | $\bar{x}_1 = 3.75$ | | | $\bar{x}_2 = 2.63$ | | | $\bar{x}_3 = 3.25$ | |

$\Sigma$ means 'total' and $\bar{x}$ means 'average'

where

    $N$ = the total number of subjects

    $n_c$ = the number of subjects in each group

    $\Sigma\dfrac{R^2}{n_c}$ = the sum of each rank total squared and divided by the number of Ss in that condition.

6. Calculate the degrees of freedom (d.f.) value using the following formula:

$$\text{d.f.} = C - 1$$

where $C$ is the number of conditions or groups of subjects

where

    $N = 24$

    $n_c = 8$

$$\Sigma\frac{R^2}{n_c} = \left(\frac{127.5^2}{8} + \frac{71.5^2}{8} + \frac{101^2}{8}\right)$$
$$= [0.02(2032.03 + 639.03 + 1275.13)] - 75$$
$$= 78.92 - 75$$

$$\therefore H = 3.92$$

6. d.f. $= 3 - 1$
     $= 2$

*Looking up the results in the probability tables*

The results of the calculations produced an $H$ of 3.92 with d.f. = 2. To find out whether this represents significant differences in initiative between the groups, the value of $H$ must be looked up in the probability tables associated with the Kruskal-Wallis test. There are two relevant probability tables here, Table C.8, which covers three conditions with up to five subjects per condition, and Table C.1, which covers more than three conditions and/or more than five subjects per group. As we have three conditions with eight subjects in each, Table C.1 is relevant. Across the top of this table are levels of significance or probability for a two-tailed test or hypothesis, while down the left-hand side are d.f. values. Taking our d.f. of 2, you will see that there are five numbers to the right of this value:

$$4.60 \quad 5.99 \quad 7.82 \quad 9.21 \quad 13.82$$

These are called 'critical values' and each is associated with the level of significance indicated at the top of its column, such that 13.82 has a level of significance (or probability) of 0.001. For our $H$ value to be significant at one of the levels indicated at the top of the table, it has to be *equal to* or *larger than* the corresponding critical value. If the obtained $H$ value is equal to the critical value, then the relevant probability is exactly the same as that of the critical value. So, had our $H$ been 7.82, the probability would have been exactly 0.02. On the other hand, had $H$ been larger, say 7.92, then the probability would have been even smaller than 0.02. This is expressed as $p < 0.02$.

Our $H$ value of 3.92 is smaller than the critical value 4.60, which means that the level of significance or probability is greater than 0.10 or 10%.

*What do the results mean?*
The probability of our results being due to random error is greater than 10%. If we use the normal cut-off probability point of 5% (or less) to claim that the results are significant, we can conclude that the results of this study are *not* significant. The experimental hypothesis has not been supported and the null (no relationship) hypothesis must be accepted. There is no relationship between the type of nurse training and initiative shown at the end of the 1st year.

There may be some questions that you wish to raise particularly if you have been calculating the Kruskal-Wallis on your own data.

> Q1: *I used three conditions in my study, with five subjects in two conditions and four in the third. How do I look up my* H *value?*

Where three conditions and five (or fewer) subjects have been used in each, Table C.8 is appropriate. This table caters for various numbers of subjects in each group, as denoted by the values under the $n_1$, $n_2$ and $n_3$ headings. As you have five subjects in two conditions and four in one, you will need to look at the section of n values labelled 5 5 4. To the right of the these ns are six critical values of $H$, each associated with the probability noted next to it, e.g. the critical value 5.6657 has a corresponding p value of 0.049. For your obtained $H$ value to be significant at one of these probability levels, it has to be *equal to* or *larger than* the critical value. If $H$ is equal to a given critical value, its probability is exactly the same as that of the critical value. So, had you obtained $H = 5.6429$, then the p value would has been 0.05 exactly. On the other hand, if your obtained $H$ is larger than a given critical value, then the associated probability is even less. So, for an $H$ of 5.7721, the probability is less than 0.049. This is expressed as $p < 0.049$.

> Q2: *My* H *value is smaller than all the relevant critical values. What does this mean?*

If your obtained $H$ is smaller than any of the critical values given, then your results are classified as 'not significant'. In this case, the null (no relationship) hypothesis would have to be accepted and the experimental hypothesis rejected.

*Exercises (answers on page 318)*

**3** Look up the following $H$ values and state whether or not they are significant and at what level.
  (a) $H = 9.84$,    $n_1 = 4$,    $n_2 = 3$,    $n_3 = 2$,    $n_4 = 5$
  (b) $H = 3.91$,    $n_1 = 4$,    $n_2 = 2$,    $n_3 = 1$

(c) $H = 7.98$,     $n_1 = 5$,     $n_2 = 5$,     $n_3 = 5$
(d) $H = 12.47$,    $n_1 = 8$,     $n_2 = 8$,     $n_3 = 8$,     $n_4 = 8$
(e) $H = 14.86$,    $n_1 = 7$,     $n_2 = 5$,     $n_3 = 8$

**4** Calculate a Kruskal-Wallis on the following data, specifying the $H$, d.f. and $p$ values. Clarify what your results mean.

**H₁:** There is a difference in the evaluations (by patients, qualified nurses and doctors) of the quality of student nurses trained on the pilot RGN scheme.

**H₀:** There is no difference in the evaluations by patients, nurses and doctors of student nurses trained on the pilot RGN scheme.

*Brief outline of study*
Taking an alternative perspective on the pilot nurse training scheme, you decide to compare the evaluations of doctors, patients and other nurses of the overall quality of nursing care provided by nurses trained on the pilot scheme. You randomly select five patients, four sisters and four doctors and ask each one to rate the quality of nursing care, provided by the pilot scheme trained student nurses, along a six point scale (1 = very poor, 6 = excellent.) The results are given in Table 11.3.

**Table 11.3**

| Subject | Condition 1: patients | Subject | Condition 2: nurses | Subject | Condition 3: doctors |
|---------|------------------------|---------|----------------------|---------|----------------------|
| 1 | 6 | 1 | 4 | 1 | 3 |
| 2 | 5 | 2 | 5 | 2 | 4 |
| 3 | 5 | 3 | 4 | 3 | 4 |
| 4 | 6 | 4 | 5 | 4 | 3 |
| 5 | 6 | | | | |

### 11.1.3 The Jonckheere trend test

The Jonckheere trend test is used under similar conditions as the Kruskal-Wallis test:

1. With three or more separate groups of subjects (a different subject design).
2. When the data are ordinal or interval/ratio.
3. When the other conditions required by a parametric test cannot be fulfilled.

In addition, however, is one very crucial condition – if a *trend* in the results is predicted in the hypothesis the Jonckheere trend test must be used to analyse the data.

So, while the Kruskal-Wallis simply establishes whether or not differences exist between the groups, without specifying the direction of these differences, the Jonckheere trend test is used when a *specific direction* to the results is anticipated. So, if you look at the hypothesis quoted in Exercise 4 you will note that it simply predicts that there will be *differences* in the evaluations made by nurses, doctors and patients, without clarifying which group will make the most positive or negative evaluation. If it has been anticipated that patients will have the most favourable views, followed by nurses and then doctors, a *trend* in the results in predicted, such that patients have more positive views than nurses, who have more positive views than doctors.

Where such a trend in the results is predicted in the hypothesis, a Jonckheere trend test should be used to analyse the results.

Because the Jonckheere trend is only used when a specific direction to the results has been anticipated, any hypotheses associated with it must be one-tailed.

When you calculate the Jonckheere, a numerical value for $S$ is found which is then looked up in the probability tables associated with the Jonckheere. This will give you the probability (or $p$ value) of your results being due to random error. If the $p$ value is 5% or less then it is unlikely that random error can account for the results, which are then said to be *significant*. The experimental hypothesis has been supported and the null (no relationship) hypothesis can be rejected. If it is more than 5%, the results could be explained by random error, and so are classified as *not* significant. The null (no relationship) hypothesis would have been supported and the experimental hypothesis must be rejected. A further point should be emphasised – the Jonckheere trend test requires *equal* subject numbers for each group.

**Example**

Anxiety neuroses can be particularly difficult to deal with, each psychiatrist or clinical psychologist having their own preferred method of therapy. Imagine you are working in an out-patients department with a large number of anxiety neurotics, who receive various treatments, which depend on the perspective of the person treating them. The efficacy of these treatments has not been compared, but you observed that those patients receiving psychotherapy appear to do better than those receiving behaviour modification, while those on pharmacological treatment appear to do worst. You decide to make a systematic study of these therapies in the hope of improving patient care.

*Hypotheses*

$H_1$: Patients suffering from anxiety neurosis make more progress on psychotherapy, followed by behaviour modification and finally pharmacological intervention.

$H_0$: There is no relationship between the type of treatment an anxiety neurotic receives and the rate of progress.

The $H_1$ is predicting a *trend* in the results and consequently is one-tailed.

*Brief outline of study*
Eighteen patients, all suffering from anxiety neurosis, are randomly selected on arrival at an out-patient's clinic. Of these, six are to receive psychotherapy, six behaviour modification and six drug therapy. The severity of the condition is comparable for all patients. After six weeks, the patients are assessed on a five point scale of progress:

| 1 | 2 | 3 | 4 | 5 |
|---|---|---|---|---|
| No progress | Minimal progress | Moderate progress | Good progress | Complete recovery |

In this way, three equal sized groups of patients (psychotherapy, behaviour modification and drug therapy) are assessed on an ordinal scale of progress, with a *trend* in the results predicted. All the conditions required by the Jonckheere trend test are therefore fulfilled.

The design is shown in Figure 11.10. (Note that there would be many constant errors that would have to be eliminated if this study were to be carried out properly.) You are interested in whether progress differs according to the type of treatment received.

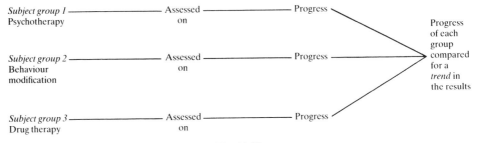

**Fig. 11.10**

*Results*
The results in Table 11.4 are obtained and set out in a format which places the condition expected to have the *lowest* scores on the *left* and that with the *highest* scores

**Table 11.4**

| Subject | Condition 1: drugs | Subject | Condition 2: behaviour modification | Subject | Condition 3: psychotherapy |
|---|---|---|---|---|---|
| 1 | **2** (11) | 1 | **3** (4) | 1 | **3** |
| 2 | **3** (6) | 2 | **4** (1) | 2 | **5** |
| 3 | **1** (12) | 3 | **2** (6) | 3 | **4** |
| 4 | **2** (11) | 4 | **3** (4) | 4 | **3** |
| 5 | **2** (11) | 5 | **3** (4) | 5 | **4** |
| 6 | **3** (6) | 6 | **4** (1) | 6 | **4** |

on the *right*. The remaining conditions should be arranged accordingly (bold type = data from study, light type = calculations of the Jonckheere).

*Calculating the Jonckheere trend test*

| PROCEDURE | WORKED EXAMPLE |
|---|---|
| 1. Calculate the mean score for each condition to give $\bar{x}_1$, $\bar{x}_2$ and $\bar{x}_3$. | 1. $\bar{x}_1 = 2.17$ <br> $\bar{x}_2 = 3.17$ <br> $\bar{x}_3 = 3.83$ |
| 2. Taking subject 1, condition 1 first, count up all the scores in the condition(s) to the right (i.e. conditions 2 and 3) which are *larger*. Do *not* count them if they are the same. Enter this figure in brackets to the side of subject 1's score. Repeat the process for the rest of the subjects in condition 1. | 2. See the figures in brackets to the right of the scores in condition 1. |
| 3. Next, taking subject 1, condition 2, count up all the scores in the condition(s) to the right (here condition 3) which are larger. Again do *not* count any scores which are the same. Enter the figures in brackets by the score for subject 1, condition 2. Repeat the process for the rest of the subjects in condition 2. Obviously, as condition 3 has no condition to its right, this process terminates with condition 2. | 3. See the figures in brackets to the right of the scores in condition 2. |

---

*Special point*

Ensure that you set out your data such that the condition expected to have the lowest scores is on the left and that expected to have the highest scores is on the right with the remaining conditions arranged accordingly.

---

| | |
|---|---|
| 4. Calculate a value for $A$ by adding up *all* the scores in brackets. | 4. $A = 77$ |
| 5. Calculate a value for $B$ using the formula $$B = \frac{C(C-1)}{2} \times n^2$$ | 5. $$B = \frac{3(3-1)}{2} \times 6^2$$ |

where $C$ is the number of conditions
$n$ is the number of subjects
in each condition

6. Calculate $S$ using the formula

$$S = (2 \times A) - B$$

where $C = 3$
$n = 6$
$B = 3 \times 36$
$\therefore B = 108$

6. $S = (2 \times 77) - 108$
$= 154 - 108$
$\therefore S = 46$

*Looking up the results in the probability tables*
A value for $S$ of 46 has been calculated which must be looked up in the probability tables associated with the Jonckheere trend test to see if it represents a significant trend in the progress of anxiety neurosis patients. Turn to Table C.9 which consists of two significance tables – the upper one for less than 0.05 levels and the lower one for less than 0.01 levels (both for one-tailed hypotheses). Across the top of each table are values of $n$ (i.e. number of subjects in each condition), while down the side are values of $C$ (i.e. number of conditions). Taking the lower table first, find our values of $C$ and $n$ (3 and 6 respectively) and locate the figure at the intersection point (59). This is called the 'critical value of $S$'. For our obtained $S$ value of 46 to be significant at the level indicated at the top of this table it must be equal to or larger than the value at the intersection point. Our value of 46 is smaller than 59 and so is not significant at the <0.01 level. In this case, the process must be repeated with the <0.05 table. The critical value at the intersection point for $C = 3$, $n = 6$ is 42. As our obtained $S$ value is larger than this, the results are significant at <0.05 level.

*What do the results mean?*
For the obtained $S$ of 46, the results are significant at less than the 0.05 level. This means that there is less than a 5% chance that random error could account for the results which, if we take the usual cut-off probability level of 5% or less, means that our results are significant. The experimental hypothesis has been supported and the null (no relationship) hypothesis can be rejected. There is a significant trend in the progress rate of patients with anxiety neuroses undergoing different therapies, with psychotherapy producing most progress, followed by behaviour modification and finally drug therapy.

---

### Exercises (answers on page 318)

**5** Look up the following $S$ values and state whether or not they are significant and at what level. (Note the < sign at the top of the tables.)
   (a) $S = 66$,     $C = 4$,     $n = 6$
   (b) $S = 92$,     $C = 3$,     $n = 8$
   (c) $S = 87$,     $C = 3$,     $n = 10$
   (d) $S = 73$,     $C = 5$,     $n = 5$
   (e) $S = 121$,    $C = 4$,     $n = 7$

**6** Calculate a Jonckheere trend test on the following results, stating the $A$, $B$, $S$, $C$, $n$ and $p$ values. Clarify what your results mean.

>   **H₁:** The severity of a heart attack is related to the smoking be-
>   haviour of the individual, with smokers having the most
>   severe attacks, followed by ex-smokers and finally non-
>   smokers.

>   **H₀:** There is no relationship between severity of heart attack and
>   smoking behaviour of the patient.

*Brief outline of study*
Randomly select fifteen male patients who have been admitted to hospital following a heart attack. Of these five are smokers, five ex-smokers and five non-smokers. The severity of heart attack is assessed on a four point scale (1 = very mild, 4 = very severe). Note that a number of constant errors would need to be eliminated if this study were being carried out properly, e.g. age, weight, stress, previous coronary heart disease. The results are given in Table 11.5.

**Table 11.5**

| Subject | Condition 1: non-smokers | Subject | Condition 2: ex-smokers | Subject | Condition 3: smokers |
|---------|--------------------------|---------|-------------------------|---------|----------------------|
| 1 | 2 | 1 | 3 | 1 | 4 |
| 2 | 3 | 2 | 2 | 2 | 3 |
| 3 | 1 | 3 | 3 | 3 | 4 |
| 4 | 2 | 4 | 4 | 4 | 4 |
| 5 | 2 | 5 | 3 | 5 | 2 |

## 11.2 Parametric tests

### 11.2.1 One-way analysis of variance (anova) for unrelated designs

The one-way analysis of variance (or *anova*) for unrelated designs is used under the following conditions:

1. Three or more separate groups of subjects take part in the study, i.e. an un-related or different subject design.
2. The data are interval/ratio.
3. The remaining three conditions required by a parametric test can be (approxi-mately) fulfilled (see Section 7.2).

This test is the parametric equivalent of the Kruskal-Wallis, in that it compares the performance of three or more separate groups of subjects to see if there are differences between them. It can only tell the researcher whether *general* differences exist and not the direction of these differences. As a result, any hypothesis associated with this anova must be two-tailed. If you want to find out whether one group (or condition) does significantly better than another, the Scheffé multiple range test should be used *after* the anova has been calculated. This test is described in the next section.

When calculating the anova test a numerical value for *F* is obtained which represents any differences between the subject groups. To see if these differences are significant, the *F* value has to be looked up in the probability tables associated with the anova. If the resulting *p* value (or probability) is 5% or less, then the results are said to be *significant* and it is unlikely that random error can explain them. The experimental hypothesis would have been supported and the null (no relationship) hypothesis can be rejected. If it is greater than 5%, then the results are *not* significant, since they could be explained by random error. The null (no relationship) hypothesis would have been supported and the experimental hypothesis must be rejected.

The formula given here for the one-way anova for unrelated designs is *only* suitable for designs which have equal numbers of subjects in each group. If you have unequal subject numbers a different formula is required (see Ferguson, 1976).

The anova involves some very long-winded calculations, which although not complex, are laborious. The process may be simplified if some detail is provided about the function of the anova. This will be done through an example.

**Example**

A great deal is now known about the effects of obesity on health. However, as many of us know, it is easier to accept the information than do anything about it. Imagine you are working as a practice nurse and are involved in running a well-woman clinic. A number of the women who attend are seriously overweight to the degree that it is affecting their health. In an attempt to encourage them to diet you decide to compare three educational approaches, to see which produces the greatest weight loss.

*Hypotheses*

> $H_1$: There is a relationship between successful weight loss and the type of health education process involved.

> $H_0$: There is no relationship between successful weight loss and the type of health education process involved.

The $H_1$ is two-tailed, since it predicts only differences in success, without specifying which health education programme is most successful.

*Brief outline of study*

Twenty-one women from a well-woman clinic are randomly selected for study. All are at least 25% overweight and are beginning to experience weight related problems (e.g. arthritis of the hips and knees, oedema, varicose veins, high blood

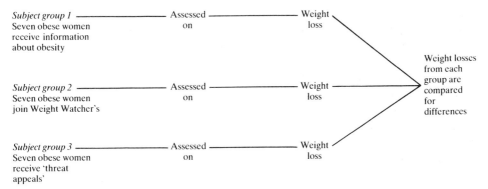

**Fig. 11.11**

pressure). All are between 40 and 50 years of age. Three sub-groups each comprising seven women are randomly chosen. One group is simply given a fact sheet about obesity and its effects; a second group is enrolled at a Weight Watcher's club and the third group is given negative information (a 'threat appeal') about the deterimental effects of obesity by a video showing interviews of women who have experienced severe health problems as a result of being overweight (coronaries, hip replacements, etc.). Therefore, each of the three subject groups experiences a different health education approach. At the end of three months, you compare the three groups on the amount of weight lost (in lbs).

Because three separate groups have been used (information, Weight Watcher's, threat appeals), the data are interval/ratio (lb weight lost) and the remaining conditions for a parametric test can be (more or less) met, the one-way anova is the appropriate test for analysing the data.

The design is shown is Figure 11.11. You are concerned whether the weight lost by the groups differs significantly.

*Results*
The results in Table 11.6 were obtained (lb lost). (It does not matter which condition is 1, 2, or 3.) You have predicted that weight loss will be different according to the type of educational process used. In other words, significant differences are expected between the conditions. This source of variation in scores is called the between conditions variance, and is represented by the value of $F$ that is calculated in the anova.

However, there is another possible source of variation in the scores – that due to random error. In other words, the transitory factors about the subjects (such as mood, personality, temperament) could also produce a variation in scores between the conditions. These variations would not reflect any genuine differences due to the educational programme used, but would result from individual differences among the subjects. This source of variation in scores is called error variance.

What we would hope to find is a significant between conditions variance which would be indicated by a probability for the obtained $F$ value of 5% or less. In order

**Table 11.6**

| Subject | Condition 1: information only | Condition 2: Weight Watcher's | Condition 3: threat appeal |
|---------|-------------------------------|-------------------------------|----------------------------|
| 1 | 5 | 10 | 7 |
| 2 | 7 | 14 | 6 |
| 3 | 3 | 8 | 8 |
| 4 | 4 | 9 | 9 |
| 5 | 1 | 5 | 10 |
| 6 | 2 | 12 | 12 |
| 7 | 5 | 11 | 9 |

to compute this $F$ value, three other values are needed first: sums of squares (SS) mean square (MS) and degrees of freedom (d.f.).

These values, once calculated, have to be set out in the format of Table 11.7. To calculate these values, take the following steps.

**Table 11.7** Sample format of an anova table (unrelated designs)

| Source of variance | Sums of squares (SS) | Degrees of freedom (d.f.) | Mean square (MS) | F ratios (F) |
|---------------------|----------------------|---------------------------|------------------|--------------|
| Variation in scores between groups or conditions, i.e. between conditions variance | $SS_{bet}$ | $d.f._{bet}$ | $MS_{bet}$ | $F$ |
| Variation in scores due to random error, i.e. error variance | $SS_{error}$ | $d.f._{error}$ | $MS_{error}$ | |
| Total | $SS_{tot}$ | $d.f._{tot}$ | | |

*Calculating the one-way anova for unrelated designs*

PROCEDURE

1. Calculate the totals for each condition to give $T_{c1}$, $T_{c2}$ and $T_{c3}$.

2. Calculate the total of all the scores by adding up the $T_c$ values to give $\Sigma x$.

3. Calculate the sums of squares (SS) for each source of variation by calculating the following values:

WORKED EXAMPLE

1. $T_{c1} = 27$
   $T_{c2} = 69$
   $T_{c3} = 61$

2. $\Sigma x = 27 + 69 + 61$
   $= 157$

3.

$\Sigma T_c^2$ the total of the squared $T_c$ values

$$\Sigma T_c^2 = 27^2 + 69^2 + 61^2$$
$$= 729 + 4761 + 3721$$
$$= 9211$$

$n$ is the number of subjects in each condition

$n = 7$

$N$ is the total number of scores ($n \times$ the number of conditions)

$N = 21$

$(\Sigma x)^2$ is the value of $\Sigma x$, squared

$$(\Sigma x)^2 = 157^2$$
$$= 24\,649$$

$\dfrac{(\Sigma x)^2}{N}$ is a constant to be subtracted from all SS calculations

$$\frac{(\Sigma x)^2}{N} = \frac{24\,649}{21}$$
$$= 1173.76$$

$\Sigma x^2 =$ is the total of each squared individual score

$$\Sigma x^2 = 5^2 + 10^2 + 7^2 + 7^2 + 14^2 +$$
$$6^2 + 3^2 + 8^2 + 8^2 + 4^2 +$$
$$9^2 + 9^2 + 1^2 + 5^2 + 10^2 +$$
$$2^2 + 12^2 + 12^2 + 5^2 +$$
$$11^2 + 9^2$$
$$= 1415$$

---

*Special point*

Note the difference between
$(\Sigma x)^2$ which is the total of all the scores, *squared* and
$\Sigma x^2$ which is the total of the squared individual scores.

---

4. To calculate the sums of squares between the conditions (SS$_{bet}$) use the following formula:

$$\frac{\Sigma T_c^2}{n} - \frac{(\Sigma x)^2}{N}$$

4.

$$SS_{bet} = \frac{9211}{7} - 1173.76$$
$$= 1315.86 - 1173.6$$
$$= 142.1$$

5. Calculate the total sums of squares (SS$_{tot}$) using the following formula:

$$\Sigma x^2 - \frac{(\Sigma x)^2}{N}$$

5.

$$SS_{tot} = 1415 - 1173.76$$
$$= 241.24$$

6. Calculate the sums of squares for the error variance (SS$_{error}$) using the following formula:

6.

$$SS_{tot} - SS_{bet}$$

$$SS_{error} = 241.24 - 142.1$$
$$= 99.14$$

7. Calculate the degrees of freedom (d.f.) values for each source of variance using the following formula:

$$d.f._{bet} = C - 1$$

where $C$ is the number of conditions

$$d.f._{tot} = N - 1$$

where $N$ is the total number of scores

$$d.f._{error} = d.f._{tot} - d.f._{bet}$$

7.

$$d.f._{bet} = 3 - 1$$
$$= 2$$

$$d.f._{tot} = 21 - 1$$
$$= 20$$

$$d.f._{error} = 20 - 2$$
$$= 18$$

8. Calculate the mean squares (MS) value for each source of variance using the following formulae:

$$MS_{bet} = \frac{SS_{bet}}{d.f._{bet}}$$

$$MS_{error} = \frac{SS_{error}}{d.f._{error}}$$

8.

$$MS_{bet} = \frac{142.1}{2}$$
$$= 71.05$$

$$MS_{error} = \frac{99.14}{18}$$
$$= 5.51$$

9. Calculate the $F$ ratio by using the formula:

$$\frac{MS_{bet}}{MS_{error}}$$

9.

$$F = \frac{71.05}{5.51}$$
$$= 12.9$$

Enter these values into the anova table; Table 11.8.

**Table 11.8**

| Source of variance | SS | d.f. | MS | F ratio |
|---|---|---|---|---|
| Between conditions variance | 142.1 | 2 | 71.05 | 12.9 |
| Error variance | 99.14 | 18 | 5.51 | |
| Total | 241.24 | 20 | | |

*Looking up the results in the probability tables*
An *F* ratio of 12.9 has been obtained. This value represents the differences in weight loss between the three subject groups. To see whether these differences are significant, $F = 12.9$ must be looked up in the probability tables associated with the anova. Turn to Tables C.6(a)–(d). Table C.6(a) is associated with the least significant probabilities and Table C.6(d) with the most significant. Each table has values for $v_1$ across the top and $v_2$ down the side; the $v_1$ value is the d.f.$_{bet}$ figure and the $v_2$ value is the d.f.$_{error}$ figure. So, starting with Table C.6(a), the d.f.$_{bet}$ of 2 and the d.f.$_{error}$ of 18, locate the figure at the intersection point. This is 3.55 and is called the 'critical value of *F*'. To be significant at the level indicated at the top of the table, the obtained *F* must be *equal to* or *larger than* the critical value at the intersection point. As 12.9 is larger than 3.55, the results are significant at <0.05. But are the results significant at an even lower probability (i.e. can we achieve a smaller/better *p* value)? Repeat the process with Table C.6(b). The critical value of *F* at the intersection point is 4.56. As 12.9 is larger than this, the results are significant at <0.025. Repeat the process with Tables C.6(c) and (d). The critical values are 6.01 and 10.39 respectively. As the obtained *F* value is larger than 10.39 the results are significant at <0.001 level.

*What do the results mean?*
The obtained *F* value of 12.9 is larger than the critical value of 10.39 on the $p < 0.001$ table. This means that the probability of these results being due to random error is less than 0.1%. Given that this is less than the usual cut-off significance level of 5%, the results can be said to be significant. The null (no relationship) hypothesis can be rejected and the experimental hypothesis accepted. There is a significant relationship between amount of weight lost and the educational approach used for obese women.

Note again that the anova only tells you whether there are overall differences between the groups and not where these differences lie. Obviously, in the example just given, it would be extremely useful to know whether one educational approach was significantly more effective than another, so that any other seriously overweight woman could be given that approach as an aid to losing weight. Where it is useful to have this information, a Scheffé multiple range test can be used (see next section). However, this test can *only* be used *after* the anova has been calculated and if the results from the anova are significant.

If you have been calculating a one-way anova for unrelated designs on your own data there may be some questions you want to raise:

*Q1:  My obtained* F *value was smaller than the relevant critical value on Table C.6(a)* (p < 0.05). *What does this mean?*

If you obtained an *F* value that was smaller than the critical value on Table C.6(a), this simply means your results are not significant and that random error could account for the data. In this case the null (no relationship) hypothesis would have been supported and the experimental hypothesis must be rejected.

*Q2: There is no exact number for my d.f. values in Tables C.6(a)–
(d). What do I do?*

In this case, you would use the next lowest d.f. value in the table, except
where the d.f.s were substantially bigger than even the largest d.f.s in the
table, in which case you would use the value for infinity ($\infty$).

---

## Exercises (answers on page 318)

**7** Look up the following $F$ values and state whether or not they are significant
and at what level:
(a) $F = 4.01$,    d.f.$_{bet} = 3$,    d.f.$_{error} = 18$
(b) $F = 7.83$,    d.f.$_{bet} = 4$,    d.f.$_{error} = 9$
(c) $F = 2.91$,    d.f.$_{bet} = 2$,    d.f.$_{error} = 16$
(d) $F = 3.32$,    d.f.$_{bet} = 2$,    d.f.$_{error} = 31$
(e) $F = 12.97$,    d.f.$_{bet} = 4$,    d.f.$_{error} = 10$

**8** Calculate a one-way anova for unrelated designs on the following data,
stating the MS, SS, d.f., $F$ and $p$ values using a format like that of Table
11.8. Clarify what the results mean.

**H$_1$:** There is a relationship between the method of administering
morphine to cholecystectomy patients and the amount of
pain experienced.

**H$_0$:** There is no relationship between the method of administering
morphine to cholecystectomy patients and the amount of
pain experienced.

*Brief outline of study*
Randomly select eighteen cholecystectomy patients, of whom six have con-
tinuous morphine delivered by a pump, six have morphine injections at regular four hour intervals and six have morphine on request. Using the visual
analogue method of assessing pain (i.e. a 10 cm line is drawn with divisions

**Table 11.9**

| Subject | Condition 1:<br>morphine pump | Condition 2:<br>regular injections | Condition 3:<br>morphine on request |
|---|---|---|---|
| 1 | 3 | 7 | 4 |
| 2 | 4 | 8 | 5 |
| 3 | 6 | 6 | 4 |
| 4 | 4 | 7 | 5 |
| 5 | 2 | 7 | 6 |
| 6 | 3 | 8 | 5 |

at 1 cm intervals and the patient is required to mark a point on that line to indicate the degree of pain experienced, 1 = none, 10 = excruciating), each patient reports their pain level on the first day after the operation.

The results are given in Table 11.9.

## 11.2.2 The Scheffé multiple range test

The Scheffé multiple range test is used under the following conditions:

1. Following the calculation of a one-way anova for unrelated designs.
2. Only if the anova produces significant results.

The anova only tells the researcher whether or not there are significant differences between the groups (or conditions) without specifying whether the performance of one group was significantly better or worse than any other. Sometimes this information would be extremely useful as in the example used in the previous section, where it was shown that significant differences in weight loss were achieved with various educational approaches. However, no indication was given as to the most effective method. Clearly if it could be demonstrated that one particular method of encouraging overweight women to diet was more successful than the others, this method would be used more widely in the campaign against obesity. The Scheffé test can be used to derive this information. In particular it can compare the information only and the Weight Watcher's groups, the information only and the threat appeal groups and the Weight Watcher's and the threat appeal groups to see whether there are significant differences between them.

When calculating the Scheffé, two numerical values are found, one for $F$ and one for $F'$. If $F$ is equal to or larger than $F'$ then the difference between the two conditions being compared is significant.

Do note that the formula given for the Scheffé for unrelated designs is the same as that given for the Scheffé for related or same subject designs (Section 9.2.2). It is included again here, because it is easier to understand a formula when it uses data you are immediately familiar with, rather than having to turn back to a different chapter.

**Example**
Let us return to the example given in the previous section for one-way anovas for unrelated designs. There it was predicted that there is a relationship between weight loss and the type of educational approach used with middle-aged obese women. Significant results were obtained ($p < 0.001$) so the conditions required by the Scheffé have been fulfilled. Using these data, we will now calculate the Scheffé to see if any single approach is significantly better than another, i.e.:

Condition 1 (information only) v. Condition 2 (Weight Watcher's)
Condition 1 (information only) v. Condition 3 (threat appeals)
Condition 2 (Weight Watcher's) v. Condition 3 (threat appeals)

*Calculating the Scheffé multiple range test for unrelated subject designs*

| PROCEDURE | WORKED EXAMPLE |
|---|---|
| 1. Calculate the mean scores for each condition to give $\bar{x}_1$, $\bar{x}_2$ and $\bar{x}_3$. | 1. $\bar{x}_1 = 3.86$<br>$\bar{x}_2 = 9.86$<br>$\bar{x}_3 = 8.71$ |

2. To compare condition 1 with condition 2 (information v. Weight Watcher's), use the following formula:

$$F = \frac{(\bar{x}_1 - \bar{x}_2)^2}{MS_{error}/n_1 + MS_{error}/n_2}$$

where
$\bar{x}_1$ is the mean score for condition 1
$\bar{x}_2$ is the mean score for condition 2
$n_1$ is the number of Ss in condition 1
$n_2$ is the number of Ss in condition 2
$MS_{error}$ is the value for the mean squares error variance from the anova results

2.

$$F = \frac{(3.86 - 9.86)^2}{5.51/7 + 5.51/7}$$

where
$\bar{x}_1 = 3.86$
$\bar{x}_2 = 9.86$
$n_1 = 7$
$n_2 = 7$
$MS_{error} = 5.51$

$$F = \frac{36}{1.58}$$

$$\therefore F = 22.79$$

3. To compare condition 1 with condition 3 (information v. threat appeals) use the following formula:

$$F = \frac{(\bar{x}_1 - \bar{x}_3)^2}{MS_{error}/n_1 + MS_{error}/n_3}$$

where
$\bar{x}_1$ is the mean score for condition 1
$\bar{x}_3$ is the mean score for condition 3
$n_1$ is the number of Ss in condition 1

3.

$$F = \frac{(3.86 - 8.71)^2}{5.51/7 + 5.51/7}$$

where
$\bar{x}_1 = 3.86$
$\bar{x}_3 = 8.71$
$n_1 = 7$

$n_3$ is the number of Ss in condition 3

MS$_{error}$ is the value for the mean squares for the error variance from the anova table

$n_3 = 7$

MS$_{error} = 5.51$

$$F = \frac{23.52}{1.58}$$

$$\therefore F = 14.89$$

4. To compare condition 2 with condition 3 (Weight Watcher's v. threat appeals), use the following formula:

$$F = \frac{(\bar{x}_2 - \bar{x}_3)^2}{\text{MS}_{error}/n_2 + \text{MS}_{error}/n_3}$$

where

$\bar{x}_2$ is the mean score for condition 2

$\bar{x}_3$ is the mean score for condition 3

$n_2$ is the number of Ss in condition 2

$n_3$ is the number of Ss in condition 3

MS$_{error}$ is the value for the mean squares for the error variance from the anova table

4.

$$F = \frac{(9.86 - 8.71)^2}{5.51/7 + 5.51/7}$$

where

$\bar{x}_2 = 9.86$

$\bar{x}_3 = 8.71$

$n_2 = 7$

$n_3 = 7$

MS$_{error} = 5.51$

$$F = \frac{1.32}{1.58}$$

$$\therefore F = 0.84$$

---

*Special point*

If you have more than three conditions which you want to compare, simply substitute the appropriate $\bar{x}$ and $n$ values in the formula.

---

5. To calculate $F'$, select the d.f.$_{bet}$ and d.f.$_{error}$ values from the anova table, to be $v_1$ and $v_2$ respectively.

5. d.f.$_{bet}$ = 2
   d.f.$_{error}$ = 18

6. Turn to Table C.6(a) ($p < 0.05$ values) and find the critical value of $F$ at the intersection point of the relevant $v_1$ and $v_2$ values. Call this value $F^0$.

7. Calculate $F'$ using the following formula:

$$F' = (C - 1) F^0$$

where

$C$ is the number of conditions
$F^0$ is the critical value at the intersection point

6. Critical value of $F$ (i.e. $F^0$) at the intersection point of $v_1 = 2$, $v_2 = 18$ is 3.55

7.

$$F' = (3 - 1) 3.55$$
$$= 2 \times 3.55$$
$$\therefore F' = 7.1$$

*Finding out whether the results are significant*

To find out whether the $F$ values derived from the three comparisons of the conditions are significant, compare each $F$ with $F'$. If $F$ is *equal to* or *larger* than $F'$, it is significant at $<0.05$ level (because the $p < 0.05$ table was used to calculate $F'$).

The obtained $F$ values were as follows:

1. Condition 1 v. condition 2 = 22.79.
2. Condition 1 v. condition 3 = 14.89.
3. Condition 2 v. condition 3 = 0.84.

Comparing each with the $F'$ of 7.1, it can be seen that the first two comparisons are significant, while comparison 3 is not.

*What do the results mean?*

Using the usual cut-off significance level of 5% or less, the results mean the following:

1. A significant difference in weight loss between the groups who experienced information only and Weight Watcher's ($p < 0.05$). If we look back at the mean scores for these conditions, we can see that the Weight Watcher's group lost significantly more weight than the information only group (9.86 lb as opposed to 3.86 lb).
2. A significant difference in the weight loss for the groups who received information only and threat appeals ($p < 0.05$) with the information only group losing less than the threat appeals group (3.86 lb as compared with 8.71 lb).
3. No significant difference in the weight loss of the Weight Watcher's group and the threat appeals group.

In this latter case, random error could account for the results.

You could conclude from these findings that Weight Watcher's and threat appeals are comparably effective in inducing middle-aged obese women to lose weight. Information only is significantly less effective.

You might by now be wanting to ask some questions about the Scheffé test:

*Q1: Why is the* $p < 0.05$ *table used to calculate* $F'$?

The Scheffé test is extremely stringent. If the smaller probability tables are used it reduces the chances of finding significant differences between conditions. If, however, you have $F$ values which are considerably bigger than $F'$, then you could recalculate $F'$ by substituting the critical value of $F$ from the smaller probability tables C.6(b)–(d).

*Q2: Although I obtained a significant result from the anova, none of my Scheffé comparisons is significant. What does this mean?*

The answer to this lies partly with the stringency of the Scheffé test, which only yields significant comparisons of means if there is a sizeable difference in the results from the conditions. However, it is also explained by the workings of the anova itself. Significant results from the anova without significant results from the Scheffé mean that all the data made a contribution to the significance of the anova without any one condition being primarily responsible.

---

*Exercise (answers on page 319)*

**9** Calculate a Scheffé on the data from Exercise 8 (page 188). State the $F$, $F'$, $F^0$ and $p$ values and clarify what the results mean.

> $H_1$: There is a relationship between the method of administering morphine to cholecystectomy patients and the amount of pain experienced.
>
> $H_0$: There is no relationship between the method of administering morphine to cholecystectomy patients and the amount of pain experienced.

*Brief outline of study*
For details of this, see Exercise 8.

*Results*
Note that these, too, are derived from Exercise 8.

$$d.f._{bet} = 2$$
$$d.f._{error} = 15$$
$$MS_{error} = 1.0$$
$$\bar{x}_1 = 3.67$$
$$\bar{x}_2 = 7.17$$
$$\bar{x}_3 = 4.83$$

# 12 Statistical tests for correlational designs

## 12.1 Introduction

The first three tests covered in this chapter are used to analyse data from correlational designs; the fourth test is used in a slightly different way – to make predictions from existing correlated data. Because less space has been devoted to these types of design throughout the book, it is probably worth recapping on their main features, and particularly how they relate to experimental designs:

1. *Like* experimental designs, correlational designs start off with an experimental hypothesis which predicts a relationship between two variables.
2. *Unlike* experimental designs, which manipulate one of these variables and monitor the effects of this on the other, neither variable is manipulated or varied in a correlational design. Instead, *whole ranges* of scores on *both* variables are typically collected, to see if there is any link or association between them. Because neither variable is manipulated, there is no independent or dependent variable in correlational designs.
3. *Unlike* the experimental design which can attribute cause and effect as a result of manipulating the independent variable, correlational designs cannot assume any causal link between the variables. All that can be concluded from these designs is that the data vary in a way which suggests a link between them. (One very nice recent example from a Sunday newspaper illustrates perfectly the problem in inferring causality from correlated data. An MP was reported as asking whether AIDS was caused by Greek yoghurt, since he noticed an increase in yoghurt supplies in the supermarket which coincided with the sudden rise in AIDS cases ...).
4. *Unlike* experimental designs which look for *differences* between sets of data, correlational designs are concerned with *similarities* or *patterns* in the data from each variable.
5. *Unlike* experimental designs which, because they can determine cause and effect, can draw definite conclusions from the results (as long as the study has been carried out properly), correlational designs do not permit this. Because they are less conclusive, they may be a less popular design with some researchers. However, because they do not involve any manipulation or 'interference' with

the subjects, they are also considered to be less problematic ethically and so more suitable for medical and paramedical research.
6. Both experimental and correlational designs are subject to sources of error. Order effects, experimenter bias, and constant and random errors all need to be controlled in either design, since their presence can distort the results.

Some of these characteristics need to be emphasised when planning a correlational design, because they can be a source of confusion.

Firstly, in order for the sort of correlational design we have been talking about to be feasible, it *must* be possible to have a *range* of scores on *both* variables. If, for example, your hypothesis predicted that the incidence of stress-related illness is greater among the unemployed than among the employed, a correlational design would not be possible. The reason for this is simply that while you could obtain a range of scores on the 'incidence of stress-related illness' variable (from zero to infinity) you cannot have a range on the 'employed/unemployed' variable, since the subjects would be either one or the other; two options only are available on that variable. If, however, you had modified your hypothesis slightly to predict a greater number of stress-related illnesses the longer the period of unemployment, then a correlational design would be possible because you could obtain a range of scores on both variables. Therefore, a salient question you should ask yourself when deciding between a correlational of the type or experimental design covered by this text, is: 'Is a range of scores on both variables possible?' If the answer is 'no' you cannot use a correlational design.

Second, in a correlational design you are interested in either of two patterns of data – either *high* scores on one variable relating to *high* scores on the other and, consequently, *low* scores on one variable relating to *low* scores on the other. An example of this is the hypothesis that high fat diets lead to high blood pressure. This is called *positive correlation* and is indicated by a plus sign in front of the correlation coefficient that derives from the calculations from the appropriate statistical tests. The closer the resulting correlation coefficient figure to $+1.0$, the stronger the positive correlation between the two variables (see Chapter 3).

The other pattern in which you may be interested concerns the relationship of *high* scores on one variable with *low* scores on the other. An example of this might be the hypothesis that high levels of exercise lead to low levels of cholesterol in the blood. This is called a *negative correlation* and is indicated by a minus sign in front of the correlation coefficient derived from the statistical test you calculate. The closer the correlation coefficient figure is to $-1.0$, the stronger the negative correlation between the two variables. From this, we can formulate another question which you should ask yourself when deciding on the type of design which would be appropriate for testing your hypothesis: 'Am I predicting that the higher the scores on one variable, the higher (or lower) the scores on the other?' If the answer is 'yes' then you have confirmation that a correlational design is appropriate for your hypothesis.

In practice, when carrying out a correlational design you would first of all identify the constant errors and deal with them as outlined in Chapter 5. Then you would collect a *range* of data on one of the variables. The relevant data on the other variable would then be obtained, giving you two sets of scores. These two sets would be analysed using the correct statistical test. This test will provide a positive or negative correlation coefficient, which will give you an idea of the strength of the relationship between the two variables. This figure is then looked up in the appropriate probability tables which will tell you whether or not the relationship between the two variables is a significant one.

So if you were interested in looking at the relationship between temperature and incidence of hypothermia among the elderly (the lower the temperature, the higher the incidence), you would select, say, twenty days during December, January and February when the temperature ranged from $-10$ to $+10\,°C$; these would constitute your range of scores on the temperature variable. You would then collect the numbers of reported incidences of hypothermia for these days; the numbers would constitute the data for the hypothermia variable. These two sets of figures would be analysed using the correct statistical test. This would probably yield a correlation coefficient around $-1.0$, suggesting a negative correlation between these two

**Table 12.1** Summary table outlining the conditions required by the relevant statistical tests

| Test | Design | Conditions for use |
|---|---|---|
| Spearman test | Correlational | Can *only* be used with: (a) two sets of scores; (b) ordinal or interval/ratio data; and should be used when the conditions required by a parametric test cannot be fulfilled. |
| Pearson test | Correlational | Can *only* be used with: (a) two sets of scores; (b) interval/ratio data; and when the remaining conditions required by a parametric test can be fulfilled. |
| Kendall's coefficient of concordance | Correlational | Can *only* be used with: (a) three or more sets of scores; (b) ordinal or interval/ratio data. |
| Linear regression | Correlational | Can *only* be used: (a) with interval/ratio data; (b) when a significant correlation between two variables has been previously demonstrated; (c) to make predictions about the scores on one variable from established scores on the other. |

variables – which was what was predicted. This figure is then looke
probability tables to see whether it represents a *significant* correlation
two sets of data or whether the result could be explained by randoı..
Section 5.6).

If you still have any questions concerning correlational designs, re-read Chapter 3.

The statistical tests covered by this chapter are all used in conjunction with
correlational designs, although each has slightly different specifications for its used.
These specifications will be outlined briefly here and in more detail under the
relevant section:

1. The Spearman test, which is non-parametric, is used to analyse two sets of
   ordinal or interval/ratio data.
2. The Pearson test, which is parametric, is used to analyse two sets of interval/
   ratio data.
3. The Kendall coefficient of concordance test, which is non-parametric, is used to
   analyse three or more sets of ordinal or interval/ratio data.
4. The linear regression test is used with two sets of interval/ratio data which has
   already been shown to be significantly correlated. Essentially it allows the researcher to *predict* scores on one variable when the scores on the other are
   already known.

These features are summarised in Table 12.1.

## 12.2 Non-parametric test for use with two sets of data

### 12.2.1 The Spearman rank order correlation coefficient

The Spearman test should be used under the following conditions:

1. When the research design is correlational.
2. When there are two sets of scores *only*.
3. When the data are ordinal or interval/ratio.

The Spearman test will tell you whether your scores are *positively* correlated (i.e.
high scores on one variable associated with high scores on the other) or negatively
correlated (i.e. high scores on one variable associated with low scores on the other).
When calculating the Spearman test, you find a value for $r_s$ or rho; this value is a
correlation coefficient and will be a figure between $-1.0$ and $+1.0$. If the $r_s$ is close
to $-1.0$ it is indicative of a negative correlation, while if it is close to $+1.0$ it is
indicative of a positive correlation. This $r_s$ figure is then looked up in the associated
probability tables to see if it represents a significant relationship between two sets
of scores. If the probability (or $p$) value for a given $r_s$ is 5% or less, then we can
conclude that there is a significant correlation between the two variables, and it
is unlikely that the results can be accounted for by random error. The experimental

hypothesis would have been supported. If it is more than 5%, then the relationship between the two variables is not significant, and the results can be explained by random error. The null (no relationship) hypothesis would have been supported.

One other point is important before proceeding. With the Spearman test, it doesn't matter whether the scores on one variable are ordinal and on the other interval/ratio, since all this test does is compare the *rankings* of the two sets of scores. So you could compare pain levels along a five point scale with heart rate, or position on an operating theatre's daily waiting list with the number of hours in the theatre. So the *levels* of measurement for your two sets of scores do not have to be the same for this test.

**Example**

The NHS is currently very concerned with consumer appraisal or how a patient evaluates the service. This is a notoriously difficult area to research, since evaluations may be dependent on a host of factors besides quality of care. Imagine you are interested in one of these factors – the amount of time the doctor spends with the patient in a GP's surgery. Your prediction is that the longer the time spent, the more positive the patient's evaluation, irrespective of the actual quality or efficacy of treatment.

*Hypotheses*

> $H_1$: The longer the time a patient spends with the GP the more positive their evaluation of the service.
>
> $H_0$: There is no relationship between the length of time a patient spends with the GP and evaluation of the service.

The $H_1$ is predicting a positive correlation, with *high* scores on the length of time variable being associated with *high* scores on the evaluation variable. Because it is making a specific prediction it is a one-tailed hypothesis.

*Brief outline of study*

Randomly select fifteen patients who have just spent varying amounts of time (from three to twenty minutes) with the GP and ask each to evaluate the service along a five point scale of quality:

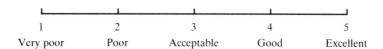

| 1 | 2 | 3 | 4 | 5 |
|---|---|---|---|---|
| Very poor | Poor | Acceptable | Good | Excellent |

We have, then, interval/ratio data on the length of time variable and ordinal data along the evaluation of service variable. (Assume that all the constant errors have been taken into account, e.g. same GP for all patients.) You are interested in whether high scores on the time variable are associated with high scores on the evaluation variable.

*Results*
The results are given in Table 12.2. (Bold type figures are the data from the study while the lighter type figures are from the calculations. It does not matter which is variable A and which B.)

**Table 12.2**

| Subject | Data from study | | Calculations from the Spearman | | | |
|---|---|---|---|---|---|---|
| | Variable A: length of time in minutes | Variable B: evaluation | Rank of A | Rank of B | d (A − B) | d² (A − B)² |
| 1 | 10 | 4 | 9.5 | 11 | −1.5 | 2.25 |
| 2 | 5 | 2 | 3 | 3.5 | −0.5 | 0.25 |
| 3 | 7 | 4 | 5.5 | 11 | −5.5 | 30.25 |
| 4 | 3 | 1 | 1 | 1 | 0 | 0 |
| 5 | 12 | 3 | 11 | 7.5 | +3.5 | 12.25 |
| 6 | 8 | 3 | 7.5 | 7.5 | 0 | 0 |
| 7 | 6 | 2 | 4 | 3.5 | +0.5 | 0.25 |
| 8 | 13 | 4 | 12 | 11 | +1 | 1 |
| 9 | 15 | 3 | 13 | 7.5 | +5.5 | 30.25 |
| 10 | 10 | 5 | 9.5 | 14 | −4.5 | 20.25 |
| 11 | 7 | 3 | 5.5 | 7.5 | −2.0 | 4.0 |
| 12 | 20 | 5 | 15 | 14 | +1 | 1 |
| 13 | 18 | 5 | 14 | 14 | 0 | 0 |
| 14 | 4 | 2 | 2 | 3.5 | −1.5 | 2.25 |
| 15 | 8 | 2 | 7.5 | 3.5 | +4.0 | 16.0 |
| | | | | | | $\Sigma d^2 = 120$ |

Σ means 'total'

*Calculating the Spearman test*

PROCEDURE

1. Rank the scores on variable A, giving a rank of 1 to the smallest score, a rank of 2 to the next smallest, etc. If two or more scores are the same use the tied ranks procedure outlined in Section A.2.
2. Rank the scores on variable B, giving the rank of 1 to the smallest scores, 2 to the next smallest etc. If two or more scores are the same, use the tied ranks procedure.
3. Take each subject's rank B score away from their rank A score to give d(A–B). Remember to put

WORKED EXAMPLE

1. See 'Rank of A' column

2. See 'Rank of B' column

3. See d(A–B) column

in the plus and minus signs as appropriate.

4. Square each score in the $d(A-B)$ column to give $d^2$. Enter these figures in the $d^2$ column above.

5. Add up the scores in the $d^2$ column to give $\Sigma d^2$.

6. Find $r_s$ from the formula

$$r_s = 1 - \frac{6\Sigma d^2}{N(N^2 - 1)}$$

where
$\Sigma d^2$ is the total of all the $d^2$ values
$N$ is the number of subjects or pairs of scores

4. See $d^2$ column

5. $\Sigma d^2 = 120$

6.

$$r_s = 1 - \frac{6 \times 120}{15(15^2 - 1)}$$

where
$\Sigma d^2 = 120$
$N = 15$

$$r_3 = 1 - \frac{720}{15 \times 224}$$

$$= 1 - 0.21$$
$$\therefore r_3 = +0.79$$

---

*Special point*

Ensure that you mark in the plus or minus sign in front of the final $r_s$ figure to indicate whether the correlation coefficient is positive or negative. Also, note how far the division lines extend in the formula. Many students continue the division line under the '1' at the beginning of the formula and so end up with wrong results.

---

*Looking up the results in the probability tables*

The results of our calculations have produced a correlation coefficient or $r_s$ of +0.79. This must be looked up in Table C.10 to find out whether this figure represents a significant correlation between length of time spent with the GP and evaluations of the service.

Turn to Table C.10, where you will see a column down the left-hand side entitled $N$ (number of subjects) and across the top levels of significance for one- and two-tailed hypotheses or tests.

Look down the $N$ column to find our $N$ value of 15. You will see that no $N = 15$ value is given, so you must take the next *lowest* number, i.e. 14. To the right of this are four numbers (called 'critical values of $r_s$'):

0.456    0.544    0.645    0.715

Each critical value is associated with the probability level indicated at the top of its column, such that the critical value 0.544 is associated with a probability of 0.05 for a two-tailed hypothesis and 0.025 for a one-tailed hypothesis.

For our $r_s$ to be significant at one of the probability levels associated with one-tailed hypotheses, it must be *equal to* or *larger than* one of the critical values to the right of $N = 15$. (Ignore the plus or minus sign in front of the $r_s$ value for the time being). If the $r_s$ value is exactly the same as one of the critical values, then the associated $p$ value is exactly the same as that of the critical value. So, for example, an $r_s$ of 0.645 ($N = 14$) has a $p$ value of exactly 0.01 for a one-tailed hypothesis. This is expressed as $p = 0.01$.

However, if the $r_s$ value is larger than one of the critical values then the associated probability is even less than that of the critical value. For example, an $r_s$ of 0.682 ($N = 14$, one-tailed) has a $p$ value of even less than 0.01, which is expressed as $p < 0.01$.

Our $r_s$ value is larger than even the largest critical of 0.715, which means then the probability of our results being due to random error is even less than 0.005 (one-tailed hypothesis).

*What do the results mean?*
The probability associated with our $r_s$ of +0.79 is <0.005, which means that there is less than a $\frac{1}{2}$% chance that our results could be due to random error. As we usually use a cut-off point of 5% or less to indicate significant results, this means that there is a significant positive correlation between length of time spent with the GP and evaluation of service. From this we can state that the longer the time spent with the GP the more positive the evaluation of the care provided. However, before we can conclude that our hypothesis has been supported it is essential that we confirm that we also predicted a positive correlation in the hypothesis. This is an important consideration since it is possible for a significant correlation to be obtained, which does *not* accord with correlation predicted in the hypothesis. As we made a positive prediction and obtained a positive correlation coefficient from the results, we can conclude that our experimental hypothesis has been supported. The null (no re-lationship) hypothesis can be rejected. The longer a patient spends with the GP the more favourable their evaluation of the quality of care.

Some queries that may arise when calculating the Spearman on your own data:

> *Q1: What happens if my $r_s$ is smaller than any of the critical values in the table?*

This simply means that your results are not significant and there is no correlation between the scores on your two variables. So, for example, had you obtained an $r_s$ of 0.421 ($N = 10$, two-tailed), this is smaller than the smallest critical value of 0.564, which means that the probability of your results being due to random error is greater than 10% (and so, greater than the usual cut-off point of 5%).

*Exercises (answers on page 319)*

**1** Look up the following $r_s$ values and state whether or not they are significant and at what level:
   (a) $r_s = -0.51$,     $N = 17$,     two-tailed
   (b) $r_s = -0.46$,     $N = 12$,     two-tailed
   (c) $r_s = +0.456$,    $N = 15$,     one-tailed
   (d) $r_s = +0.61$,     $N = 20$,     two-tailed
   (e) $r_s = -0.44$,     $N = 25$,     one-tailed

**2** Calculate a Spearman test on the following data and state the $r_s$, $N$ and $p$ values. Clarify what the results actually mean.

> **H$_1$:** The longer it takes an individual to travel to work, the greater the degree of stress experienced.
>
> **H$_0$:** There is no relationship between length of journey to work and the degree of stress experienced.

*Brief outline of study*
Select twelve nurses whose travelling times vary across a range from five to sixty minutes. Ensure that all subjects have been making this journey for at least three years. Ask each subject to mark on a six point scale what sort of stress category they believe themselves to be in:

| 1 | 2 | 3 | 4 | 5 | 6 |
|---|---|---|---|---|---|
| Never stressed | Very infrequently stressed | Infrequently stressed | Frequently stressed | Very frequently stressed | Permanently stressed |

**Table 12.3**

| Subject | Variable A: travel time (in minutes) | Variable B: normal stress level |
|---|---|---|
| 1 | 45 | 4 |
| 2 | 20 | 4 |
| 3 | 15 | 3 |
| 4 | 30 | 2 |
| 5 | 60 | 5 |
| 6 | 10 | 3 |
| 7 | 5 | 3 |
| 8 | 15 | 2 |
| 9 | 18 | 2 |
| 10 | 12 | 5 |
| 11 | 25 | 6 |
| 12 | 22 | 2 |

Assume all other constant errors have been eliminated. The results obtained are given in Table 12.3.

## 12.3  Parametric test for use with two sets of data

### 12.3.1  The Pearson product moment correlation coefficient test

The Pearson test is used under the following conditions:

1. When a correlational design has been used.
2. When there are two sets of data only.
3. When the data are of an interval/ratio type.
4. When the remaining three conditions required by a parametric test can be reasonably fulfilled (see Section 7.2).

The Pearson test will tell you whether your two sets of scores are either positively correlated (i.e. high scores on one variable being associated with high scores on the other) or negatively correlated (high scores on one variable being associated with low scores on the other).

When you calculate the Pearson test you find a value for $r$ which is a correlation coefficient. This figure will be somewhere between $-1.0$ and $+1.0$; the closer to $-1.0$ it is, the stronger the negative correlation between the two sets of figures, while the closer to $+1.0$ it is, the stronger the positive correlation between the two sets of figures. The correlation coefficient figure is looked up in the appropriate probability tables to ascertain whether the correlation between the data is a significant one or whether it could be the result of random error. If the probability (or $p$ value) associated with the obtained $r$ is 5% or less, then we can conclude that there *is* a significant correlation between the two variables, and it is unlikely that the results can be accounted for by random error. The experimental hypothesis would have been supported. If it is more than 5%, then the relationship between the two variables is not significant, and the results can be explained by random error. The null (no relationship) hypothesis would have been supported.

When using the Pearson test, it does not matter what type of measurement you have on each variable, as long as both are interval/ratio. In other words, you can compare time in minutes with weight in kilograms, or IQ with heart rate. Providing the data on both variables are interval/ratio, it can be of any measurement whatsoever.

The Pearson test has quite a complicated formula, producing large numbers. Do not be put off by this, since if you use a calculator and follow the instructions a step at a time, there should be few problems.

**Example**

There has been a drive in many schools of nursing recently to recruit mature students onto the RGN course. Imagine you are a nurse tutor in one of these schools and you are interested in monitoring the progress of these students. In particular, you have noticed that while many of them seem to have some initial difficulty in returning to study, they appear to do very well on the wards. You decide to look at the examination performance of a group of mature students to see if there is a link between maturity (as defined by age on entry) and examination results.

*Hypotheses*

$H_1$: There is a relationship between age of student nurse on entry to the RGN course and performance on first-year examinations. (The older the student, the better their performance.)

$H_0$: There is no relationship between age of student nurse on entry to the RGN course and performance on first-year examinations.

The $H_1$ is predicting a positive correlation (high scores on the age variable being related to high scores on the performance variable), and because it is making a definite prediction about the scores, it is a one-tailed hypothesis.

*Brief outline of study*
Randomly select twelve students from the RGN course, whose ages range from 18 to 42 upon entry. Collect their marks (in percentages) from the first-year examinations. (Assume that all the constant errors such as attendance levels and intellectual ability have been taken into account and controlled for.) You therefore have two sets of interval/ratio data (age in years and marks in percentages) and can fulfil the most important parameter required by the Pearson test. You are interested in whether high scores on the age variable are related to high scores on the examinations.

**Table 12.4**

| Subject | Data from study | | Calculations from Pearson | | |
|---|---|---|---|---|---|
| | *Variable A (age on entry)* | *Variable B (exam. mark)* | $A \times B$ | $A^2$ | $B^2$ |
| 1 | 25 | 62 | 1550 | 625 | 3844 |
| 2 | 18 | 63 | 1134 | 324 | 3969 |
| 3 | 21 | 58 | 1218 | 441 | 3364 |
| 4 | 23 | 69 | 1587 | 529 | 4761 |
| 5 | 30 | 72 | 2160 | 900 | 5184 |
| 6 | 34 | 69 | 2346 | 1156 | 4761 |
| 7 | 19 | 50 | 950 | 361 | 2500 |
| 8 | 42 | 71 | 2982 | 1764 | 5041 |
| 9 | 25 | 55 | 1375 | 625 | 3025 |
| 10 | 26 | 60 | 1560 | 676 | 3600 |
| 11 | 20 | 47 | 940 | 400 | 2209 |
| 12 | 38 | 67 | 2546 | 1444 | 4489 |
| | $\Sigma A = 321$ | $\Sigma B = 743$ | $\Sigma A \times B = 20\,348$ | $\Sigma A^2 = 9245$ | $\Sigma B^2 = 46\,747$ |

$\Sigma$ means 'total'

*Results*
You obtain the sets of data given in Table 12.4. (Figures in bold type are the raw data from the study, while those in light type are the calculations from the Pearson test. It does not matter which variable is A and which B.)

*Calculating the Pearson test*

PROCEDURE

1. Add up all the scores for variable A to give $\Sigma A$.
2. Add up all the scores for variable B to give $\Sigma B$.
3. For each subject, multiply their variable A score by their variable B score to give $A \times B$ e.g. for subject 1, $A \times B = 25 \times 62$. Enter these figures in the '$A \times B$' column.
4. Add up all the figures in the column $A \times B$ to give $\Sigma A \times B$
5. Square each score in the variable A column to give $A^2$ e.g. for subject 1, $25^2 = 625$. Enter these scores in the $A^2$ column.
6. Add up all the scores in the $A^2$ column to give $\Sigma A^2$.
7. Square each score in the variable B column to give $B^2$, e.g. for subject 1, $62^2 = 3844$. Enter these scores in $B^2$ column.
8. Add up all the scores in the $B^2$ column to give $\Sigma B^2$.
9. Find the value of $r$ from the formula:

$$r = \frac{N\Sigma A \times B - \Sigma A \times \Sigma B}{\sqrt{\{[N\Sigma A^2 - (\Sigma A)^2][N\Sigma B^2 - (\Sigma B)^2]\}}}$$

Where:
$N$ is the number of subjects or pairs of scores
$\Sigma A \times B$ is the total of the scores in the $A \times B$ column
$\Sigma A$ is the total of the variable A scores
$\Sigma B$ is the total of the variable B scores

WORKED EXAMPLE

1. $\Sigma A = 321$

2. $\Sigma B = 743$

3. See $A \times B$ column

4. $\Sigma A \times B = 20\,348$

5. See '$A^2$' column

6. $\Sigma A^2 = 9245$

7. See '$B^2$' column

8. $\Sigma B^2 = 46\,747$

9.

$$r = \frac{12 \times 20\,348 - 321 \times 743}{\sqrt{\{[12 \times 9245 - 103\,041][12 \times 46\,747 - 552\,049]\}}}$$

Where:
$N = 12$

$\Sigma A \times B = 20\,348$

$\Sigma A = 321$

$\Sigma B = 743$

$\Sigma A^2$ is the total of the scores in the $A^2$ column

$$\Sigma A^2 = 9245$$

$\Sigma B^2$ is the total of the scores in the $B^2$ column

$$\Sigma B^2 = 46\,747$$

$(\Sigma A)^2$ is the total of the variable A scores, squared

$$(\Sigma A)^2 = 103\,041$$

$(\Sigma B)^2$ is the total of the variable B scores, squared

$$(\Sigma B)^2 = 552\,049$$

$\sqrt{\phantom{x}}$ is the square root of all the calculations from under this sign

$$r = \frac{244\,176 - 238\,503}{\sqrt{\{[110\,940 - 103\,041][560\,964 - 552\,049]\}}}$$

$$= \frac{5673}{\sqrt{(7899 \times 8915)}}$$

$$= \frac{5673}{8391.64}$$

$$\therefore r = +0.68$$

10. Calculate the d.f. (degrees of freedom) value using the formula

$$\text{d.f.} = N - 2$$

10. $\text{d.f.} = 12 - 2$
    $= 10$

---

*Special point 1*

Do make sure you are clear about the difference between $\Sigma A^2$ and $(\Sigma A)^2$ and $\Sigma B^2$ and $(\Sigma B)^2$. $\Sigma A^2 =$ the total of the *already squared* scores from variable A, while $(\Sigma A)^2$ is the total of the scores from variable A, which is then squared. The same distinction applies to $\Sigma B^2$ and $(\Sigma B)^2$.

*Special point 2*

Do remember to put in the appropriate plus or minus signs in front of the value for *r*. This will indicate whether the correlation coefficient is positive or negative.

---

*Looking up the results in the probability tables*

The calculation of the Pearson test has produced a correlation coefficient (or *r*) of +0.68. This now has to be looked up in Table C.11 to see whether this figure represents a significant correlation between student's age on entry and practical exam mark.

If you now turn to Table C.11, you will see down the side d.f. or degrees of freedom values, while across the top there are various levels of significance for one- and two-tailed tests (or hypotheses). If you look down the d.f. column to find our d.f. = 10, you will see five figures to the right of this:

0.4973    0.5760    0.6581    0.7079    0.8233

These are called 'critical values of $r$' and each is associated with the level of significance indicated at the top of its column, such that 0.5760 has a probability of 0.025 for a one-tailed test and 0.05 for a two-tailed test.

Ignoring the plus or minus sign in front of $r$ for the moment, for our $r$ to be significant at one of these levels, it must be *equal to* or *larger than* one of the critical values to the right of the relevant d.f. value. If the obtained $r$ is equal to one of the critical values it would have exactly the same probability level so that value. So, for d.f. = 10, $r$ = 0.6581, two-tailed hypothesis, the $p$ value is 0.02 exactly. This is expressed as $p$ = 0.02 (or 2%). If the obtained $r$ is *larger* than one of the critical values to the right of a given d.f., then the associated $p$ value is even less than that for the critical value. So, for d.f. = 10, $r$ = 0.6961, two-tailed hypothesis, the probability value is *even less* than 0.02. This is expressed as $p < 0.02$.

Our $r$ of +0.68 (d.f. = 10, one-tailed hypothesis) is *larger* than 0.6581 and so has a $p$ value of less than 0.01. This means that the probability of our results being due to random error is even less than 1%.

*What do the results mean?*
The probability level associated with $r$ = +0.68 (d.f. = 10, one-tailed hypothesis) is less than 0.01 or 1%. This means that there is less than a 1% chance that random error could account for the results. Using the usual cut-off point of 5% or less for concluding that results are significant, this suggests that there *is* a significant correlation between the two sets of data in our study. As the correlation coefficient of 0.68 was preceded by a plus sign, this indicates that this correlation is a positive one, i.e. high scores on one variable being associated with high scores on the other. If we translate this to the present study, we can conclude that the *older* the student on entry to the RGN course, the *higher* the practical exam mark at the end of the first year. As this is exactly what was predicted in the original hypothesis, we can state that the experimental hypothesis has been supported and that the null (no relationship) hypothesis can be rejected.

You must always check that the type of correlation (positive or negative) between the sets of data is in accord with the prediction made by the experimental hypothesis. This is essential, since it would be possible to obtain a significant correlation which was the opposite of the one anticipated and so would *not* support the hypothesis.

You may have some queries about the Pearson test if you have been analysing your own data:

> *Q: My r value was smaller than any of the critical values to the right of my d.f. What does this mean?*

This means that you do *not* have a significant correlation between your two sets of data, and that random error could explain your results. So, if you obtained an *r* of 0.3210 (d.f. = 18, two-tailed hypothesis) this is smaller than the smallest critical value of 0.3783 and so the associated probability would be *greater* than 0.10 (or 10%). As the usual cut-off point for claiming significant results is 5% or less, you would have to conclude that the correlation between your data is not significant and could be explained by random error.

> *Q2: I have a d.f. of 24, but there is no corresponding d.f. value in the table. What do I do?*

In such cases where there is no equivalent d.f. value in the table, you should take the next smallest. Here, then, you would use the d.f. = 20 value.

---

## Exercises (answers on page 319)

**3** Look up the following *r* values and state whether or not they are significant and at what level:
(a) *r* = 0.5386,    d.f. = 20, one-tailed
(b) *r* = 0.3423,    d.f. = 26, two-tailed
(c) *r* = 0.7781,    d.f. = 13, one-tailed
(d) *r* = 0.3232,    d.f. = 28, two-tailed
(e) *r* = 0.4575,    d.f. = 12, one-tailed

**2** Calculate a Pearson test on the following data and state the *r*, d.f. and *p* values. Clarify what the results actually mean.

**H₁:** There is a relationship between the length of time a woman

**Table 12.5**

| Subject | Variable A: length of time on Pill (years) | Variable B: length of time to become pregnant (months) |
|---|---|---|
| 1 | 4 | 6 |
| 2 | 2 | 5 |
| 3 | 7 | 7 |
| 4 | 10 | 4 |
| 5 | 1 | 5 |
| 6 | 11 | 10 |
| 7 | 9 | 12 |
| 8 | 5 | 18 |
| 9 | 3 | 6 |
| 10 | 6 | 9 |
| 11 | 7 | 3 |
| 12 | 10 | 7 |

has been on oral contraceptives and the time it takes to become pregnant subsequently.

$H_0$: There is no relationship between length of time on oral contraceptives and length of time to become pregnant.

*Brief outline of study*
Randomly select twelve women from a pre-conceptual counselling clinic in a health centre who have been on oral contraceptives for varying periods. Monitor the length of time between coming off the Pill and conception for each patient (taking account of all constant errors). The results obtained are given in Table 12.5.

## 12.4  Non-parametric test for use with three or more sets of data

### 12.4.1  Kendall's coefficient of concordance

Kendall's coefficient of concordance is used under the following conditions:

1. When the design is correlational.
2. When there are three or more sets of data.
3. When the data are either ordinal or interval/ratio.

The Kendall test is particularly valuable when analysing data from studies which are concerned with how far people's judgements on a particular issue agree.

However, two factors associated with the Kendall test are important. Firstly, this test is used when at least three sets of data have been derived from the study and correlations are predicted between them. Because of this, the Kendall test can only tell you whether or not the data are *positively* correlated; it gives no information about negative correlations. The reason for this becomes apparent if we look more closely at what the Kendall test actually does.

It was said earlier that the Kendall test compares at least three sets of data to see if there is any correlation between them. Imagine that we have obtained three sets of scores (I, II and III). It is possible for all three sets to *agree* completely in terms of the direction of the results, such that high scores on set I are associated with high scores on sets II and III. This is, of course, a positive correlation. But is not possible for the scores to be negatively correlated since high scores on set I could be associated with low scores on set II, but what about the scores for set III? If the scores for set III are also low, then they will concur with those from set II, thus providing some measure of agreement (or positive correlation) between these two sets of scores. On the other hand, if the scores on set III are high, then they will accord with those from set I, and so will provide some measure of agreement between those two sets. Therefore, where there are three or more sets of data, it is not possible to achieve a negative correlation. So, the Kendall test will only tell us whether the results show no correlation or a positive correlation. As a result the correlation coefficient derived from the Kendall will be somewhere between 0 and +1.0.

One other point emerges from this. Because the Kendall test can only say whether or not results are *positively* correlated, the experimental hypotheses used with this test must be one-tailed (since the results can only go in one direction).

Two further points should be outlined before proceeding. Firstly, the raw data for this test must be ordinal or interval/ratio. However, if you *do* have interval/ratio data, then you must rank order this before calculating the Kendall (giving a rank of 1 to the lowest score and so on as outlined in Section A.2) since the test can only be performed on rank ordered data. Secondly, there is a limit to the amount of data that can be used with the Kendall. While you can use three or more *sets* of data, the number of scores within each set should not exceed seven. This limit should not be a major inconvenience since seven scores per set should be sufficient for most purposes. Should you ever be in a situation where more scores are required, a different formula should be used and this is outlined in Siegel (1956).

Essentially, the Kendall test calculates the theoretical *perfect* agreement between the sets of scores from your study and then compares the *actual* sets of data with this, to determine how far they deviate from perfection. When calculating the Kendall test you find the value of $W$, which, roughly speaking, is the index of the divergence of the data from the maximum agreement possible. This will be a figure between 0 and +1.0 and will give you an idea of the strength of agreement between the sets of data. A value for $s$ is also calculated which is looked up in the relevant probability tables, to see whether or not there is a significant positive correlation between the sets of scores. If the probability (or $p$ value) associated with $s$ is 5% or less, then there is significant agreement between the sets of data; it is unlikely that random error can explain the results, and the experimental hypothesis would have been supported. If it is more than 5%, then there is no agreement between the sets of data and the results could be explained by random error. In this case, the null (no relationship) hypothesis would have been supported.

**Example**

Everybody believes they can identify a good nurse, though not everyone may agree as to what good nursing constitutes. However, as a senior nurse tutor in a large school of nursing you and four other senior tutors are involved in selecting the annual award winner from among the trainees. But because of inability to agree on the winner, there is conflict every year. You decide that it might be more productive initially to obtain tutors' opinions on which qualities they consider are most important in a good nurse. If some agreement on that could be achieved, then it might serve as a guideline for selecting the award winner and so reduce the degree of conflict.

*Hypotheses*

**H₁:** There is a positive correlation between the opinions of senior tutors about essential qualities of a good nurse.

$H_0$: There is no relationship between the opinions of senior tutors about the qualities of a good nurse.

The $H_1$ is, of necessity, one-tailed.

*Brief outline of study*
After looking through the literature you identify six qualities which appear to be central to a good nurse:

1. Intelligence.
2. Sensitivity.
3. Compassion.
4. Detachment.
5. Practical skill.
6. Theoretical knowledge.

These qualities are numbered from 1 to 6.

You and the four other senior tutors rank order each of these qualities according to how important the tutor considers the quality to be. A rank of 1 is given to the one considered to be most important, etc. Try to ensure that the people carrying out the ranking do not assign too many equal ranks to the qualities or events being ranked, since this has the effect of reducing the significance of your results. The five sets of rankings are then subjected to statistical analysis to determine the level of agreement between the tutors. The conditions for the Kendall test have been met, since we have five sets of data (from the five tutors) with six scores in each set (from the rankings of the six qualities). The qualities have been rank ordered so that the data are now ordinal (Table 12.6).

---

### Special point

It is essential that you set out your data in the way outlined below such that the rank orderings go *across* the page.

---

**Table 12.6**

| Senior tutor | Tutors' rankings of nursing qualities | | | | | |
| | Quality 1 | Quality 2 | Quality 3 | Quality 4 | Quality 5 | Quality 6 |
|---|---|---|---|---|---|---|
| I | 5 | 1 | 4 | 2 | 3 | 6 |
| II | 1 | 2 | 6 | 3 | 4 | 5 |
| III | 4 | 1 | 2 | 6 | 3 | 5 |
| IV | 5 | 3 | 4 | 2 | 6 | 1 |
| V | 5 | 6 | 4 | 2 | 1 | 3 |
| Rank totals | 20 | 13 | 20 | 15 | 17 | 20 |

*Calculating the Kendall test*

| PROCEDURE | WORKED EXAMPLE |
|---|---|
| 1. Add up the rank assigned to each quality to give the rank totals. | 1. See 'Rank Totals' row |
| 2. Add up all the rank totals to give $\Sigma R$. | 2. $\Sigma R = 105$ |
| 3. Divide $\Sigma R$ by the number of qualities being ranked to give the average rank $\bar{x}_R$. | 3. $\bar{x}_R = 105 \div 6$ <br> $= 17.5$ |

4. Take each rank total from $\bar{x}_R$ and square the results to give $d^2$.

4. (a) $d^2 = (17.5 - 20)^2$
$= 6.25$
(b) $d^2 = (17.5 - 13)^2$
$= 20.25$
(c) $d^2 = (17.5 - 20)^2$
$= 6.25$
(d) $d^2 = (17.5 - 15)^2$
$= 6.25$
(e) $d^2 = (17.5 - 17)^2$
$= 0.25$
(f) $d^2 = (17.5 - 20)^2$
$= 6.25$

5. Add up the $d^2$ values to give $\Sigma d^2$. $\Sigma d^2$ is the value of $s$.

5. $\Sigma d^2 = 6.25 + 20.25 + 6.25 + 6.25$
$+ 0.25 + 6.25$
$= 45.5$
$\therefore s = 45.5$

6. Find $W$ from the formula

$$W = \frac{s}{1/12 \ n^2(N^3 - N)}$$

where
$s$ is the value of $\Sigma d^2$
$n$ is the number of *sets* of data (i.e. number of judges carrying out the ranking)
$N$ is the number of scores in each set of data. (i.e. the number of objects or features being ranked)

6.

$$W = \frac{45.5}{(1/12 \ 5^2(6^3 - 6)}$$

$s = 45.5$
$n = 5$

$N = 6$

i.e. $W = \dfrac{45.5}{1/12 \times 5^2(6^3 - 6)}$

$= \dfrac{45.5}{0.08 \times 5250}$

$\therefore W = 0.108$

*Looking up the results in the probability tables*
The Kendall test has yielded a $W$ value of 0.108 which is the correlation coefficient, and a value of $s$ of 45.5. To see whether this represents a significant agreement between the senior tutors' rankings, you need to look up the value of $s$ (not $W$) in the probability tables.

To look up the value of $s$, you also need the values of $n$ (the number of sets of data) and $N$ (the number of scores per set of data). Turn to Table C.12. This table is divided into two parts: the top part refers to probabilities or significance levels of 5%, while the bottom part refers to probabilities or significance levels of 1%. Down the left-hand side of each table, you will see values of $n$ (number of judges), while across the top are values of $N$ (number of items being judged). Starting with the top table first, locate our $n = 5$ down the side and $N = 6$ across the top. At their intersection point is the figure 182.4. This is called a 'critical value of $s$'. For a given $s$ value to be significant, it must be equal to or larger than the critical value at the intersection point of the $n$ and $N$ values. If $s$ is equal to the critical value, it is significant at exactly the 5% level (since we are dealing with the 5% probability table). Therefore, had we obtained an $s$ of 182.4 it would have a probability of *exactly* 5%. This is expressed as $p = 0.05$ (or 5%).

However, if the obtained $s$ is *larger* than the critical value at the intersection point, then its associated probability is *even less* than 5%. In this case you should repeat the process with the 1% probability table underneath. If the obtained $s$ is the same as the figure at the intersection point then the results are significant at the 1% level exactly and $p$ would then equal 1% (or 0.01). If, on the other hand, the obtained $s$ is larger than the relevant critical value, then the associated probability value is even less than 1%. So, had we obtained an $s$ of 243.6 ($n = 5$, $N = 6$) then since this is larger than the critical value 229.4, the associated probability is less than 1%. This is expressed as $p < 0.01$.

Our $s$ of 45.5 is smaller than the critical value 182.4, for $n = 5$, $N = 6$. This means that there is no significant correlation between the data because the chances of random error accounting for the results is greater than 5%.

*What do the results mean?*
The probability of our results being due to random error is greater than 5%, which is the standard cut-off point for claiming results as significant. This means that there is no agreement among senior tutors as to the qualities of a good nurse. The experimental hypothesis must be rejected since the null (no relationship) hypothesis has been supported.

You may have some questions about the Kendall test if you have been using it to analyse your own data.

    Q1: *Why must I calculate* W, *since it is not used to look up the probability levels?*

W is the correlation coefficient and tells you on a scale of 0 to +1.0 how strong the agreement is between your sets of data. Although it is not used in determining the significance of your results, it does give you a good idea of the strength of the correlation, since a $W$ close to $+1.0$ is indicative of a strong positive correlation, while a $W$ close to 0 is indicative of no agreement. In other words, although $s$ will give you the significance of your results, the $W$ value can be informative as to the absolute degree of relationship between your sets of data.

> Q2: I have twelve sets of scores. How do I look this up, given that $n = 12$ is not listed in the tables?

When there is no equivalent $n$ value, use the next *lowest* figure in the table. Therefore, here you would need to use $n = 10$.

---

## Exercises (answers on page 320)

**5** Look up the following $s$ values and state whether or not they are significant and at what level.
(a) $s = 230.5$,    $n = 4$,    $N = 7$
(b) $s = 99.4$,    $n = 8$,    $N = 4$
(c) $s = 74.6$,    $n = 9$,    $N = 3$
(d) $s = 309.1$,    $n = 10$,    $N = 5$
(e) $s = 450.2$,    $n = 14$,    $N = 6$

**6** Calculate a Kendall coefficient of concordance on the following data and state the $W$, $s$, $n$, $N$ and $p$ values. Clarify what the results mean.

**H$_1$:** There is significant agreement among nurses on a coronary care ward as to how the gift of a large sum of money should be spent.

**H$_0$:** There is no agreement between nurses on a coronary care

Table 12.7

| Nurse | Ways of spending the money | | | | |
|---|---|---|---|---|---|
| | *1* | *2* | *3* | *4* | *5* |
| 1 | 4 | 1 | 2 | 3 | 5 |
| 2 | 3 | 2 | 1 | 4 | 5 |
| 3 | 4 | 3 | 2 | 1 | 5 |
| 4 | 5 | 3 | 1 | 2 | 4 |
| 5 | 5 | 1 | 2 | 3 | 4 |
| 6 | 5 | 1 | 2 | 3 | 4 |
| 7 | 4 | 2 | 1 | 3 | 5 |
| 8 | 4 | 2 | 3 | 1 | 5 |
| 9 | 5 | 1 | 4 | 2 | 3 |
| 10 | 4 | 1 | 2 | 3 | 5 |

ward as to how the gift of a large sum of money should be spent.

*Brief outline of study*
A grateful patient has donated a sum of money to the coronary care ward on which you are the sister. You decide to ask the views of the nursing staff as to how they think this money should be spent. You itemise five possible ways of disposing of this sum and ask the ten nurses on the ward to rank order these in order of preference. You obtain the results in Table 12.7.

## 12.5  Making predictions from correlated data

### 12.5.1  Linear regression

Linear regression is a particularly useful tool in research, since it allows the researcher to make predictions from correlated data. More precisely, if a significant correlation has been established between two variables ($A$ and $B$), then a score on variable $B$ can be predicted from any known score on variable $A$, and vice versa. Let us take an example. Supposing you had found a significant correlation between the amount of time a community psychiatric nurse spends with a discharged manic depressive patient, and the number of subsequent manic episodes, such that the *greater* the amount of time spent, the *fewer* the relapses. From this knowledge you could use linear regression to predict the number of manic relapses for any given patient providing you knew the amount of time the CPN spent with him or her. This would obviously be a useful prediction, since it would enable closer monitoring of at-risk patients and the possible provision of back-up support for patients who had limited contact with the CPN.

One or two or examples might further demonstrate the value of this technique. Nocturnal enuresis in adulthood is a particularly distressing condition. Recently a drug has been introduced which slows down the activity of the bladder, allowing the sufferer more time in which to get to the lavatory. The drug has been shown to be effective, but has a major side-effect for multiple sclerosis victims for whom enuresis is a considerable problem. The drug, when given in quantity, increases their paralysis. Now if a significant correlation could be demonstrated between the amount of drug given and the severity of paralysis, then for any known dosage of the drug, the level of subsequent paralysis could be predicted using linear regression. This would be invaluable knowledge for those patients for whom certain increased levels of paralysis would be more unacceptable than the enuresis. Essentially what would happen in such cases is that the drug dosage for each patient could be altered to a level whereby the level of subsequent paralysis would be less problematic than the incontinence.

Similarly, in nurse training, if it could be demonstrated that a significant correlation existed between scores on the first theory assessment completed after the first six weeks of the RGN course, and clinical skill during subsequent ward experience, then predictions could be made about students' clinical performance as soon as the theory test scores were known. This would allow the nurse tutor to give additional help and support to those students who are not predicted to do well during ward experience.

Essentially, then, linear regression allows the researcher to make predictions about future performance on one variable from known scores on the other which in nursing research could mean that pre-emptive intervention could be instituted in cases where problems are predicted.

Certain conditions are required before linear regression can be used:

1. The two variables that you are interested in must be shown to be significantly correlated *before* the linear regression technique can be applied.
2. This correlation must be established using a valid statistical test, i.e. the Spearman or the Pearson.
3. Linear regression can only be used on interval/ratio data or on ordinal data where equal intervals along the scale have been assumed (see Section 6.4).

Linear regression involves the calculation of an equation, which is then used to make predictions about the unknown variable. When using linear regression, the variable whose score is known is called $X$, while the score that is to be predicted is called $Y$. So in the previous three examples:

1. Amount of time the CPN spends with the patient is *known* and is therefore called $X$, while the number of relapses is to be predicted and so is called $Y$.
2. Drug dosage is *known* and therefore would be called $X$, while degree of paralysis is to be predicted and so is called $Y$.
3. Theory exam mark is *known* and is therefore called $X$, while ward performance is to be predicted and so is called $Y$.

In order to explain how linear regression works, it may be useful to recap briefly on correlations. From Chapter 4, you will remember that there are two types of correlation. Firstly, there are positive correlations, where high scores on one variable are related to high scores on the other (and so low scores on one variable are related to low scores on the other). If scores from *positive* correlations are plotted they produce a graph, called a scattergram, with an *upwards* slanting slope. Negative correlations occur when high scores on one variable are associated with low scores on the other. If the scores from *negative* correlations are plotted, they will produce a scattergram with a *downwards* slanting slope. The scattergrams are of particular interest in linear regression, and so will be discussed more fully.

If two sets of scores are perfectly correlated they will produce an absolutely straight line scattergram. Let us take the first example given, i.e. that the *more* time a CPN spends in following up a manic depressive patient, the *fewer* the number of subsequent manic relapses that will occur. This is a *negative* correlation, and so would produce a downward sloping scattergram. Imagine this study has been

**Table 12.8**

| Subject | Amount of time spent with patient (in hours) | Number of subsequent manic relapses |
|---------|------------------|------------------|
| 1 | 75 | 0 |
| 2 | 70 | 2 |
| 3 | 65 | 4 |
| 4 | 60 | 6 |
| 5 | 55 | 8 |
| 6 | 50 | 10 |
| 7 | 45 | 12 |
| 8 | 40 | 14 |

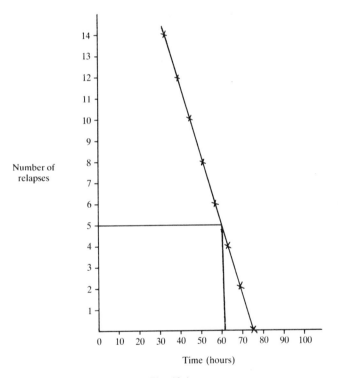

**Fig. 12.1**

carried out and the results in Table 12.8 obtained. If these scores are plotted on a scattergram then the graph in Figure 12.1 is obtained.

A line drawn through all the dots is perfectly straight, representing a *perfect* negative correlation. A perfect correlation is highly unlikely ever to occur, of course, but it suits the purposes of explanation. Now, supposing you know of a

CPN who spends sixty-two and a half hours with a patient. If you find $62\frac{1}{2}$ along the horizontal axis and draw a straight line across to meet the slope, and then another straight line down to meet the vertical axis (see dotted lines on graph) you will find that the number 5 is encountered. This means that from these particular set of results, a patient who spends sixty-two and a half hours with the CPN could be expected to experience five manic relapses. In other words, you have made a prediction about the number of relapses, based on your knowledge of the amount of time the CPN spends with the patient. This is the essence of linear regression.

But (and it is a big but) you are very, very unlikely to achieve a set of perfectly correlated scores. For this study, the sorts of results you might achieve would be more likely to look like the ones in Table 12.9. If these scores are plotted, the scattergram looks like Figure 12.2. This slope of scores is still roughly downward

**Table 12.9**

| Subject | Amount of time in hours | Number of relapses |
| --- | --- | --- |
| 1 | 73 | 2 |
| 2 | 68 | 1 |
| 3 | 65 | 4 |
| 4 | 59 | 4 |
| 5 | 51 | 6 |
| 6 | 48 | 5 |
| 7 | 20 | 9 |
| 8 | 12 | 10 |

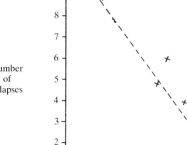

**Fig. 12.2**

sloping, but a straight line cannot be drawn through all the points. If you were asked to draw in a straight line, you would have to opt for a line which was the 'best fit' for all points. A line of best fit is defined as that line which, when drawn in, allows every point on the graph to be as close to it as possible.

It is extremely difficult to draw in this line of best fit by hand. I have put one in the graph above, but there may be as many variations to this as there are readers of the book. If there is no agreement as to where the line of best fit should be drawn, how can any predictions be made? For example, what would the relapse rate be now, if the CPN had spent sixty-two hours with a patient? The answer to this must be imprecise if we cannot be sure our line *is* the best fit.

The linear regression procedure is simply a formula or equation for calculating that line of best fit, such that every point on the graph is as close to the line as possible. Using this formula, accurate predictions can be made about one variable, from the knowledge of any given value on the other variable. This formula or equation is:

$$Y = bX + a$$

where  $Y$ is the variable value to be predicted

 $X$ is the variable value which is already known

a and b are constants which have to be calculated

**Example**

Imagine you are interested in the third of the examples quoted earlier, i.e. the relationship between theory test marks and clinical performance. Obviously it would be extremely useful to be able to predict a student's ward performance from his or her mark on the theory exam taken after six weeks' training, since those students who are expected to do poorly during ward experience could be given extra tuition and support. In order to make these predictions you must initially collect students' theory and clinical marks to find out whether or not there is a significant correlation between them. If there is, you can proceed with the linear regression; if there is no correlation the linear regression should not be performed.

You obtain the results in Table 12.10.

**Table 12.10**

| Subject | Variable X: theory mark (out of 20) | Variable Y: ward assessment (out of 20) |
|---|---|---|
| 1 | 18 | 15 |
| 2 | 17 | 15 |
| 3 | 15 | 13 |
| 4 | 12 | 14 |
| 5 | 10 | 7 |
| 6 | 13 | 12 |
| 7 | 14 | 14 |
| 8 | 9 | 6 |

*Calculating the linear regression equation*

PROCEDURE

1. Calculate a Spearman or Pearson on the two sets of scores from your study to establish whether or not they correlate.

WORKED EXAMPLE

1. Using a Pearson test on the data in table $r = +0.87$
$p < 0.01$ (two-tailed)
∴ the scores are significantly correlated. The higher the theory mark, the better the clinical assessment.

---

### Special point 1

If you do not get a significant correlation, you should not proceed with the linear regression equation.

---

2. Set out the raw scores in a table under the headings; Subject, Variable $Y$, Variable $X$, $X \times Y$, $X^2$.

2. See Table 12.11.

**Table 12.11** Initial linear regression calculations

| Subject | Variable Y: ward assessment | Variable X: theory exam. mark | $X \times Y$ | $X^2$ |
|---------|---------|---------|---------|---------|
| 1 | 15 | 18 | 270 | 324 |
| 2 | 15 | 17 | 255 | 289 |
| 3 | 13 | 15 | 195 | 225 |
| 4 | 14 | 12 | 168 | 144 |
| 5 | 7 | 10 | 70 | 100 |
| 6 | 12 | 13 | 156 | 169 |
| 7 | 14 | 14 | 196 | 196 |
| 8 | 6 | 9 | 54 | 81 |

---

### Special point 2

Remember that $Y$ is the variable score to be predicted and $X$ is the known variable score.

---

3. To obtain the values in the column $X \times Y$, multiply each variable $X$

3. See $X \times Y$ column

score by the corresponding variable $Y$ score, e.g.

$$15 \times 18 = 270$$

4. To obtain the values in the $X^2$ column, square each variable $X$ score, e.g. $18^2 = 324$.

    4. See $X^2$ column

5. Add up all the scores for variable $Y$ to give $\Sigma Y$.

    5. $\Sigma Y = 96$

6. Add up all the scores for variable $X$ to give $\Sigma X$.

    6. $\Sigma X = 108$.

7. Add up all the values in the $X \times Y$ column to give $\Sigma X \times Y$.

    7. $\Sigma X \times Y = 1364$

8. Add up all the values in the column $X^2$ to give $\Sigma X^2$.

    8. $\Sigma X^2 = 1528$

9. Calculate the constant b from the formula:

    9.

$$b = \frac{N\Sigma X \times Y - (\Sigma X)(\Sigma Y)}{N\Sigma X^2 - (\Sigma X)^2}$$

$$b = \frac{8 \times 1364 - 108 \times 96}{8 \times 1528 - 11\,664}$$

where

$N$ is the total number of subjects or pairs of scores

$\Sigma X \times Y$ is the total of the values in the $X \times Y$ column

$\Sigma X$ is the total of the values in the variable $X$ column

$\Sigma Y$ is the total of the values in the variable $Y$ column

$\Sigma X^2$ is the total of the values in the $X^2$ column

$(\Sigma X)^2$ is the total of the scores in the variable $X$ column, squared

where

$N = 8$

$\Sigma X \times Y = 1364$

$\Sigma X = 108$

$\Sigma Y = 96$

$\Sigma X^2 = 1528$

$(\Sigma X)^2 = 11\,664$

$$= \frac{10\,912 - 10\,368}{12\,224 - 11\,664}$$

$$= \frac{554}{560}$$

$$\therefore \ b = 0.97$$

---

*Special point 3*

Do discriminate between $\Sigma X^2$, which is the total of the already squared variable $X$ scores and $(\Sigma X)^2$ which is the total of the variable $X$ scores, squared.

10. Find $a$ from the formula

$$a = \frac{\Sigma Y}{N} - b \times \frac{\Sigma X}{N}$$

where

$N$ is the number of subjects or pairs of scores

$\Sigma Y$ is the total of scores in the variable $Y$ column

$\Sigma X$ is the total of the scores in the variable $X$ column

$b$ is the constant already calculated

11. Substitute the values of a and b in the formula

$$Y = bX + a$$

where $X$ is any known value or variable

10.

$$a = \frac{96}{8} - 0.97 \times \frac{108}{8}$$

where

$N = 8$

$\Sigma Y = 96$

$\Sigma X = 108$

$b = 0.97$
$\quad = 12 - 13.1$
$\therefore\ a = -1.1$

11. $Y = 0.97 \times X + (-1.1)$

where $X$ is any known theory test mark

*What do the results means?*
We have calculated the linear regression formula for the correlated data between theory test mark and ward assessment. We can now substitute any theory test mark for $X$ in the formula in order to predict ward performance ($Y$). So, imagine you have collected the students' marks from their first theory exam after six weeks' training. Some marks look rather borderline and you decide to see which of these might be associated with substandard ward performance. In particular, you are interested in student A with 11 for the theory test, student B with 8 for the theory test and student C with 6 for the theory test.

If we substitute student A's theory test mark in the linear regression formula we obtain

$$Y = 0.97 \times 11 + (-1.1)$$
$$= 9.57$$

This student, then, is likely to obtain a ward assessment mark of 9.57, which is a marginal fail. With a bit of extra tuition and help, she could probably achieve a reasonable pass.

If we substitute student B's theory test mark in the linear regression formula, we get

$$Y = 0.97 \times 8 + (-1.1)$$
$$= 6.66$$

This is clearly a poor fail, and you would have to decide whether to give considerable extra coaching or ask her to rethink her chosen career.

If student C's theory mark is substituted, we get

$$Y = 0.97 \times 6 + (-1.1)$$
$$= 4.72$$

This is an extremely low mark and it is conceivable that this student may never make the grade. You decide to consult with the other nurse tutors about whether or not to request this student's withdrawal from the course.

This formula, then, provides a method of making predictions about performance from a known score, and as such is an invaluable technique in a variety of research contexts.

---

## Exercises (answers on page 320)

7 You have already established, using a Pearson test, that the more information a patient has about their impending bowel resection operation, the
more adapted they are to the situation post-operatively. (Both variables
were measured along a seven point scale which assumed equal intervals.)
The results were as given in Table 12.12.

Using a Pearson test, the results were $r = 0.7, p < 0.025$ (for $N = 8$, one-
tailed hypothesis). On your ward, there are three men each of whom is
awaiting bowel surgery. You ascertain the extent of their knowledge about
the operation in order to make predictions about their subsequent adaptation levels. If these can be established, the services of the specialist nurse
adviser can be called upon to counsel those patients who are not expected
to adapt to the situation after their operation.

You obtain the following knowledge about the operation scores for the
three patients:

**Table 12.12**

| Subject | Knowledge | Adaptation |
|---------|-----------|------------|
| 1 | 7 | 6 |
| 2 | 4 | 3 |
| 3 | 5 | 3 |
| 4 | 5 | 4 |
| 5 | 6 | 7 |
| 6 | 1 | 3 |
| 7 | 1 | 2 |
| 8 | 4 | 5 |
| 9 | 4 | 6 |
| 10 | 6 | 5 |

Patient I    = 3
Patient II   = 2
Patient III  = 4

Calculate the linear regression formula using the above data and state the values of a and b. Using these values, calculate the predicted adaptation levels for the three patients.

# 13    Estimation

## 13.1  An introduction to some basic issues and terms

The concept of **estimation** was introduced briefly in Chapter 2. Like hypothesis testing, estimation allows the researcher to infer that the information derived from a small group of people (known as a **sample**) also applies to the larger group from which the sample was drawn. This larger group is called the **population**. The terms 'sample' and 'population' are discussed in more detail in Section 2.1.

However, estimation as a branch of inferential statistics, differs from hypothesis testing in that it does not involve testing a prediction or hypothesis by experimental methods. Instead it involves collecting data from a small sample of people on a particular characteristic and then using an appropriate statistical formula to make estimates about this characteristic for the population from which the sample comes. The characteristic that we want to estimate is called a *parameter*. Thus, the technique of estimation can be seen as a scientific 'best guess'. Let us illustrate this by an example. Supposing you were interested in finding out how many children in Birmingham under the age of 5 had been vaccinated against measles, in a city of over a million people, it would be too costly and time consuming to ask every family about their children's vaccination history. So instead, you would select a small sample of children for study and from their responses you could *estimate* the number of children in Birmingham as a whole who had been vaccinated. This technique is a very useful one in nursing and has applications in a number of areas. For example, estimates could be made of the number of people in a given area or group who are HIV positive, or the number of people likely to require treatment for secondary tumours following carcinoma of the bowel, or the number of nurses who leave the profession within five years of qualifying. Estimation is an important approach in nursing research, especially where planning and directing resources is involved.

In order to carry out estimates of this sort the following steps have to be taken:

1. Firstly, you must identify your area of interest. This involves selecting the relevant population and defining it carefully. Vague terms such as 'patients', 'children', etc., may be inappropriate. For example, if you are interested in estimating the number of children in a large city who have had whooping cough, you must define the terms carefully, taking into account the age of the children, what area constitutes 'the city', etc.

Identifying the area of interest also means specifying the relevant population parameters in which you are interested. In other words, you must state clearly what it is you wish to estimate. For example, it may be the drop-out rate of student nurses, the number of incontinence appliances required for a given area's multiple sclerosis sufferers, the number of people over the age of 65 who require treatment for varicose ulcers, etc.

2. The second step involves selecting a representative sample of the population in which you are interested. This process involves the **random sampling** procedure described in Section 4.1.2. Stated simply, random sampling means that every member of a given population has an equal chance of being selected for study. In practice, the procedure typically involves putting the names of that population into a hat and pulling out the required number, or using random number tables.

It is important to note that the sample should be of an adequate size. The term 'adequate size' is not an easy one to define since it can be altered to suit the subject under investigation. Certainly, larger samples are more likely to yield representative and unbiased subjects than are small samples (see Section 4.1.1). So, try to obtain as many subjects in your sample as time and resources will allow.

3. The next stage in estimation is the data collection. This may take the form of simple yes/no questions and answers, direct measurement or observation. Whatever method is selected, it *must* be a suitable measure of the parameter. For example, to estimate the number of children under 5 who had received the measles vaccine, it would *not* be appropriate to find out whether or not the children in the sample had had measles. It would be much better to find out from medical records or from the families directly whether or not the children in the sample had received the vaccination.

4. The final stage in the estimation process involves the application of an appropriate statistical formula to the data collected from the sample. From the calculation of this formula, the population's characteristics or parameters can be estimated.

## 13.2  Point estimations and interval estimations

There are two types of estimation – **point estimation** and **interval estimation**. A point estimate is a single figure derived from the sample which is used as an estimate for the population. This single figure may be the *average* of the sample which is used to estimate the population average. For example, if you were interested in estimating the number of days lost per annum due to sickness amongst trained nurses within a particular regional health authority, you would select a sample of trained nurses, and calculate the average number of days lost. If the average derived from the sample was, say, 17.3 days per annum you could estimate that an average 17.3 days per annum would be lost across the region as a whole.

This simple figure of 17.3 days derived from the sample is a point estimate of the average number of days lost within the population of trained nurses within the region.

On the other hand, the point estimate could be a single *percentage* figure. So if your focus of interest was the percentage of women between the ages of 50 and 55 who had had a cervical smear within the preceding twelve months, you would select a sample of women between 50 and 55 and, on the basis of their responses, the percentage of women who had obtained smears could be calculated. If this figure was 32% you could estimate that the percentage of the total population of women between the ages of 50 and 55 who had received cervical smears was also 32%.

This technique of point estimation is a simple one, but its accuracy depends on the quality of the random sampling that took place when selecting the subjects. Even if the sampling procedure was impeccably carried out, it is still very likely that the point estimate could be wrong, if only marginally, and sometimes even marginal errors can be disastrous when planning resources or budgets. In these cases it is preferable to know the degree of error surrounding the point estimate. Calculating the degree of error is called interval estimation and this technique simply involves stating two figures (instead of one), between which the researcher can be reasonably confident the population parameter lies. So rather than calculating a single point estimate of 17.3 days lost due to sickness amongst trained nurses, it might be better to specify an upper and lower number of days within which the average days lost by the population of nurses within the region can confidently be expected to fall. So, an interval estimate from the sample might produce the figures 16.8–18.1 days lost due to sickness and the researcher would feel reasonably confident that the average number of days lost for the nurse population would fall between these two figures.

The phrase 'reasonably confident' is an important one, and is an essential difference between point and interval estimates. The degree of confidence a researcher has in the estimate is expressed in a percentage. To illustrate this, let us compare point and interval estimates. With both techniques the researcher is asking the question 'How does the population rate on this particular parameter?' With point estimates the answer is 'I don't know exactly, but my guess is . . . (a single figure)'. With interval estimates the answer is 'I don't know exactly, but I'm 95% confident that the answer is between these two figures . . . (x and y).'

If we translate these questions and answers into the example of days lost per annum due to sickness amongst trained nurses, the question the researcher is asking is: 'What is the average number of days lost through illness p.a. amongst trained nurses within the regional health authority?' The point estimate answer is: 'I don't know precisely, but my guess is 17.3 days'. The interval estimate answer is: 'I don't know for certain, but I'm 95% confident that the average lies between 16.8 and 18.1 days'.

Some definitions and explanations of terms are now required. The upper figure in an interval estimate is called the **upper confidence limit**, while the lower figure is called the **lower confidence limit**. Thus, 18.1 days is the upper confidence limit and 16.8 days is the lower confidence limit. The difference between these two figures is called the **confidence interval**. The degree of confidence expressed (e.g. 95% confident) can be varied but is derived from the calculation of the interval estimate. This procedure will be described in detail shortly.

However, confidence intervals are frequently misinterpreted which can lead to all sorts of incorrect assumptions and possibly dangerous conclusions. For this reason, some theoretical background to interval estimates will be provided. Although I said in the Preface that I would not indulge in statistical theory, it is so important in interval estimation that here I shall break my promise. This theoretical explanation is rather long-winded and may appear to be hopelessly irrelevant at times, but as it is so critical, do bear with it.

## 13.3 Normal distribution

Two concepts are crucial to the understanding of interval estimation: **standard deviation** and **normal distribution**. Take the normal distribution first: the term refers to a particular shape of a graph. Let me explain by going back to some basic principles of data analysis. One way of presenting sets of figures or data from a research project is to draw a graph. There are many ways of drawing graphs (see Clifford and Gough 1990, for example) but one type of graph is of particular relevance here – the **frequency polygon**. We have all drawn frequency polygons in school. They are graphs which represent how often, or the frequency with which, a particular event or characteristic occurs. So, you might have collected data on the percentage of women who are taking the oral contraceptive within certain age groups (Table 13.1). These data could be more clearly represented in a graph; Figure 13.1.

When drawing a frequency polygon, the categories of event are represented along the horizontal axis, and the frequency of each category's occurrence along the vertical axis.

Obviously, the shape of the frequency polygon depends upon the data being plotted. However, one shape that occurs time and again is the normal distribution curve. This curve has a number of important qualities. Firstly it is bell-shaped;

**Table 13.1**

|     | Age groups (years) | Percentage on Pill |
|-----|--------------------|--------------------|
| 1.  | 15–20              | 14                 |
| 2.  | 21–25              | 29                 |
| 3.  | 26–30              | 36                 |
| 4.  | 31–35              | 42                 |
| 5.  | 36–40              | 32                 |
| 6.  | 41–45              | 23                 |
| 7.  | 46–50              | 11                 |

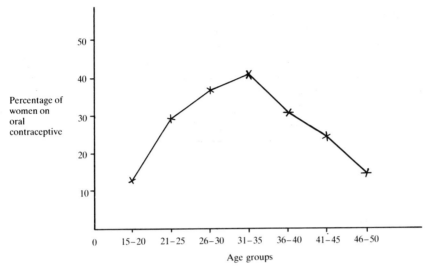

Fig. 13.1

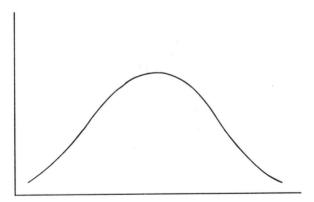

**Fig. 13.2** The normal distribution curve.

second it is symmetrical; third the tails of the curve never touch the horizontal axis. The normal distribution looks like Figure 13.2.

If we were to plot frequency data from a wide range of subject areas, we would find that they are normally distributed. For example, height, weight and IQ are all normally distributed throughout the population.

The normal distribution has another important quality which is central to interval estimation and which is tied up with the concept of the standard deviation. The concept of the standard deviation will be explained first and then the link between the normal distribution and standard deviation will be outlined in relation to interval estimates.

## 13.4  Standard deviation

The standard deviation (SD) refers to the degree of variability within a set of scores. For example, supposing you had collected some first-year examination results from a set of ten student nurses from the 1988 intake and a further set of ten students from the 1989 intake. You might have obtained the scores in Table 13.2.

**Table 13.2**

| Student | 1988 | 1989 |
|---------|------|------|
| 1 | 54 | 30 |
| 2 | 50 | 82 |
| 3 | 53 | 26 |
| 4 | 58 | 74 |
| 5 | 55 | 39 |
| 6 | 50 | 96 |
| 7 | 57 | 15 |
| 8 | 59 | 85 |
| 9 | 54 | 23 |
| 10 | 50 | 70 |
| Total | 540 | 540 |
| Average | 54 | 54 |

The total and average scores for each set of students are exactly the same, but if we inspect the data more carefully, we can see that all the examination results for the 1988 intake are in the 50s while those for the 1989 intake swing about wildly, from 15% at one extreme to 96% at the other. Clearly, if someone just compared the mean scores of the two groups, they might conclude that the students from the two sets were of similar ability, and yet the actual *individual* scores clearly indicate that this is not the case. The students in the 1988 intake are fairly homogeneous with regard to examination performance, while those in the 1989 intake are very different. The point that emerges from this is that average scores can be very misleading, and really do not provide enough information to draw conclusions from a set of data. However, if some statement about the **variability** of the scores could be made, then the researcher can glean much more information about a set of results. This is where the standard deviation comes in. The standard deviation is a figure which gives a statement about how variable a set of scores is. It is calculated by comparing each individual score with the average score using the following formula:

$$\sqrt{\frac{\Sigma(x - \bar{x})^2}{N - 1}}$$

where  $\sqrt{\phantom{x}}$  is the square root of the result of all the calculations under this sign

$\Sigma$  is the total or sum of the calculations to the right of this sign

$x$   is the individual score
$\bar{x}$   is the average score
$N$   is the total number of scores in the set of data.

Therefore, if we calculate the standard deviation for the examination marks for the 1988 intake we have first to take the average score (54%) away from each individual score and then square the result (Table 13.3).

**Table 13.3** Initial calculations of the standard deviation

| Subject | Score | $x - \bar{x}$ | $(x - \bar{x})^2$ |
|---------|-------|---------------|-------------------|
| 1 | 54 | 54 − 54 | 0 |
| 2 | 50 | 50 − 54 | 16 |
| 3 | 53 | 53 − 54 | 1 |
| 4 | 58 | 58 − 54 | 16 |
| 5 | 55 | 55 − 54 | 1 |
| 6 | 50 | 50 − 54 | 16 |
| 7 | 57 | 57 − 54 | 9 |
| 8 | 59 | 59 − 54 | 25 |
| 9 | 54 | 54 − 54 | 0 |
| 10 | 50 | 50 − 54 | 16 |

All the squared differences between each score and the mean are added together to give $\Sigma(x - \bar{x})^2$:

$$0 + 16 + 1 + 16 + 1 + 16 + 9 + 25 + 0 + 16 = 100$$

This figure is then divided by the total number of scores in the set of data (i.e. 10):

$$\frac{100}{10} = 10$$

and the square root of this figure is then taken to give the standard deviation. Therefore

$$\sqrt{\frac{100}{10}} = 3.16$$

Therefore the standard deviation of the examination results for the 1988 intake is 3.16. You will probably have realised that the standard deviation is really an average figure for how far a set of scores varies about the mean. It is a very useful statistic, in that a small standard deviation indicates that a set of scores does not vary much, while a large standard deviation suggests a wide range of scores. This can be demonstrated by calculating the standard deviation for the examination results of the 1989 intake. The standard deviation for these scores is 28.69, which indicates a much greater degree of variation in the scores than the SD of 3.16 for the 1988 intake.

## 13.5 The relationship between standard deviations and the normal distribution

Now, how do standard deviations and normal distributions relate to each other? In any set of normally distributed data, a fixed percentage of the scores *always* falls within a given area under the curve. These 'given areas' are determined by the standard deviation of the set of scores. This means that as long as we know the mean and the standard deviation of a set of normally distributed scores then we can also find out just what percentage of those scores are likely to fall within a given range.

Put more specifically, the following hold:

1. 68% of the scores will fall within *one* standard deviation either side of the mean.
2. 95% of the scores will fall within *two* standard deviations either side of the mean.

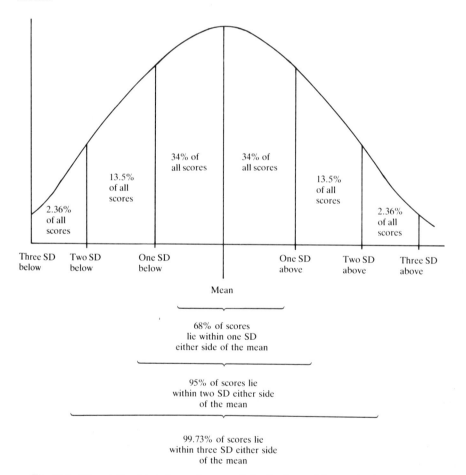

**Fig. 13.3**    The relationship between the normal distribution and the standard deviation.

3. 99.73% of all the scores will fall within *three* standard deviations either side of the mean.

This information can also be shown as in Figure 13.3. (If you have added up the figures represented under the curve of the graph, you will have arrived at 99.72%, and not 99.73% as stated at the bottom of the graph. The figure 99.73% is normally used because of the correction to a standard number of decimal places.)

Let us translate all this theory into a practical example. Supposing a nurse manager was interested in the durability of hospital mattresses in a large regional health authority, with a total of 1500 beds. She or he randomly selects 100 new mattresses and monitors how long each lasts before having to be replaced. Assume the mean life-span is calculated as being 30 months, and the standard deviation of the scores as 2.5 months. On plotting the scores derived from the sample of mattresses, they are found to be normally distributed. The frequency graph looks like Figure 13.4.

Because the scores are normally distributed and the SD is 2.5 months, the nurse manager would know that 68% of all the new mattresses would last between 27.5

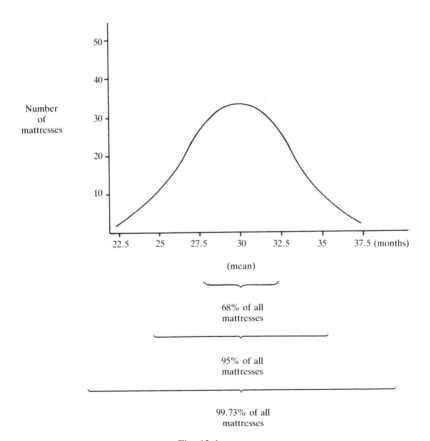

**Fig. 13.4**

and 32.5 months (One SD either side of the mean), 95% of all new mattresses would last between 25 and 35 months, and 99.73% of all new mattresses would last between 22.5 and 37.5 months.

Now all this has been calculated on a *sample* of new mattresses. If the nurse manager wants to make an interval estimate with 95% confidence for the life expectancy of the whole *population* of mattresses in the region, she or he simply has to compute the average and SD for a sample of mattresses. To the average life expectancy of the sample is added two SDs to give the upper confidence limit, and two SDs are subtracted from the average to give the lower confidence limit (because 95% of the scores fall within two SD either side of the mean). In the present example, this produces a confidence interval of 25–35 months. Because 95% of all scores lie within two SD of the mean, the nurse manager can say that she or he is 95% confident that the average life expectancy of the whole population of mattresses in the region fall within the interval 25–35 months.

This 95% figure is called the **level of confidence**, and can be varied to suit the topic being researched. For example, if the researcher was carrying out potentially dangerous research, she or he might want more assurance that the population parameter was included in the confidence interval. In this case, a 99% confidence level might be used. So, for example, if the researcher wanted to estimate the average length of time a drug was able to inhibit the growth of a spinal tumour, a sample of say, seventy-five patients might be selected and the mean length of time the drug was effective in suppressing the tumour could be calculated. If the SD of these scores was also computed, three SD could be *added* to the mean and three SD could be subtracted, to give the confidence interval. So if 47 months was the mean, and 7.8 months the standard deviation, some estimates of the total population of spinal tumour patients could be made. Because 99.73% of all scores are known to be contained within three SD either side of the mean, the researcher could say that she or he is 99.73% confident that the average length of time the drug would suppress a spinal tumour in the whole population of spinal tumour patients is between 23.6 months and 70.4 months. (Note that these calculations have been based on the assumption that time, in this case, would be normally distributed. In reality, this is unlikely to be so; the example is hypothetical.)

By the same token, if the researcher does not need such a high level of assurance, the confidence level can be reduced to, say, 90% or lower. The mechanisms for making these alterations to the confidence levels are given in the sections that follow.

One very important point to note is that the figures of 95%, 99%, etc., are *not* statements of the probability that the population parameter will fall within the confidence interval, but a statement of *belief* that the population parameter is contained within the confidence interval. Put another way, if a number of random samples were selected from the parent population, the mean of less than 5% of the samples would fall outside the confidence limits defined by the 95% level, the mean of less than 10% of the samples would fall outside using the 90% level, the mean of less than 1% of the samples would fall outside using the 99% level. This point is an essential one in interpreting estimates and confidence intervals.

**Summary of key points**

1 Estimation is a statistical technique which allows the researcher to make a scientific 'best guess' about a given characteristic or parameter in a population from information derived from just a small sample of that population.

2 There are two types of estimation. The first – *point estimation* – provides a single figure (which is typically an average or percentage) for the characteristic from the sample. This figure is then assumed to apply to the parent population from which the sample was drawn.

   The second type of estimation – *interval estimation* – involves the calculation of two figures, called confidence limits, for the characteristic. These are derived from the sample, but they allow the researcher to state with some degree of confidence that the parent population's characteristic also falls between these two figures.

3 Estimation is a useful technique in nursing, particularly at a managerial and planning level.

## 13.6  Calculating interval estimates

The formula you should choose for calculating interval estimates varies according to a number of factors. Firstly, is the sample size greater or less than thirty? For example, if you are looking for patients with a rare condition, the sample selected is very likely to be less than thirty. On the other hand if you are concerned with people with blood group O+, then your selected sample will probably be greater than thirty. You should note, however, that confidence limits can *only* be calculated on samples of less than thirty, if the parent population is *known* to be normally distributed on the characteristic in question.

Second, are you interested in estimating the population *average* for a particular parameter (e.g. average length of time in hospital following prostatectomy) or are you interested in proportions or percentages (e.g. the percentage of the female population which has had rubella)?

And finally, how much assurance do you require for your confidence level? A standard rule-of-thumb is 95%, but if your research requires greater confidence, this can be altered to 99%, or alternatively reduced to 90% if that degree of assurance is unnecessary. This last point will be dealt with within each of the sections that follow.

### 13.6.1  Samples larger than thirty

**Estimating confidence limits for the population mean**

The formula for estimating the population mean of a parameter from a sample with 95% confidence is:

$$1.96 \times \frac{SD}{\sqrt{n}}$$

where  SD    is the standard deviation of the parent population
       $n$     is the sample size
      $\sqrt{\ }$    is the square root of the sample size
     1.96   is a constant to be used with the 95% confidence level.

Note that if the standard deviation of the parent population is unknown (as it often is) then the standard deviation of the *sample* can be used as an estimate instead.

Taking an example, let us suppose you are interested in estimating with 95% confidence the average age on entry into nurse training over a five year period in the United Kingdom. To make this estimate, you randomly select 100 new students and work out the average age when starting training. Therefore, you have a sample larger than 30 and your interest is in estimating the *mean*; the following formula is consequently appropriate. You find that the mean of the sample is 22.3 years, with a standard deviation of 1.75 years. As you do not know the SD of the parent population, you have to use the SD of the sample when calculating the formula. Therefore, if we substitute our figures in the formula

$$1.96 \times \frac{SD}{\sqrt{n}}$$

we obtain

$$1.96 \times \frac{1.75}{\sqrt{100}} = 0.35$$

This figure is then

(a) *added* to the sample mean to obtain the *upper confidence limit*, i.e.

$$22.3 + 0.35 = 22.65$$

(b) *subtracted* from the sample mean to give the *lower confidence limit*, i.e.

$$22.3 - 0.35 = 21.95$$

This provides us with the confidence interval of 21.95–22.65 years. From this we can state that we are 95% confident that the average age on entry for the total population of student nurses falls between 21.95 and 22.65 years.

Should you wish to increase your confidence level to 99%, the formula becomes

$$2.58 \times \frac{SD}{\sqrt{n}}$$

and if you want to reduce it to 90% it becomes

$$1.64 \times \frac{SD}{\sqrt{n}}$$

**Estimating confidence limits for proportions of the population**
The formula for estimating with 95% confidence the confidence limits of the proportion or percentage of the population which possesses a particular characteristic is

$$1.96 \times \sqrt{\frac{pq}{n}}$$

where 1.96   is a constant to be used with confidence levels of 95%
    $p$    is the *proportion* of the sample which possesses the characteristic
    $q$    is the *proportion* of the sample which does not possess the characteristic
    $n$    is the sample size
    $\sqrt{\phantom{x}}$    is the square root of all the calculations under this sign

Note that the proportion of a sample is calculated using the formula $a/n$, where $a$ is the actual *number* of the sample who possess the characteristic and $n$ is the sample size.

Imagine you are interested in estimating the proportion of diabetics in the United Kingdom who have retinopathy, and randomly selected a sample of 175 diabetics. As the sample is larger than 30 and your interest is in the *percentage* of diabetics with retinopathy, this formula is suitable. Imagine you found that 49 of the sample had developed retinopathy. This means that 49 possess the characteristic in question, and 126 do not. To calculate these figures as a proportion:

$$\text{The proportion of diabetics with retinopathy} = \frac{49}{175} = 0.28$$

and

$$\text{The proportion of diabetics without retinopathy} = \frac{126}{175} = 0.72$$

Therefore, the formula for the confidence limits is

$$1.96 \times \sqrt{\frac{0.28 \times 0.72}{175}} = 1.96 \times 0.03$$
$$= 0.06$$

This figure of 0.06 is

(a) *added* to the proportion of the sample which possess the characteristic to give the upper confidence limit, i.e.

$$0.28 + 0.06 = 0.34$$

(b) *subtracted* from the proportion of the sample which possess the characteristic to give the lower confidence limit, i.e.

$$0.28 - 0.06 = 0.22$$

This gives the confidence interval of $0.22 - 0.34$.
The researcher can then estimate with 95% confidence that the proportion of the

total population of diabetics who have retinopathy is between 0.22 and 0.34 or 22–34%. Should different degrees of assurance be required for this estimate, the formula

$$1.64 \times \sqrt{\frac{pq}{n}}$$

will provide a 90% confidence level and

$$2.58 \times \sqrt{\frac{pq}{n}}$$

will provide a 99% confidence level.

### 13.6.2 Samples of less than thirty

**Estimating the confidence limits for the population mean**
To estimate with 95% confidence the confidence interval for the population mean from the sample mean, the following formula is used:

$$t \times \frac{SD}{\sqrt{n}}$$

where $t$ is a value derived from the probability table (see Table C.3 page 292)
     SD is the standard deviation of the sample
     $n$ is the number of scores in the sample
     $\sqrt{}$ is the square root of the number under this sign.

Let us use an example. Imagine you are a nurse manager and are interested in the average length of time qualified nurses remain in your hospital after qualifying. You select a sample of eleven newly qualified SRNs and monitor them over a period of years. (This could, of course, be done retrospectively.) You find that the average length of time they stay is 3.4 years. To derive confidence intervals from this, you need to take the following steps:

1. Calculate the standard deviation for the sample using the formula

$$SD = \sqrt{\frac{\Sigma(x - \bar{x})^2}{N - 1}}$$

where $\sqrt{}$ is the square root of all the calculations under this sign
     $\Sigma$ means the total of all the calculations to the right of this sign
     $x$ is the individual score
     $\bar{x}$ is the average score
     $N$ is the total number of scores in the sample.

Look back to Section 13.4 if you are unsure about how to calculate the SD. Imagine that the SD in this sample is 0.85 years.

2. Calculate $N - 1$, which is the number in the sample minus 1. This is the d.f. value.
3. Turn to Table C.3. This is the probability table associated with the $t$ test. Here you will see down the left-hand side numbers from 1 to 120 under the letters d.f. Across the top you will two lines of figures, one line under the heading 'Level of significance for a one-tailed test' and a second under the heading 'Level of significance for a two-tailed test'. These terms are explained in more detail in Section 7.4, but do not need to concern us here. Look down the column of figures under the letters 'd.f.' until you find your $N - 1$ value. Ours is $11 - 1 = 10$. To the right of 10 are six figures:

$$1.372 \quad 1.812 \quad 2.228 \quad 2.764 \quad 3.169 \quad 4.587$$

Only the figure under the column '0.05 Level of significance for a two-tailed test' is of interest to us here, i.e. 2.228. This will provide us with a 95% confidence level.
4. Using the figure you have just derived from the tables, calculate the confidence interval estimate from the formula

$$t \times \frac{\text{SD}}{\sqrt{N}}$$

Substituting our values

$$\text{the confidence interval estimate} = 2.228 \times \frac{0.85}{\sqrt{11}}$$
$$= 2.228 \times 0.26$$
$$= 0.58$$

5. *Add* this figure to the average score to give the *upper confidence limit*, i.e. $3.4 + 0.58 = 3.98$.
6. *Subtract* this figure from the average score to give the *lower confidence limit*, i.e. $3.4 - 0.58 = 2.82$.

   Thus the confidence interval estimate for the length of time newly qualified RGNs stay in the hospital's employment is between 2.82 and 3.98 years. The degree of credibility that can be placed in this estimate is 95%.

   Should you require a 99% confidence level, then the column of figures under '0.01 Level of significance for a two-tailed test' in Table C.3 should be used. And should you require a 90% confidence level, then the column of figures under '0.10 Level of significance for a two-tailed test' in Table C.3 should be used. From all this, you will have noted that values of $t$ replace the constants of 1.64, 1.96 and 2.58 given in the formula for large samples. Apart from this, the formula is the same.

   You will have noticed that no formula has been given for estimating confidence limits for proportions of the population from samples of less than thirty. The reason for this is that these estimates are unreliable unless the proportion derived from the sample is close to 50%. For this reason, estimates of proportion from small samples are not usually calculated.

*Exercises (answers on page 320)*

1 The DNS at a large general hospital wants to alter the hours of the day shifts. Before doing so, the opinions of a sample of thirty-eight nurses are sought. Of these, seventeen are in favour of the change. Calculate the 95% confidence limits for the proportion of the population (i.e. all the nurses in the hospital) who would also support the change. (Remember to calculate the proportions for and against the change first, e.g. 17 as a proportion of 38 = 17/38 = 0.45.)

2 A DNS at a psychiatric hospital wants to estimate the average length of time schizophrenic patients are able to function in the community before having a relapse which requires hospitalisation. A sample of twenty-five patients is selected and a mean time of 4.8 months between discharge and readmission is calculated. Calculate the confidence limits for the population with 90%, 95% and 99% degrees of confidence. The standard deviation of the example is 1.76 months.

3 In order to estimate the number of A rhesus negative patients in a regional health authority, a senior nursing officer selects a sample of a hundred patients. Of these sixteen are A-. Estimate the proportion of the population (i.e. patients within the regional health authority) who are also A-, using the 95% and 99% confidence limits. (Again remember to convert to a proportion first.)

## 13.7  Key words and terms

Confidence interval

Estimation

Frequency polygon

Interval estimation

Level of confidence

Lower confidence limit

Normal distribution

Point estimation

Population

Random sampling

Sample

Standard deviation

Upper confidence limit

Variability

# 14 Practical application of the theory: carrying out your research

You might be interested in carrying out a piece of research for a whole host of reasons. Perhaps you have a dissertation to complete as part of a higher education course; or maybe you have been asked by your health authority to look at a particular topic. You may even want to do some research on your own initiative, because you have a burning interest in a specific issue. But how do you get started?

The preceding chapters may have left you feeling somewhat bewildered, without clear guidelines as to how to start and carry out your research in practice. It might, therefore, be useful at this stage to recap on some of the main points of research design and the order in which they should be considered.

## 14.1 Planning and preparing the research

### 14.1.1 Defining your area of interest

A vague notion that you would like to do some research might be a powerful motivating force, but it needs to be defined *specifically* before you can begin. In other words, once you have decided on the broad topic area you wish to research, you must narrow the field to a specific *testable* hypothesis. So, your main focus of concern may be infection control on surgical wards, but before you can proceed further with your research you must devise a hypothesis in accordance with the principles outlined in Chapter 3. So perhaps after some thought you decide to compare the contamination levels of plastic *v.* metal washbasins. As your hypothesis must predict a relationship between two variables, the $H_1$ might look like this:

> **$H_1$:** There is a relationship between the type of washbasin and the level of bacterial contamination.

From this you can identify the null hypothesis, and the independent and dependent variable, if applicable.

Ensure at this stage that your hypothesis is testable, by which I mean that it is within your skills as a researcher to carry out and analyse, and that it is logically, ethically and practically feasible. Do you have the time, energy, commitment and resources to carry it through? For instance, a hypothesis which requires major changes in patient care, staffing ratios or hospital policy may be beyond your scope. So do think carefully at this stage about what would be involved.

Second, once you have clearly stated your hypothesis, you must decide how the dependent variable is to be measured, or alternatively if the hypothesis warrants a correlational design, how *both* variables are to be measured. General terms like 'progress', 'improvement' or 'recovery', while acceptable in the hypothesis, must be clarified at this stage so that you know what sorts of scoring procedure and data collection are required.

So the first steps in the research process are as follows:

1. Define your area of interest.
2. State the experimental hypothesis, ensuring that a relationship between two variables is predicted.
3. State the null hypothesis.
4. Identify the variables.
5. Define the terms used in the hypothesis.

## 14.1.2 Reviewing previous studies

Once you have specified your hypotheses, you must look through all the relevant journals in order to find out exactly what research has been carried out in the area before. This **literature search** has a number of functions. Firstly, it allows you to find out whether or not your study has been carried out before. If it has, then you may wish to modify your hypothesis so that your project is an original one. (Original studies often stand a better chance of funding and publication.) However, it can be just as valid to replicate an existing study to see whether or not its results are reliable and generalisable, since if you end up with comparable results, the original research has gained additional support and credibility and may influence whether or not its conclusions are adopted into nursing practice. If you end up with different results, some doubt has been cast on the value of the original study and its conclusions. In such cases there may be merit in repeating the study on two or three separate occasions to establish which results are more reliable.

So the literature search will establish the originality of a proposed piece of research.

Second, by studying other projects in a particular area the researcher can glean a lot of ideas about the most appropriate methodology to be used, as well as which constant errors are important. However, if your aim is to replicate an existing study, then its methodology should be followed exactly, so that a direct comparison of the two sets of results can be made.

Third, a thorough review of the literature will provide a full understanding of the theoretical issues surrounding your chosen topic, which, besides alerting you to relevant constant errors, should also provide you with some possible explanations and interpretations of any results you obtain.

And fourth, if you wish to carry out a totally original piece of work, say for a dissertation, a complete literature search can guard against considerable wastage of time and effort. It is not unknown for a student to discover immediately prior to the submission of a doctoral thesis that the study has been carried out previously.

How do you go about carrying out a literature search? The first step is to visit the librarian of your hospital, school of nursing or medical library and state your requirements.

The librarian will be able to tell you which journals the library has access to, how to use the journal indexes, how to search the abstract system for any studies in your area of interest and how to borrow books and journals from other libraries on the inter-library loan scheme.

In addition, many libraries have computing facilities, which will provide a computer search of the literature. While this takes all the effort out of the task, you do have to pay for this service. Talking to the library staff will establish the price and feasibility of this facility.

If you do not have access to reasonable library facilities, then the library of the Royal College of Nursing may be able to help, particularly as they will reply to written enquires. The address is:

The Librarian
Library of Nursing
Royal College of Nursing
Henrietta Place
Cavendish Square
London W1M 0AB

All this may look rather daunting at first, but do remember that you can get a considerable amount of help and guidance from your library. However, to help you on your way, the following examples of journals all have research articles:

*Nursing Times*
*Nursing Research*
*Journal of Advanced Nursing*
*International Journal of Nursing Studies*
*British Journal of Hospital Medicine*

A number of other journals covering specialist areas of nursing also publish research articles. Some examples are:

*Cancer Nursing*
*Geriatric Medicine*
*Journal of Paediatric Nursing*
*AORN* (Association of Operating Room Nurses)

In addition a number of **abstracting** and **indexing systems** are available which cite research both by subject and author. This enables you to look up the particular subject you are researching to find out what research articles have previously been written on the topic. Details such as title, author and journal are given for each entry, and from this you can then locate the original article to find out how pertinent it is to your study. The most relevant and comprehensive systems are as follows:

*International Nursing Index*
*Nursing Bibliography*, published by the RCN and consisting of a current awareness service
*Cumulative Index of Nursing and Allied Health Literature*
*Health Service Abstracts*
*Hospital Abstracts* } all published by the Department of Health
*Quality Assurance Abstracts*

*Nursing Research Abstracts* (Department of Health) also publishes details of ongoing research, although it is not comprehensive since it relies on voluntary submissions by the researchers themselves.

Other indexes and abstracts which may have relevant information include:

*Index Medicus*
*British Humanities Index*
*British Reports, Translations and Theses* } British Library publications
*Current Research in Britain*

It should be noted that some health authorities also offer current research awareness services.

The abstract systems are especially useful as they provide a short summary of the aims, methods and findings of the research article. By reading this you can determine whether it is potentially useful or not. If it is, then you should find the original article and read it in full. However, if you decide from the abstract it is not useful, you can reject it without wasting further time and energy reading the full report.

Learning how to use these systems may take a little time but the effort is well worth while. The reader is referred to Clifford and Gough (1990) for a more detailed description of their usage.

So, the second major stage in your research is to review the available literature in order to find out the following:

6. Whether your study is original.
7. The best approach to testing the hypothesis.
8. The background theory to the topic area.

### 14.1.3 Keeping a card index file

Allied to the literature search is the necessity of keeping a **card index file** of all the research articles that are relevant to your study. For each article use a separate card and write on it the following:

1. The *full* name(s) of the author(s).
2. The title of the article.
3. The title of the journal in which the article was published.
4. The year of publication.
5. The volume and part number of the journal.
6. The first and last page numbers of the article.
7. A brief outline of the article's contents.

Should the relevant information be found in a book rather than a journal ensure that you make a note on your index card of the following:

1. The editor(s) or author(s) of the book.
2. The title of the book.
3. The author and title of the relevant section if the book is edited.
4. The year of publication.
5. The publisher.
6. The place of publication.
7. A brief outline of the essential information.

Ensure that the information is complete. Do not, for instance, simply make a note of the author's surname, without the forenames, or omit the page or volume numbers, since this will only make extra work for you at a later date. I speak from experience when I say that it is profoundly irritating to spend hours looking for an article the details of which are committed only to memory and not to paper.

What is the point of this rather laborious clerical exercise? Its main function is apparent when you write up your research for publication, dissemination at research seminars or for a dissertation. It is *essential* that any research to which you have referred in your report is listed in detail at the end. This is described more fully in the next chapter, but suffice it to say that a comprehensive card index system is an essential ingredient to efficient research.

A second function of the card index file is as a store of information for any future research you might wish to do on the topic. The results from your first study might generate a host of ideas you want to test out on related issues. The card index file of relevant articles will save you the effort of having to repeat the literature search.

The third step, then, in the research process is the following:

9. Keep a card index system citing all the details of relevant research articles on your selected topic.

## 14.1.4 Designing the research

This stage involves making a number of decisions about how to design a research project which best tests your hypothesis. The decisions, while not always obvious and clear-cut, must be *informed*, i.e. reasoned choices based on your understanding of research methodology and on the information gained in the course of your literature search.

The first decision you have to make when deciding how to test your hypothesis is whether a correlational or experimental design is more appropriate. The issues surrounding the basic differences between these designs are outlined in chapters 3 and 12. However, a brief overview will be presented at this stage.

In your hypothesis are you anticipating that the higher the scores on one variable, the higher (or lower) the scores on the other? If you are, then a correlational design is required. You can double-check this by asking yourself whether a whole *range* if scores is possible on *both* variables. So, for example, suppose you had hypothesised the greater the amount of highly saturated fats eaten, the higher the blood pressure. Here, you are anticipating that the higher the scores on the saturated fats variable, the higher the scores on the blood pressure variable, and thus a correlational design would be appropriate. This can be confirmed by the fact that a whole range of scores on both intake of saturated fats and blood pressure is possible.

However, a slight modification of this hypothesis would make an experimental design appropriate. Remember that in an experimental design, the researcher manipulates one variable and measures the consequent effects on the other to see if there are any *differences* in the results. So, had the hypothesis been:

$H_1$: People who eat large amounts of saturated fats have higher blood pressure than those who do not.

the variable concerned with intake of fats would be manipulated by comparing one group who ate a lot of saturated fat with another group which did not and comparing their blood pressure for *differences*.

In this way, it can be demonstrated that the manner in which predictions are made in the hypothesis will influence the choice of an experimental or correlational design.

The next phase requires consideration of the subjects, the procedure and the analysis. The following questions may provide a framework to help you clarify some of these details.

## Subjects
1. Who are your subjects going to be?
2. How many will you need?
3. How are they to be selected?
4. Are there any constant errors concerning your subjects that need to be controlled?

## Procedure
1. Do you need any apparatus or equipment?
2. If so, can you obtain it and do you know how to use it?
3. Do you need any materials, such as score sheets, pencils, paper, temperature graphs?
4. How are you going to conduct your study, if you have opted for an experi-

mental design, i.e. will you be using a same, different or matched subject design?

5. How are you going to measure your dependent variable (or both variables if you are using a correlational design)?
6. What are the constant and random errors and how will they be controlled?
7. Are order or experimenter bias effects going to be a problem? If so, how will they be dealt with?
8. Have you ensured that the procedure and instructions are identical for each subject?
9. Is the usual 5% significance level appropriate for your study? If not, what level of significance would be better?
10. Is your hypothesis one- or two-tailed?
11. Where will the research be carried out? Do you need to obtain permission?

**Analysis**
1. Do you know which statistical test should be used to analyse your data? (You may only be able at this stage to narrow down the field to the relevant parametric and non-parametric tests. The final decision should really be made when the data have been collected.)
2. Can you calculate this test or do you have a computer program that will do it for you?

Stage 4 involves a number of decisions about the research design:

**10.** Is a correlational or experimental design more appropriate?
**11.** The nature and number of subjects.
**12.** How the project will be executed and analysed.

## 14.2  Writing a research proposal

A **research proposal** is essential if you are intending to carry out a study which involves patients, or patients' relatives or colleagues, since you will need to present a written outline of your intended research to the relevant personnel in order to obtain permission to go ahead. Even the most apparently innocuous and simple project can have far-reaching and unforeseen ethical implications and so you *must* seek clearance and advice from senior nursing personnel before you begin.

I cannot stress this issue of obtaining permission too strongly. Before any research can be carried out within a health authority, agreement must be obtained from all the professionals likely to be involved. The best way of ensuring this is by writing first to the relevant senior nursing personnel explaining what you would like to do and perhaps arranging a meeting. At this juncture you present your research proposal. However, do not quibble if your request is turned down. There may be a host of reasons unknown to you concerning the undesirability of carrying out your project. However, if possible, try to find out what these reasons are, since it is conceivable that with minor amendments to your design, permission might be granted. As a preliminary guide to the ethical issues involved, Royal College of Nursing (1977) and relevant chapters in Downie and Calman (1987) might be helpful.

If you present a clear research proposal then all the major problems can be discussed and ironed out with the nursing officers (or other personnel) which can eliminate any risk to your subjects.

A research proposal is also essential if you are applying for **funding** to carry out the project. While many nurses may be prepared to foot the bill for any minor expenditure such as paper, postage, literature searches, etc., many studies incur substantial costs both in time and money and, to cover this, grants or written permission to use various resources may be necessary. To help you decide how costly a research project is going to be, the following items should be considered:

1. What are the costs of stationery, postage, telephone, photocopying, etc.?
2. Are any secretarial services required and, if so, can you estimate the approximate number of hours and costs involved?
3. Do you need to buy any equipment, apparatus or special materials, or alternatively, will a charge be levied for using existing equipment?
4. Will you require any computing facilities and, if so, what is their cost?
5. Are any travel or subsistence costs involved for your subjects and/or your experimenters?
6. If you are intending to employ either full- or part-time research workers, what are the costs of the salaries, National Insurance, superannuation and overheads? (A chat with the personnel or staffing office should help you with these points.)
7. If the research is proposed to continue over a period of months, have you built in an allowance for inflation?

Costing a research project is a complex business and anyone new to the task would be advised to seek the help of someone who is familiar with the process either because they have first-hand experience or because they are trained to deal with these issues. A telephone call to any of the senior nursing personnel in the District may help you identify such a person. Do be careful not to undercost the project, though. While an apparently cheap study may look appealing to the funding body, they will be less than thrilled if the project has to be discontinued half way through, owing to lack of money. This all-too-common a problem simply wastes time, patience and resources.

Many would-be researchers are amazed by the expense of even a small project. Costing the items in this way will help you decide whether the cost is justified, or alternatively where costs can be cut.

The third function of a research proposal is for your benefit – it focuses your attention on all the relevant aspects of the design and procedure, and enables you to plan the details of your research. Furthermore, if you are intending to write up your research at a later date then a detailed proposal is invaluable, since it serves as a reminder to you of what you did and why. It is amazing how easy it is to forget the reasoning behind a set of decisions taken months earlier.

The mechanics of writing up a research proposal are similar to preparing an article for publication. As this is covered in the next chapter, only outline headings will be given in this section, and the reader will be referred to the next chapter where appropriate. Where this happens, the instructions given in the next chapter should be followed closely. The following information should be provided in the research proposal:

1. The title of the research project (see next chapter).
2. The background to the study – the extent of the problem, previous research in the area and its conclusions. (See Section 15.2.4.)
3. Why is your topic worth studying? What implications does it have? i.e. what is the rationale for your research? (See Section 15.2.4.)
4. A clear statement of the experimental and null hypotheses.
5. What design are you going to use, correlational or experimental? Same, different or matched subjects if experimental? (See Section 15.2.5.)
6. Who are you subjects to be? How many are needed and how will they be recruited? (See Section 15.2.5.)
7. What apparatus, equipment and materials are required? (See Section 15.2.5.)
8. What is the research procedure, i.e. how will the study be carried out in detail? (See Section 15.2.5.)
9. Where will the research be carried out? Has permission been obtained?
10. What safeguards will be taken to ensure confidentiality of data, of subjects' personal details, ethical issues, etc.?
11. What test will be used to analyse the results?
12. What is the anticipated value of the study? What are the expected implications and how might they improve patient care, hospital policy, nursing standards, etc.?
13. How long will the study take to complete? While accurate predictions cannot be made, guestimates can, although you should take care not to *under*estimate the time required at this stage. Problems almost always arise in the course of research, such as equipment breaking down, subjects failing to turn up, etc., all of which will delay you. Hence it is better to overestimate the time needed in order to take account of these obstacles. A chart which notes the approximate timing of each major stage in the project can be useful particularly since it may encourage the researcher to stick to a timetable.
14. A list of all the major personnel who will be involved in the project, together with their qualifications and relevant experience should be provided. Typically the personnel fall into three categories:
    (a) The research director who takes overall responsibility for the project and any major decisions. This person should have a thorough understanding of research methods and all the issues involved.
    (b) The research workers who typically collect the data, administer the treatments, procedures, etc., and are responsible for the day-to-day running of

the project. These people should have some knowledge of research methodology and should have been thoroughly appraised of their functions before starting the project.

(c) The clerical, secretarial, technical and computing staff who provide a support and back-up system.

15. List any published articles which are related to the subject area in alphabetical order. (See Section 15.2.9.)

It is worth noting at this point that the written style of the proposal is important. While this is covered in greater detail in the next chapter, it is important to bear in mind that it should be as clear, concise and objective as possible, without reference to anecdotes and personal opinions. More information on appropriate literary styles is given in Section 15.1.

If you are applying for funding (or formal permission) then you should consult the funding body about their preferred layout. Many organisations have their own forms or guidelines for research proposals and you should follow these to the letter. Variations, modifications and sheer disregard for stated requirements may ensure that your proposal does not even get considered!

Two final points at this stage. It may be worth while having a meeting with any hospital personnel who may be involved in the research, however minimally, in order to explain your intentions. At this point, a resumé of your research proposal could be circulated. Not only is this courteous, it may also keep working relations smooth and avoid problems along the way. Arising out of this point is an important reminder: patients and colleagues do *not* have to co-operate with you in your project. Do not coerce them if they wish to opt out and do not go in for re-criminations. Research should only be entered into willingly!

Stage 5, then, involves the following:

**13.** Writing a research proposal, covering all the relevant details of aims and proposed methodology.

**14.** Making an estimate of the costs involved.

## 14.3 Obtaining funding

Even small research projects can be costly of time, resources and money. To offset at least part of these costs you may decide to apply for some funding, although finding a suitable source of money is not always easy. So where do you apply? The following list of possible funding bodies should be considered:

1. Your own hospital's research fund. Many hospitals have large sums of money earmarked for research activities and which are usually only tapped by doctors. Applications concerning nursing research are therefore often welcomed.

2. Your local health authority. A number of health authorities commit money to research activities and may be willing to consider support for a nursing project.

Your senior nurses in the district should be able to provide more detailed information on this.

3. The Department of Health has a Nursing Research Fellowship Scheme which provides financial backing for a small number of research activities. Enquiries can be made to The Principal Nursing Officer (Research), Department of Health, Alexander Fleming House, London SE1 6BY.

4. Drug or equipment manufacturers can be a fecund source of funding, particularly if the research relates to their products. They are less likely, however, to back research which is totally unrelated. So, a research project which focused on incontinence might well be favourably received by the manufacturers of incontinence equipment, but rejected by a company producing insulin.

If none of these sources seems appropriate then the reader is referred to Royal College of Nursing Personal Advisory Service (1981). This provides a review of funding available to nurses wishing to undertake research.

Once you have decided whom to approach for financial support it is worth having an informal discussion with them to confirm that your project is the sort of thing they would be willing to consider backing. If the project is aligned to their interests, obtain any official application forms and relevant information they may provide for prospective applicants and complete these in accordance with the guidelines laid down in the accompanying information. If no standard forms are available, prepare a research proposal along the lines indicated in the previous section, remembering to stress the practical value and implications of your proposal. Projects which are dangerous, mundane or lacking in apparent usefulness are very unlikely to receive sponsorship. The more you can relate your study to the funding organisation's particular interests, the more likely it is to be favourably considered.

Make sure, too, that the research proposal is well-presented. A type-written report, devoid of spelling, grammatical or other errors will add to your overall credibility.

If the funding agency decides to award you a grant, a number of courses of action are advisable. Firstly, you could invite someone from the organisation to be on a working party for the project. In this way your sponsor can be kept fully informed at all stages of everything that goes on. If this does not seem appropriate, then regular progress reports should be sent instead. These reports should recount how far the research has progressed and also acknowledge any obstacles that have been encountered (and overcome) along the way.

In addition to this, it is your responsibility as the grant holder to ensure that you fulfil your side of the contract. In other words, you should endeavour to complete the project within the time and budget allowed. Requests for more money may well be rejected and half-completed research is of limited value. If you want to keep your sponsors favourably disposed towards you, you must meet your side of the deal.

And finally, present, *on time*, a final report of the completed project outlining clearly the findings, conclusions and implications. The format for this should be similar to that for articles prepared for publication and can be found in the next chapter.

Of course, your grant application may be turned down. If you can discover the reasons why, it can provide useful feedback for any future applications you might wish to make. It is also conceivable that the funding body would be prepared to reconsider if modifications were made. Do not be too despondent, though, if you fail to obtain financial support. A large amount of research goes on without this backing, so unless your own study is very expensive, it would be worth thinking about how to reduce its costs and then proceeding on your own.

Stage 6, therefore, involves the following:

**15.** Applying for funding if appropriate.

## 14.4  Working out the final details of the research procedure

Even if you have prepared a research proposal you may still find that there are a number of details that need to be sorted out before your study can begin in earnest. Some of these are outlined below:

1. Do you need any score sheets, temperature/blood pressure/urine output forms, etc., for recording your data? Ensure you have sufficient copies, taking account of wastage.
2. Are you planning to give your subjects any instructions or information about the project? If so, this may need to be standardised in advance, and written down.
3. Are you intending to preserve the anonymity of your subjects? If so, how do you propose to do this? Will a master record sheet containing all the necessary details about your subjects be required and, if so, where can this be kept securely?
4. Do you envisage having to contact the subjects at some future date, either to provide information about the project's outcome or to enlist their help again? If so, have you got their addresses?
5. If you need to allocate subjects or conditions randomly, have you got the details of how you are going to do this sorted out? Ensure that the techniques of random selection and allocation and of counterbalancing are sorted out *in advance.*
6. Do you need to enlist the help of anyone else to collect your data, either because you are unable to do so or to counteract experimenter bias? If so, have they been briefed and trained appropriately? Do they need written instructions concerning their role?
7. Do you need any equipment or apparatus? If so, check that it works and that *you* know how to use it.
8. Write out an instruction sheet for yourself and any other experimenters stating

what should happen, when and where *in detail*, so that everyone is familiar with the experimental procedure.

9. If you are going to send out questionnaires or forms for your subjects to complete and return by post, enclose a stamped addressed envelope as this will increase the number of replies you receive.

10. If you have recruited your subjects a long time before you actually require them to take part in the study, it is worth while sending them a reminder a few days before the start of the project. This will guard against inadequate numbers of subjects turning up.

Two cardinal rules of research should be emphasised at this stage Firstly, *never, ever* go straight into your main study without running some **pilot trials** first. A pilot run highlights all the pitfalls and problems of your experimental procedure and allows you to iron them out in advance. In this way you can establish whether the methods are suitable, the tasks too hard or easy, the time allowed adequate, whether the equipment works and can be operated smoothly and whether any instructions you have devised are intelligible. Pilot trials make for smooth efficient running of the main project and should *always* be carried out beforehand.

Secondly, most of the research you do will involve human subjects. Treat them with courtesy at all times. Never ask them to do anything embarrassing or dangerous and keep them informed of the goals and practical details of your project when appropriate. Write and thank them for their co-operation when the data collection is complete and send them the results of your project as soon as it is feasible to do so. Try to interfere with their lives and routine as little as possible and remember that they are the linchpin of your research, without whom your study would not be possible.

Stage 7 of your research preparation involves the following:

**16.** Finalising the details of your experimental procedure.
**17.** Running some pilot trials.

## 14.5  Presenting your research findings

Carrying out research is pointless unless the findings are disseminated to those in a position to use it. There are a number of channels through which this can be achieved, the most effective of which (in terms of the numbers of people reached) is through journal articles. Writing up research for publication is covered in detail in the next chapter. However, less daunting possibilities include research seminars, discussion groups or simply distributing a written report of your research to those professionals who might find it useful and interesting. Whichever approach you choose, the principles and format are similar and are covered in Chapter 15.

Do not be put off by the prospect of going into print. If research is to have any value it must be made available to any interested party. Remember, a patient's well-being may be enhanced by your research findings, so it is your duty to ensure that the information is disseminated.

If, by now, you are still keen to pursue some research but lack the confidence to start, there are a number of support systems available. Short courses on research methods are run all over the country by hospitals, health authorities, schools of nursing and universities. Enquiries to any of these bodies or a perusal of the nursing press should provide further information. Similarly there are many local and national research interest and discussion groups – for example the RCN Research Society holds regular meetings. All these may inspire and fortify the intending researcher.

Stage 8 of the research programme involves the following:

**18:** Writing up the research for dissemination.

If you now re-read the eighteen points throughout this chapter, you will have an outline framework within which you can start planning your own research project.

## 14.6  Other useful sources of information

*Sources of Nurse Knowledge*
*Identifying and Defining Questions for Research*
*The Experimental Perspective*

All three titles are written by E. Clark (1987) for the Royal College of Nursing. Available from The Distance Learning Centre, South Bank Polytechnic, Borough Road, London, SE1.

## 14.7  Key words and terms

Abstracting systems
Card index file
Funding
Indexing systems
Literature search
Pilot trials
Research proposal

# 15    Presenting your research findings

There is little point in spending months in data collection and analysis if the results are to be kept a closely guarded secret. If research is to have any real value, its findings and conclusions should be disseminated to all those who might be interested.

Many people believe that research is only worth publishing if it yields significant results. This is not necessarily true and in fact a lot of purists would claim that the only way to ultimate scientific truth is through support of the *null* hypothesis. While this philosophical debate need not concern us here, it should be stressed that non-significant results that by implication refute long-held assumptions in nursing or which fail to support existing research findings are particularly useful and should be considered for publication.

Undoubtedly, with respect to dissemination, publishing research findings in a professional journal reaches the widest audience and may have the maximum impact. However, many beginning researchers find the prospect of going into print at a national or international level rather daunting. For them, a less traumatic initiation process in the form of a presentation to a specialist research interest group is a preferable alternative, at least to begin with. Whichever avenue you choose, though, the *general* principles of presenting your research are similar. These general guidelines will be outlined first, and then more specific details concerning format and layout will be given.

## 15.1  General guidelines

*Always* bear in mind that the object of the dissemination exercise is to convey ideas and facts. It is your job to ensure that the audience understands the object, method, results and conclusions of your research, *without ambiguity*. To this end you should write clearly, simply and coherently.

Writing in a clear and simple literary style means the following:

1. Only convey one idea per sentence. Do *not* go in for complex sentence structures with half a dozen clauses. Make sure each sentence is simple and straightforward in structure and content. Four or five short sentences convey more meaning than a single sentence of thirty or so words.

2. Use simple words. Use 'start' rather then 'commence', 'finish' or 'end' rather than 'terminate'. Infrequently used polysyllabic words will merely confuse the listener or reader who will probably lose attention and interest.
3. Do not use esoteric abbreviations or jargon unless there is some clarification of their meaning. So, should you decide to shorten 'coronary heart disease' to CHD, do it in the following way when the term is *first* referred to in the report: 'A study of fifty middle age men suffering from coronary heart disease (CHD) was carried out.'
4. Write in the third person and not the first person. Remember that your research was an objective scientific investigation, with no place for personal opinions and views. Therefore, phrases like 'I think', 'I administered the drugs', 'I asked the patients' are all inappropriate, and should be replace by 'It is suggested', 'Drugs were administered' and 'The patients were asked'.

   This might sound rather pretentious, but a report written in the first person looks subjective and amateurish and lacks credibility – qualities definitely to be avoided when attempting to publish research!

Remember that your article or report will not always be read by specialists in the field who are familiar with the topic and terminology. If you write the report on the assumption that the audience is new to the area then these problems should be avoided. If, after completing the article, you are unsure whether it is sufficiently clear, ask someone other than a nurse (e.g. parent, spouse or friend) to read it. If they remain in a fog, then you will have to consider rewriting. Do not forget that complexity of written style impresses no one. Clarity and simplicity do.

Do not include any jokes or anecdotes, however pertinent they may seem at the time. What happened to your neighbour's son may have a direct relevance to the content of your report, but *do not* include it! Personal experience simply detracts from the scientific message.

Ensure that the report is succinct and concise, yet comprehensive. While this is often easier said than done, it may be helpful to remember the report should be so clear and precise that anyone randomly selected from the population could understand what you did, why you did it and what you found. The procedure you used should be sufficiently detailed that the study could be replicated to the letter without questions having to be asked about how and when something was done. This may seem to open the floodgates to excess verbiage, but you should also bear in mind that whatever is included in the report should be directly relevant to the study and its outcomes. For example, did the shift times of the staff nurses in your study have any impact on the procedure or results? If not, leave it out of the report. Was the patients' hair colour salient? If not, leave it out of the report.

The report should be comprehensive enough to inform the audience of exactly what was done and found. It should also be succinct enough to maintain interest and not bore the audience to tears.

You should quote relevant existing research extensively throughout your report (see next section for how to do this). This adds credibility and support to any statement you make about theories and practices in nursing. So the statement:

'Cervical cancer is more common amongst multiparity women than amongst child-less women' is an unsubstantiated fact as its stands. This is unacceptable in research articles as it allows the author to make any assertion or statement without having to produce evidence – a potentially dangerous situation. However, if the same statement is followed by the names of the researchers who made the finding together with the date of the research, then the statement gains credibility. Clearly, a piece of research which can be attributed to someone has more impact than one which is unclaimed.

Furthermore, quoting apposite research throughout your report informs the audience that you are fully appraised of the topic you have researched and that in itself promotes your credibility as a researcher.

Finally, remember that clearly drawn graphs of the results can often be more informative to the audience than a table of raw data. Details of how to draw graphs are given in Clifford and Gough (1990).

## 15.2  How to prepare an article for publication

The following points should only be used as a model for writing up an article for publication, since all journals have their own specified format and requirements.

If you want to publish your research, you should initially consider which journal would be the most suitable. This decision should be based primarily on the subject of your research. For example, if your study was concerned with leukaemia in the under-5s, then *Journal of Paediatric Nursing, Paediatric Nursing, Cancer Nursing* plus other less specialist journals may all be suitable. One you have narrowed the field on this basis, you should consult each journal to see what level and type of article it publishes. This information can be gleaned in two ways: firstly by looking at the notes for contributors (usually contained on the front or rear inside cover) and second by perusing the articles contained in two or three issues of the journal. In this way, you can find out whether your research is appropriate, as well as the specific format and requirements of the journal.

These requirements vary from journal to journal. For example, some journals require a summary of the research and this may come at the beginning *or* end of the article. Other journals specify that references should be listed alphabetically, while others want them listed in order of appearance. Section headings and minor points of style vary too; some journals do not find abbreviations such as e.g. and i.e. acceptable, requiring that they are written in full. In addition, only one copy of the article needs to be submitted to certain journals, while others request three or four. The cardinal rule here is, *consult the journal* concerned. Do follow their requirements to the letter, too, since a layout totally at odds with the one required creates the impression of sloppiness – after all, if you cannot follow the guidelines set out by the journal, can you really follow the principles of scientific methodology and analysis?

That having been said, most research articles in professional journals follow a similar broad format and it is this that will be described here. Modifications can then be made as necessary to suit any given journal. On the left-hand side of the page are the general principles involved in writing a report. On the right-hand side is an example of how these principles should be applied in practice using a fictitious research study by way of illustration. (Do note that everything on the right-hand side is a total fabrication and does not, as far as I know, refer to any actual research or data.)

### 15.2.1 Title

The **title** should be succinct and informative such that anyone browsing through the journal could decide immediately whether the article was relevant to their purposes. The easiest way of achieving conciseness in the title is by referring back to the experimental hypothesis, and using the relationship it predicted as the basis of the title. So, supposing your study had been concerned with testing the hypothesis:

$H_1$: Healing of varicose ulcers takes longer in women who are more than 10% overweight as opposed to women who are of normal weight.

The relationship predicted here is between weight and healing rate of varicose ulcers. Therefore your title might be:

A Study of the Relationship between Patients' Weights and Rate of Healing of Varicose Ulcers

### 15.2.2 Authors

List the **author(s)**, their occupation and place of work. The order in which these names appear in a multi-authorship paper depends on the contribution of the researchers. Typically the person who initiates and carries out

Alice Payne
Senior Nursing Officer
St Varicose General Hospital
London.
Marion Hartswell
Sister

the bulk of the study takes first authorship, with any subsidiary researchers ranked according to their input. However, sometimes the researcher who holds the most senior professional position may take first authorship. Since first authorship is a prized position, trouble can arise when a senior colleague demands first authorship, having made a minimal contribution to the project. This sort of issue should be sorted out prior to the start of the research.

Note that the title and authorship should be on a separate page.

### 15.2.3 Abstract or summary

Although not required by all journals, this **abstract** or **summary** section is a short resumé of the research project. It should provide sufficient information that anyone browsing through the journal could read the abstract and make an immediate decision as to whether the article was likely to be of interest to them.

The abstract must be on a separate page and is usually about 100–200 words in length, although the journal will specify exactly what is required. It typically includes the following information:

1. The hypothesis under investigation.

2. A brief outline of the research design and procedure.

St Varicose General Hospital London.

*Abstract*

A study was carried out to investigate the hypothesis that healing of varicose ulcers takes longer in women who are more than 10% overweight than in women who are of normal weight.

Twenty overweight women and twenty normal weight women who had been referred for treatment of varicose leg ulcers were selected for study. All the subjects were prescribed 'Varicol' ointment and compressive bandages. After three months, the percentage healing for each group was compared using an unrelated $t$ test.

3. A summary of the results.

The results were significant ($t = 2.63$, d.f. $= 38$, $p < 0.01$, one-tailed)

4. A short discussion of the conclusions and implications of the research findings.

suggesting that leg ulcers take longer to heal in overweight women than in normal weight women. Weight reduction programmes might therefore be considered alongside the standard treatments when dealing with overweight patients with varicose ulcers.

## 15.2.4 Introduction

The **introduction** (on a new page) to a research report covers the following:

### Introduction

1. The general background to the research topic, thereby 'setting the scene' for your project.

Varicose ulcers are notoriously difficult to heal because of the reduced circulation to the affected area. While they cause the patient considerable pain and distress, no single method of treatment has been shown to be universally effective in the promotion of healing.

2. A review of all the relevant research literature on the topic, quoting brief findings (and method if appropriate), author(s) and date of research.

A considerable amount of research has been carried out in the area. For example, Legge and Foote (1984) found that the average healing time for varicose ulcers in the over-60s was 9.3 months, while Head and Hands (1987) reported an average time of 10.7 months.

A number of factors are known to be associated with retarded healing rates. For instance, Mann and Boyes (1986) found that gender was an important variable, with women experiencing slower healing than men. Similarly, age and smoking are also

3. A brief theoretical explanation link-
ing the findings.

related, with older patients and
smokers showing retarded recovery
(Weekes, 1981; Fagg, 1983). A full review of the literature on factors associated with slow healing may be found
in Vane and Hart (1985).

Clearly, what emerges here is the
implication that any variable which
further reduces circulation to the affected area is likely to retard the
healing process.

4. A reason why your project was
worth carrying out. Did it fill any
gaps in the existing research? Was
it useful because it replicated an
existing piece of research using a
different treatment, subject sample
or research methodology?

However, no work to date has been
carried out on the effects of obesity on
the healing of varicose ulcers, and yet
obesity has been clearly demonstrated
to be linked with poor circulation in
the body's extremities (Vane and Hart,
ibid.). This study, then, was concerned
with with looking at this issue.

5. State your experimental hypothesis
clearly.

The particular hypothesis under investigation was that healing of varicose
ulcers takes longer among women who
are more than 10% overweight than in
women who are of normal weight.

### 15.2.5 Method

This section is usually subdivided, although you should check the journal's
format and sub-titles first. However,
the main function of the **method** section
is to provide the reader with a detailed
account of the what, when, why and
how of your research procedure. It
should be sufficiently comprehensive
to allow anyone to replicate your study
to the letter, without them needing to
ask questions about what should be
done next. If you bear in mind that
some researchers may want to copy
your research in order to support or
challenge your findings, it will be clear

that to do this they need to be able to repeat your procedure *exactly*.

However, do not make detail an excuse for superfluous information. The following sections are usually included.

### Design

1. The **design** section should clarify the independent and dependent variables, if appropriate.

2. It should state whether a correlational or experimental design was used. If the latter was selected then a statement of whether a same, different or matched subject design was chosen and the reasons why (unless obvious).
3. It should also state whether any measures were taken to overcome possible sources of bias (e.g. counterbalancing for order effects, double-blind procedures).

### Subjects
In this section you should provide all the relevant details of your **subjects**. Typically this will cover the following:

1. Number of subjects used.
2. Their sex.
3. How they were selected (randomly, voluntarily, paid, etc.).
4. Age range.

5. Medical condition.

### Design
The independent variable in the study was weight of the patient while the dependent variable was the degree of healing.
An experimental, different subject design was selected in order to compare the healing rate of two groups who differed in weight.

In order to eliminate any possible experimenter bias effects, a nursing sister who was unaware of the hypothesis under investigation measured the degree of healing.

### Subjects
Two groups, each comprising twenty women were randomly selected from GP's clinics and district nursing registers. All were between 55 and 65 years of age, and all had been referred for treatment of varicose leg ulcers.

However, all the subjects in Group O (obese) were at least 10% overweight by comparison with standard height/weight charts, while the subjects in Group N (normal weight) were all within the average weight range for their height.
None of the subjects was diabetic, or

6. Any other relevant details (occupation, medical history, etc.).

had high blood pressure and all had between two and four children. No subject smoked.

## Apparatus

Here you should give details of any mechanical **apparatus** or equipment that was used in the course of your study, including type, make, manufacturer, etc. If the equipment was designed and built solely for the purposes of your study then you should describe it fully, providing an annotated diagram if relevant, so that the design could be copied by other researchers if they so wished.

## Apparatus

A Lichfield LP 4 camera was used to photograph the ulcers.

## Materials

This section should cover all equipment and **materials** that were used throughout the project. Examples include temperature and blood pressure charts, score sheets, questionnaires, etc.

## Materials

The materials used included centimetre graph paper for plotting the size of each patient's ulcers at the start and end of the project.

## Procedure

The **procedure** part of the method is critical. It should provide the reader with a step-by-step account of what was done in the study and the order in which it was carried out. Remember that it should contain sufficient detail for the reader to repeat your study exactly. It is not an easy task initially to provide just the right amount of information. A simple way of checking what you have written in this section is to ask someone totally unaware of the aims, methods and background of your project to read it. If they are not fully satisfied that they could replicate the procedure after reading it, then you should make the necessary alterations.

The following is sort of information you should include in this section:

1. How the dependent variable was measured (or both variables if a correlational design was used).

2. Any treatment the patient received and for how long.

3. Order of presentation of conditions or tasks.
4. Method of allocating subjects to conditions.
5. Timing of the study.
6. Any instructions given to the subjects or experimenters (reproduced verbatim if appropriate).

**Procedure**

Each subject in the two groups had their ulcers photographed immediately on referral to the GP or district nurse. The area of ulceration was then plotted on centimetre graph paper by an naive experimenter in order to counteract experimenter bias effects. Plotting the ulcer size at the beginning of the study provided a baseline measure against which to compare future healing.

Each patient was then prescribed 'Vari-col' ointment to be administered locally twice daily, and was instructed to use compressive bandages day and night. Every subject was instructed to rest the affected leg whenever possible.

After three months, the patients' ulcers were again photographed and plotted on graph paper by the naive experimenter. The pre- and post-test areas of ulceration were compared for size and the difference in percent noted for each subject.

## 15.2.6 Results

The raw data from your experiment are not normally required in the **results** section. If you feel they are an essential part of your report, then include them in a separate appendix.

What *is* required in this section though are the following:

1. Graphs (if used) which should be clearly labelled, ideally in Letraset or similar.
2. The average scores for each group or condition.

3. A statement of what statistical test was used.
4. The results of the test (including the level of significance).
5. A brief outline of what the results mean.

Do *not* include any workings from your statistical analysis.

## 15.2.7 Discussion

The **discussion** section should *discuss* your findings in relation to other research and theories in the area.

It typically deals with the following points:

1. A restatement of your findings.

2. Your results should be related to other research findings in the area, showing how they confirm or oppose them. Some explanation is required, however, if your results are at odds with those from other studies.

### Results

The mean percentage healing for Group O was 32%, while for Group N it was 69%.

Using an unrelated $t$ test on the data, the results were found to be significant at the $p < 0.01$ level ($t = 2.63$, d.f. = 38, one-tailed). These results suggest that healing of varicose leg ulcers is slower amongst overweight women than amongst normal weight women, thereby supporting the experimental hypothesis.

### Discussion

The results of the present study suggest that healing of varicose leg ulcers takes longer in overweight women than in normal weight women. These results accord with those of Heale (1983) who found that any wound recovery in limbs was slower amongst obese patients. Brake (1986) similarly found retarded healing of leg amputations in the overweight. However, Ketone (1985) found no difference between normal weight and obese patients in rate of healing of foot ulcers. This discrepancy in the findings may be due to the fact that all Ketone's subjects had diabetes, which would reduce healing anyway.

3. Some theoretical explanation of your findings should be offered.

The present results could be explained by the reduced circulation known to affect many overweight people. This would restrict the healing of any lower limb wound and especially varicose ulcers. In addition, it is conceivable that the degree of exercise taken by the subject groups might have influenced recovery, since it has been shown that the obese take less exercise in the course of the day than the normal weight (Idle, 1985). This would exacerbate any circulatory problems.

4. Any additional analysis of the results should be discussed. For example, you might have compared sub-groups of your subjects such as those under 40 years of age with the over 40s; males v. females; second year v. first year nurses, etc.

A further comparison of the subjects suggested that those women who had arthritis of the hip and knee joints experienced retarded varicose ulcer healing when compared with those women without arthritis, *irrespective of body weight* ($t = 2.1$, d.f. $= 38$, $p < 0.05$). Since arthritis is more common amongst the obese, it is further support for the notion that circulatory problems are a major factor in varicose ulcer recovery.

5. Any defects in your research design and procedure should be identified, and recommendations for improvements made.

While this latter finding is interesting it does highlight arthritis as a constant error which was not controlled, since it could cause restricted circulation in the lower limbs. In this way the results of the main study could be attributable to arthritis rather than obesity. Further studies are needed to eliminate this factor as a source of error.

6. Similarly, any problems encountered in the actual running of the study should be acknowledged and their implications for the results discussed. Examples of these problems include equipment breaking down, subjects not carrying out activities, treatments, etc.

Acknowledging errors and faults will not invalidate your research, unless major design flows are pre-

Furthermore, it was impossible to ascertain how far each subject complied with the treatment procedure over the three-month study period. While it is unlikely, it is nonetheless conceivable that the obese subjects were less conscientious in their treatment procedure, which might retard the healing process. Some check of patient conformity on this issue needs to be considered. On reflection, it might

sent. Recognising imperfections and recommending solutions does not detract from your credibility as a researcher, nor does it reduce the possibility of publication. Research is a process whereby informed decisions are made to achieve the optimum design for a given hypothesis. While a researcher can aspire to perfection, few achieve it.

7. Point out the practical implications of your research.

have been preferable to compare the two groups on time taken to full recovery of the ulcers, rather than relying on percentage healing after a fixed time period, which does not take account of any subsequent breakdown in the ulcerated area.

The results from the present study suggest that traditional methods of treatment for varicose ulcers in overweight women may be insufficient. What seems to be needed in addition is a weight-reduction programme.

8. Any ideas for future research which your study has revealed should be mentioned. Remember that good research should generate ideas for other research.

The efficacy of combining traditional treatments and diet might be the focus of subsequent study, using obese males as well as females.

## 15.2.8 Acknowledgements

Always **acknowledge** the help given in the course of research. This might be a statistician for analysing the results, or a colleague for helping in the data collection.

*Acknowledgements*
The author would like to thank Dr T. Test for his invaluable help in the statistical analysis and Sister A. Ward for her assistance in the data collection.

## 15.2.9 References

Any research that you have cited in your article should be fully referenced at the end of the paper (on a new page). Usually these are listed in alphabetical order; where one author has appeared several times, these **references** should be listed chronologically under the author's entry. Occasionally, the journal requires that the references are listed in the order they appear in the article. The moral here is check the prospective journal to see what is required. Typically the references should take the following form:

1. References which are *articles*: author's surname, initials, date of publication, title of article, journal in which it appears (underlined), volume and part numbers and the pages.

2. References which are *books*: author's surname, initials, date of publication, title of book (underlined), place of publication and publisher.

3. References which are independently written chapters in a book: author's surname, initials, date of publication, title of chapter, surname and initials of editor, title of book (underlined), where the book was published and by whom.

*References*

(only examples and not a fully detailed list will be given here)

Brake, L. (1986) The effect of obesity on the healing of lower limb amputations. *Journal of Major Surgery*. Vol. 3 (2) 113–116.

Ketone, D. (1985) Foot ulcers and rate of healing: a comparison of normal weight and obese diabetic patients. *Journal of Therapeutic Nursing*. Vol. 12 (4) 96–98.

Head, B. & Hands, C. (1987) *Circulatory Disorders*. London, Somerset Open Press.

Heale, O. (1983) *Limb Trauma*. New York, Oregon Pine Inc.

Idle, M. (1985), etc.

Legge, P. & Foote, S. (1984) Leg Ulcers in the Ageing Population. In Old, T. (ed.) *Problems of Ageing*. Edinburgh, Batcombe Press.

Mann, A & Boyes, S. (1986), etc.

Vane, W. & Hart, G. (1985), etc.

Weekes, L. (1981), etc.

If you now re-read everything on the right-hand side of the page, you should get a rough idea of how to construct a research article. Do not feel too despondent if your first draft is not satisfactory. Do as many of the great painters did – hide it away for a while and do something else. When you come to look at it again, some of the problems will be apparent. If, after modifications, you are still not satisfied with the product, repeat the process until you are.

Some final points about submitting an article for publication:

1. Always type the manuscript – never submit a hand-written one.
2. Ensure there are no errors and omissions and that the references are provided in full.
3. Always check that you have conformed to the format required by the journal.
4. Never send an article off to more than one journal at a time. If your first choice rejects it, then submit it to another.

And finally – *do not* despair if you get some rejection slips: everyone does!

## 15.3  Key word and terms

Abstract
Acknowledgements
Apparatus
Authors
Design
Discussion
Introduction
Materials
Method
Procedure
References
Results
Subjects
Summary
Title

# 16    Reading research articles critically

If you do not want to carry out any research of your own, it is still important to be able to evaluate the research of others, in order to assess whether or not its conclusions and recommendations should be incorporated into nursing practice.

There are a number of reasons why some nurses feel unable to examine published research articles critically. For one thing, an unquestioning attitude towards authoritative statements has been inculcated into many nurses, particularly at the training stage. And for another, critical analysis requires a sound grounding in research methodology – an area of study which eluded many nurses, at least until recently.

Yet the ability to evaluate research is crucial for two major reasons. Firstly, there is a powerful suggestion that before anyone embarks on any research themselves, it is essential that they are able to judge the attempts of others. But perhaps an even more cogent argument centres on the fact that if a nurse can identify the strengths and weaknesses in a piece of research then she or he is in a better position to assess whether or not the findings are worthy of being implemented at a practical level. Clearly, since the improvement of patient care is the ultimate aim of most nursing research, an ability to judge the quality of a piece of published research must be an essential prerequisite to implementation.

In order to evaluate research reports, the nurse should have some background knowledge of the topic under investigation and some understanding of research techniques. This chapter is an attempt to provide a general framework within which a nurse can operate when assessing research articles. Each section of a typical research report will be considered and the salient questions the reader should be asking about that section will be highlighted in bold type.

It is worth pointing out that even though you might not wish to undertake or publish any research yourself, it might still be worth reading Chapter 15 on presenting research for publication, since it outlines in some detail what is typically expected of a research report. This might be a useful aid in deciding whether or not any article you want to evaluate meets the standard requirements.

## 16.1 Title

The first stage in the evaluation process is to identify what the report was about, i.e. the topic under study. A broad idea can be gleaned from the title providing it is clear and pithy. However, many research reports fail at this first hurdle, providing esoteric, metaphorical or allegorical titles which leave the reader in confusion.

So the first question the reader might ask of the report is as follows:

*Does the title provide a clear and unambiguous statement of the topic under investigation?*

## 16.2  Abstract or summary

If the title is not illuminating, then the abstract or summary should be, since it typically includes a statement of the hypothesis under investigation, and the method, results and conclusions of the study. Having said this, not all journals provide an abstract. Where this is the case, it may be useful for the reader to produce a 150-word summary of the article, following the format outlined in Section 15.2.3. This activity has the function of focusing the reader's mind on the essential issues of the study.

If an abstract is supplied which does not include the features already discussed, then this may be a criticism, since the whole point of the abstract is to summarise what was done, why and how, as well as what was found. Omission of such salient factors is a major flaw.

The questions which should be asked about the abstract or summary are as follows:

*Is the abstract a succinct summary of the research, giving the reader a clear idea of the salient features?*

*Does the abstract state the aims or hypothesis, methods, results and conclusions?*

## 16.3  Introduction

Perhaps the most vital piece of information that can be derived from either the abstract or, failing that, from the introduction, is the hypothesis under investigation. If no clear statement of the hypothesis is made then it is impossible to assess whether or not the methodology was an adequate test of the hypothesis, or whether the conclusions the author draws are valid.

So the reader should ask the following:

*Is the hypothesis clearly defined both in the abstract and the introduction?*

*Does it predict a relationship between two (or more) variables and if so can these variables be easily identified?*

EASTBOURN.
SCHOOL OF NURSING

If the answers to these questions are 'no' then this is another major flaw, since no adequate evaluation of the rest of the project can be made.

Knowing what the hypothesis is allows the reader to assess the relevance of the rest of the introduction – in particular the literature reviewed and the rationale for carrying out the research. The literature review should describe the relevant research in the area under investigation. While it should be thorough, it does not have to be exhaustive as long as the major studies are covered. It should also be up to date. Too many articles provide only a cursory glimpse of other studies, while others report outdated findings. Obviously there may be seminal works in a given area which may be rather ancient, but these should be presented in conjunction with more recent studies. Likewise, the word 'relevant' is important here. Too often research reports cover literature which is either only tenuously related to nursing or to the particular topic in question. So, the next questions concerning the introduction are as follows:

> Is the literature review thorough?
> Is it up to date?
> Is it relevant to nursing and to the topic in question?

Severe doubt is cast on the researcher's credibility if the answers to these questions are negative, since it is likely that the author was not fully appraised of the topic area and therefore not adequately equipped to research it.

Similarly, the way in which the introduction and literature review are organised is important. A clearly logical structure which presents a thorough coverage of the literature coupled with a statement of the value of the research reported is essential. If no good argument is presented for conducting the study it is conceivable that its worth was limited or that the researcher was inadequately aware of its implications. Neither explanation does the research much credit. It is important that the problem under investigation was a significant one and that its objectives were worth while. Therefore, that next questions which should be asked of the introduction are as follows:

> Is the introduction cogently structured?
> Does it provide a good reason why the research was undertaken?
> Is the problem under investigation a significant and important one?
> Is the aim of the study worthwhile?

## 16.4 Method

The method section of a research report is typically divided into five sub-sections: design, subjects, apparatus, materials and procedure. Even if no sub-sectioning is used the method should cover a number of important issues (see Section 15.2.5).

## 16.4.1 Design

The first of these sub-sections deals with the design of the study and should describe the sort of design used (correlational or experimental, same, different or matched subjects) and the reasons for its selection. The reader should be able to assess how appropriate the design was for testing the hypothesis, as well as evaluating the validity of the author's decision-making process. The design section should also describe any procedures which were adopted to overcome potential sources of bias, e.g. random allocation of subjects to conditions, counterbalancing the order of presentations. If these precautions were not taken, should they have been? Did the researcher leave sources of constant error? It is also usual to identify (if it is appropriate) the independent and dependent variables in this section, so that the reader is aware of which variable was manipulated and which was measured.

So the questions which should be addressed to the design section are as follows:

*Is the design of the study described?*

*Are adequate reasons given for its selection?*

*Is the design an appropriate test of the hypothesis or is an alternative more suitable?*

*Are potential sources of error acknowledged and precautions taken to overcome them?*

*Are the independent and dependent variables specified (if appropriate)?*

## 16.4.2 Subjects

The second part of the method section usually describes the subject sample. Given that only a limited number of subjects can participate in a study, the researcher has to make a selection from those available. This selection has to take into consideration factors such as representativeness, number, constant errors, method of selection and suitability (see Chapter 4). If the researcher overlooks any of these points then the research findings could be invalidated. For instance, a study which only uses half a dozen subjects may have limited generalisability. This section should describe *all* the important details of the subject sample, such as age, sex, medical condition, etc. It is inadequate simply to report that twenty subjects took part in the study without saying who they were and why and now they were chosen.

This section is an important one. The questions which are relevant here are as follows:

*Is the size of the subject sample adequate?*

*Are the subjects representative of the population from which they are drawn?*

*How were the subjects selected: randomly, voluntarily, by request?*

*Is this likely to have any impact on the outcome?*

*Was any bias apparent in the selection procedure?*

*Were all the constant errors that the subjects might have introduced recognised and controlled for?*

*If not, what exactly were the omissions and how influential were these likely to be on the results?*

*Were the subjects* suitable *for the project under investigation?*

*Is the subject sample fully and adequately described, giving all the relevant information?*

### 16.4.3 Apparatus

The next section of the method deals with any mechanical apparatus which was used in the course of the study. Its type, make and other details should be mentioned. The apparatus should also be suited to the purpose for which it was intended. Out-moded equipment may not give such accurate and reliable results as more recent innovations. So the questions which are raised at this point are as follows:

 *Was any mechanical apparatus which was used clearly identifiable by type, make and model?*

*Was the apparatus appropriate for its purpose?*

### 16.4.4 Materials

Any materials such as questionnaires, patient record sheets, score cards, temperature graphs, etc., should be listed in this section.

If questionnaires or standard assessment techniques (say of attitudes, personality, etc.) were used, were data on their reliability and validity quoted? Researchers often compile a list of questions and call it a questionnaire. This is inadequate, since questionnaire design is a complex science in its own right, requiring thorough testing before it is used (see Oppenheim 1966 for a thorough review of questionnaire construction). Therefore, if a questionnaire was designed exclusively for the study was it properly tested beforehand? Is it included in the appendix for the reader's examination? Were any written instructions for the subjects or co-experimenters clear and unambiguous? The questions concerning this section, then, are as follows:

*Were the materials used suitable for their purpose?*

*Were any specially constructed questionnaires adequately tested before the start of the project?*

*Were the reliability and validity figures for questionnaires, personality inventories, etc., provided?*

*Are any specially designed materials included in the appendix?*

*If so, are they fit for the job?*

*Are the instructions clear and unambiguous?*

## 16.4.5 Procedure

The final part of the method section describes the research procedure, i.e. what was done, when and how. It should be sufficiently detailed that anyone, however unfamiliar with nursing or research, could follow the instructions to the letter, and thereby replicate the study. It should cover such details as what was actually carried out to whom and by whom and in what order; the measures that were taken to eliminate sources of constant error (e.g. randomisation, double-blind procedures, etc.); what instructions were given to the subject; how the subjects' rights were protected. This latter point is of particular relevance to nursing research, since human subjects are normally involved. They should have been appraised of the purposes and potential value of the research at some stage and their performance or responses should be carefully guarded to prevent disclosure to unauthorised personnel. Furthermore, it is rarely, if ever, acceptable to withhold treatment. If treatment *was* withheld for a control group, satisfactory explanations should be given.

Another crucial issue here is whether or not the data collection procedure and the type of data collected were adequate measures of the dependent variable. For example, the researcher might have used a nominal measure for temperature (high, average, low) when an interval/ratio measure giving the exact temperature might have been better. Or the number of days from operation to discharge might have been used as a measure of a treatment's efficacy, when a more direct method of assessing healing or recovery would have been preferable.

The questions relating to this section are as follows:

*Is the description of the steps taken clear, thorough and logical?*

*Is the procedure sufficiently comprehensive to allow a naive reader to repeat it?*

*Were all the sources of constant error controlled for? If not, can others be identified which might bias the results?*

*Were any instructions given to the subject or co-experimenter unambiguous?*

*How were the data collected and recorded?*

*Was the method of collecting the data a satisfactory measure of the dependent variable (or both variables in the case of a correlational design)?*

*Was the type of data collected a satisfactory measure of the dependent variable (or both variables in the case of a correlational design)?*
*Were the subjects' rights fully protected?*
*If a control group was used, was this ethically acceptable?*

Overall, the method section should provide sufficient detail for any reader to be able to replicate it *exactly*. Was this the case?

## 16.5 Results

While raw data are not always recorded in a research report, graphs (if used) and the results of statistical tests are. Any graph should have a self-explanatory title or label and it should be apparent what the source of data was. It should also be sufficiently large that it can be easily interpreted. The axes should be clearly labelled and the size of the intervals along each axis should be appropriate, so that trends and patterns can be readily identified.

Any tabulated data should also be clearly labelled and its source noted. It should have some direct relevance to the study and should be interpreted and discussed in the results and/or discussion section. Too often graphs and tables appear out of the blue in research reports, and have little apparent meaning and relevance.

Where statistical tests were used to analyse the data, it is *imperative* that the correct test was used for the research design. Many published articles have been shown to have used a completely inappropriate statistical test, thereby invalidating all the findings and conclusions. Given that a patient's health and well-being may ultimately be affected by a piece of nursing research, it is essential that the correct method of data analysis is used in order to avoid erroneous conclusions. Sometimes, too, a researcher uses a parametric test when a non-parametric one would have been more suitable (and vice versa). Since the working out of a statistical test is rarely recorded, the reader has to take on trust the researcher's mathematical competence. This is easier if the rest of the report has been thorough and sound. However, a sloppy and poor piece of work may (justifiably) cast doubt on the accuracy of the statistical calculations too. Even if the computations cannot be checked, a number of issues relating to the analysis can. For example, is the level of significance suitable for the subject under investigation or should it have been lower or higher? Does the researcher specify if the probability of the results was assessed according to whether the hypothesis was one- or two-tailed and, if so, does that accord with the hypothesis as originally stated? The results should also be clearly, accurately and explicitly interpreted, since if a cloak of mystery hangs over the results the reader will be unable to evaluate any interpretations or conclusions that the author may have drawn. Far too many research articles simply state the results of the statistical calculations and their significance without clarifying what these results mean and whether they support the original hypothesis. Similarly, any subsidiary analyses of the data should be presented in perspective and without dominating the main findings.

The questions which arise about this section are as follows:

*Are the graphs clearly drawn and labelled?*

*Are their sizes and the intervals used suitable?*

*Is the source of the graph data obvious?*

*Are any tables clearly labelled?*

*Are the tables of relevant data?*

*Are they interpreted and discussed elsewhere?*

*Were the statistical tests used to analyse the data the correct tests for the design?*

*Did the data warrant a parametric or non-parametric test?*

*Is the selected level of significance appropriate for the subject matter?*

*Is the probability of the results related to the nature of the experimental hypothesis, i.e. whether it is one- or two-tailed?*

*If this is specified in the results section does it accord with the experimental hypothesis as originally stated in the introduction?*

*Are the results interpreted?*

*If so, is the interpretation accurate, given the outcome of the analysis?*

*Does the author clarify whether or not the experimental hypothesis has been supported?*

## 16.6 Discussion

The discussion section of a research report is quite simply a forum for interpreting, explaining and discussing the results, together with their relevance and implications. It typically relates the current research findings to other work in the area, in order to put the results in a broader context. Theoretical explanations of the outcome are usually put forward, together with an overview of any recommendations about the practical application of the results. It is important to note whether the research interprets the findings as presented or whether the interpretations go beyond the factual evidence. Are over-generalisations made beyond the data and the subject sample used? Many researchers only present partial discussions of their findings, omitting evidence that fails to corroborate their own ideas and hypotheses. Incomplete interpretation of the results may be as misleading as erroneous interpretations.

Occasionally the findings in a research report are totally at odds with all the other work in the area. There should be a satisfactory reason for this, otherwise doubts about the adequacy of the methodology might creep in. Satisfactory and cogent theoretical explanations should be given for the findings, since theory and practice should go hand-in-hand.

It is also important to take note of whether the researcher acknowledges the limitations and flaws in the study. Failure to do so may mean that the conclusions are adopted totally and without reservation, which, should they be the outcome of an inadequate study, could jeopardise patients' well-being. The conclusions should be substantiated by *all* the findings of the study.

Finally, a piece of research should be a fecund source of other hypotheses. Whether it is should be clear in the process of reading the article.

So for this section, the questions that need to be asked are as follows:

> *Are the results related to other work in the area?*
>
> *Are adequate reasons given for any discrepancies?*
>
> *Is any plausible theoretical explanation of the results provided?*
>
> *Are the results interpreted properly on the basis of the evidence provided?*
>
> *Are extravagant interpretations and conclusions drawn which go beyond the data?*
>
> *Are all the results discussed or only those which accord with the researcher's personal viewpoint?*
>
> *Are limitations of the methodology acknowledged?*
>
> *If so, are the results interpreted in the light of these?*
>
> *Are recommendations made for improvements on these issues?*
>
> *Are the practical implications of the results presented?*
>
> *Do any hypotheses emerge as a basis for future research?*

## 16.7 References

The reference section should include all the research and work cited throughout the course of the article. It should be comprehensive in its citations and the references should be complete, covering authors' names, title and date of the article, the journal or book in which it appeared, volume and part number if relevant and the pages covered. Therefore the questions arising about the reference section are as follows:

> *Are all the articles and pieces of research quoted in the body of the article cited in the reference section?*
>
> *Are the references fully documented, stating all the relevant information?*

Finally, some general questions about the article as a whole should be asked:

> *Has the project solved a fundamental issue in nursing?*
>
> *Has it contributed to nursing knowledge?*

*Has it collected information which may be useful to nursing?*
*Is the terminology used clearly defined and consistently used?*
*Is it devoid of jargon?*
*Are abbreviations explained?*
*Is the article clearly presented?*
*Is it logical, unbiased, impartial and scientific?*
*Is the literary style clear and appropriate for a piece of empirical work?*

While it may seem as though there are a multitude of questions to be asked, these will become automatic over time, enabling the reader to make sound judgements of research before putting the results into practice.

## Appendix A
# A reminder of the basic mathematical principles

This section is divided into two parts. The first deals with the principles fundamental to mathematical calculations, such as multiplication and division. If you are unsure of these rules then you are strongly urged to read this section. The second part covers rank ordering of data, which is a component of a number of statistical tests. It is recommended that every reader familiarises themselves with this process, since errors in ranking will produce erroneous results from the tests and in consequence, incorrect conclusions.

## A.1 Basic maths

Although I suggest that either a calculator or computer is used in the analysis of the results from your research, it is still imperative that the researcher is familiar with the principles of basic maths, for two reasons. Firstly, some mathematical calculations have to be performed in a set order. So even if you are using a calculator, you must know which sequence to use when entering the numbers and instructions. Failure to observe these rules produces incorrect solutions. The second reason for understanding the basic rules of maths relates to your ability to make a quick and rough assessment of the answer produced by the calculator to see whether it 'looks' right. This process of 'eyeballing' the data is crucial, since calculators and their users are not infallible. However, it is a skill that is difficult to teach, usually only developing when the researcher has more experience of computations and more of a 'feel' for figures.

Nonetheless, you should always check your entries into the calculator and consider each calculation to see if it seems reasonable.

### A.1.1 *Order of calculation*

There are six rules which relate to the *order* of calculation.

1. If the formula contains brackets, all the calculations inside the brackets *must* be performed first. So, should you encounter

$$17 - (14 + 2)$$

The answer is $17 - 16 = 1$.

However, had you ignored the brackets and simply worked from left to right, the answer would be 5, which is incorrect.

2. Some formulae have brackets inside brackets. Where this is the case, perform the calculations on the figures in the innermost brackets first and then work outwards to the next set of brackets. So, if you found:

$$29 - [10 - (4 + 2)]$$

the answer would be

$$29 - [10 - 6]$$
$$= 29 - 4$$
$$= 25$$

On the other hand, if the brackets are ignored and the calculations carried out just from left to right the answer is 17. This is quite different and quite wrong.

3. Where there are no brackets, do all the multiplications and divisions *first*, followed by additions and subtractions. So

$$21 - 7 \times 2$$
$$= 7$$

But if this rule is ignored and the calculations are done from left to right, an incorrect answer of 28 is obtained. Similarly, with $34 + 12 \div 2$, the correct answer is 40, but if the rules are ignored the wrong answer of 23 is obtained.

4. However, any additions and subtractions above and below a division line *must* be done *before* the division

$$\text{i.e.} \quad \frac{19 + 21}{3 + 5} = \frac{40}{8}$$
$$= 5$$

5. Where there are *only* additions and/or subtractions in a formula without brackets, simply work from left to right. Therefore the answer to

$$11 + 4 + 19 - 10$$

is 24.

6. Where there are *only* multiplications and/or divisions in a formula without brackets, work from left to right. So, the answer to

$$10 \times 14 \div 4$$

is 35.

## A.1.2. Positive and negative numbers

A major hurdle for many people – even competent mathematicians and researchers – is positive and negative numbers, whose sole purpose seems to be to tie the mind in knots. While in theory the principles behind them seem quite comprehensible, in practice they can be problematic, particularly when adding and subtracting. Adding and subtracting positive and negative numbers can be construed as a banking system, with the left-hand figure representing your account. If this figure is preceded by a minus, you are in the red, while if there is no minus figure, you are in credit. The right-hand figure can be thought of as your transactions via your cheque book. If the right-hand figure has a plus before it you are paying money into your account, while if it has a minus, you are withdrawing money.

Therefore, if you have

$$-250 + 75$$

you have an overdraft of £250, but you are paying in £75. This will partially offset your debt and reduce it to £175.

On the other hand, had you been *withdrawing* £75 instead of paying it in, you would have increased your overdraft to £325. This can be expressed as

$$-250 - 75$$

There are then a number of preliminary rules to consider when adding and subtracting positive and negative numbers:

1. If no sign is indicated before a number, then the number is positive, i.e. $50 + 7$ means $+50 + 7$.
2. To add numbers with the same sign, add the numbers and prefix the common sign, e.g.

$$7 + 12 = 19$$
$$-9 + (-17) = -26$$

3. To add two numbers which have different signs, take the smaller number away from the larger one and prefix the sign of the larger number, e.g.

$$12 + (-5) = +7$$
$$-29 + 13 = -16$$

4. To subtract a positive number, subtract in the normal way, e.g.

$$51 - (+14) = 37$$

5. To subtract a negative number, change its sign to positive and *add* the two figures, e.g.

$$23 - (-9) = 32$$

Multiplying and dividing positive and negative numbers are rather more straightforward, and six rules are given for these calculations:

1. If two positive numbers are multiplied together, the answer is always positive, e.g.

$$(+4) \times (+7) = +28$$

2. If two negative numbers are multiplied together, the answer is always positive, e.g.

$$(-4) \times (-7) = +28$$

3. If a positive number and a negative number are multiplied together, the answer is always negative, e.g.

$$(-4) \times (+7) = -28$$

4. If two positive numbers are divided, the answer is always positive, e.g.

$$+60 \div +15 = +4$$

5. If two negative numbers are divided, the answer is always positive, e.g.

$$-60 \div -15 = +4$$

6. If a positive number is divided by a negative number or vice versa, the answer is always negative, e.g.

$$-48 \div +16 = -3$$
$$+48 \div -16 = -3$$

If more information on positive and negative numbers is required, the reader is referred to Bradley and McClelland (1978).

### A.1.3 Square roots and squared numbers

The 'square' of a number is that number multiplied by itself. It is expressed as $^2$ by the number to be squared. So $6^2$ means $6 \times 6$, which is 36. Remember that the answer will always be positive since a negative number multiplied by a negative number always produces a positive result.

The square root of any given number is in a way the reverse of a squared number, since it is that number which when multiplied by itself gives the number you have already. It is denoted by the sign $\sqrt{}$ which is called the *radical*. So the square root of 36, i.e. $\sqrt{36}$, is 6, since $6 \times 6$ is 36. The square root can be quite hard to compute if your calculator does not have a square root function. Try to ensure that you have access to a calculator which has a $\sqrt{}$ sign on it.

Occasionally, you will find formulae that have more than one number under the square root sign, e.g.

$$\sqrt{94} \times 3 \quad \text{or} \quad \sqrt{(94 \times 3)}$$

In these cases you should carry out all the calculations under the square root sign *before* finding the square root. So the next two rules of basic maths are as follows:

1. The square of a number means that number multiplied by itself and is expressed by $^2$.
2. The square root of a given number is the number which when multiplied by itself yields the number you already have. It is expressed by the symbol $\sqrt{\phantom{x}}$.

## A.1.4  Rounding decimal places

When working with decimals, it is quite common to end up with several figures to the right of the decimal point. To proceed with these long numbers makes the calculations unnecessarily complex, so typically the numbers to the right of the decimal point are reduced by *rounding* them up. For example, throughout this text, we have been working to two decimal places, i.e. two figures to the right of the decimal point.

To round up decimal places, the following steps are taken. If the figure on the extreme right is 5 or more, then the number to its left is increased by 1. So

$$1.78647$$

becomes

$$1.7865$$

If, however, the figure on the extreme right is less than 5, then the figure to its left stays the same. So

$$1.785644$$

becomes

$$1.78564$$

This process is repeated with the figure which is now on the right, until you have the required number of places. So

$$1.78565$$

becomes

$$1.7857$$

which becomes

$$1.786$$

which becomes

$$1.79$$

i.e. rounded up to two decimal places.

The process was straightforward with the number in the example, but what would have happened if the number had been 2.996571?

Rounding to two decimal places means

$$2.996571$$

becomes

$$2.99657$$

which becomes

$$2.9966$$

which becomes

$$2.997$$

which becomes

$$3$$

because the 9s were each increased to 10 which resulted in a final figure of 3 – and no decimal places.

So the rule pertaining to rounding up decimal places is as follows:

1. *When rounding up decimal places, start with the extreme right-hand number. If this is 5 or greater, add 1 to the number to its left. If the extreme right-hand number is less than 5, the number to its left stays the same.*

## A.2 Rank ordering data

The rank ordering procedure is an inherent part of several statistical tests. It is a simple procedure, but mistakes can be made, which will then invalidate the overall results of the statistical test.

Essentially, rank ordering data involves assigning a rank of 1 to the smallest score, a rank of 2 to the next smallest and so on until all the data have been ranked. Therefore, the rank orderings for the following figures are given in Table A.1.

**Table A.1**

| Score | Rank |
|-------|------|
| 5 | 4 |
| 3 | 2 |
| 2 | 1 |
| 4 | 3 |
| 7 | 5 |
| 9 | 6 |
| 10 | 7 |

**Table A.2**

| Score | Rank |
|-------|------|
| 2 | 2 |
| 3 | 4 |
| 2 | 2 |
| 4 | 5.5 |
| 5 | 7 |
| 2 | 2 |
| 4 | 5.5 |

Because 2 is the smallest score, it gets a rank of 1, 3 is the next smallest score and so gets a rank of 2 and so on up to the largest score of 10 which gets a rank of 7.

Sometimes a score of 0 is achieved. When this is the case, it counts as the smallest score and so gets the rank of 1. However, frequently two or more scores are the same. When this occurs, the 'tied rank' procedure has to be used. So, supposing we had obtained the data in Table A.2. The smallest score is 2, but there are three scores of 2. So here we have to add together the *ranks* that the scores of 2 would have been assigned had they been different, i.e. ranks $1 + 2 + 3$, and divided the result by 3 because there are three tied scores of 2:

$$\frac{1 + 2 + 3}{3} = 2$$

This procedure gives us the *average* rank which is then assigned to each score of 2.

The next lowest score is 3. Because we have already used up ranks 1, 2 and 3, a rank of 4 has to be given to the score of 3.

The next lowest score is 4, but there are two scores of 4. As ranks 1, 2, 3 and 4 have been used, these scores of 4 would have been assigned ranks 5 and 6 had they been different. Therefore we have to add 5 and 6 together and divide by 2 because there are two tied scores of 4: $(5 + 6)/2$. This give us an average rank of 5.5 which is assigned to each score of 4. The next lowest score is 5. Because ranks $1 - 6$ have been used up, a rank of 7 is given to the score of 5.

Where long lists of numbers have to be rank ordered it is very easy to omit one by mistake, ending up 'one out'. The easiest way to avoid this pitfall is by writing down the numbers of the ranks you will be using (which will always equal the number of scores that have to be ranked) and crossing these off as they are used. So, if you were ranking the scores in Table A.3 there are eight scores to be ranked. Write the numbers 1–8 on a piece of scrap paper and cross them off as you use them. In practice this looks like this:

$$\cancel{1} \quad \cancel{2} \quad \cancel{3} \quad \cancel{4} \quad \cancel{5} \quad 6 \quad 7 \quad 8$$

The first five ranks have been crossed off as they were used in the ranking procedure in Table A.3.

**Table A.3**

| Score | Rank |
|-------|------|
| 6 | |
| 5 | 4.5 |
| 5 | 4.5 |
| 4 | 3 |
| 7 | |
| 3 | 2 |
| 2 | 1 |
| 9 | |

The principle of applying tied ranks is exactly the same wherever you may encounter it in this book.

However, sometimes you will be required to rank the *differences* between pairs of scores, rather than the scores themselves. Where this happens, differences of zero are ignored, as are any plus or minus signs in front of the differences. Where this occurs, you will be reminded of what to do and detailed instructions will be provided.

**Summary of key points**
1 When ranking scores, give a rank of 1 to the lowest score, a rank of 2 to the next lowest, etc.
2 When two or more scores are the same, assign the *average* rank, which is calculated by adding together the ranks the number would have attained had they been different, and dividing the result by the number of scores that are the same.
3 Where *differences* between scores, rather than the scores themselves are being ranked, any differences of 0 are ignored, as are plus and minus signs in front of the differences.

*Exercise (answers on page 320)*

1 Rank the following data, giving a rank of 1 to the lowest score and so on.

Score
2
3
4
3
5

3
7
8
8

## Appendix B

# Some symbols commonly found in statistical formulae

$\div$, — and / mean 'divide by', i.e.

$$8 \div 2 \qquad \frac{8}{2} \qquad 8/2$$

$\times$ and parentheses mean 'multiply by', i.e.

$$8 \times 2 \qquad 8(2)$$

$\Sigma$ means the total of all the calculations following the symbol.

$x$ means an individual score.

$\bar{x}$ means the average or mean score.

$\sqrt{}$ means the square root of the figure or calculations under the sign, i.e. $\sqrt{9} = 3$.

$^2$ means the number in front of the $^2$ sign multiplied by itself, i.e. that number squared, $3^2 = 3 \times 3 = 9$.

$n$ means the number of subjects or scores in a condition or sub-group.

$c$ means the number of conditions or sets of scores in an experiment.

$<$ means 'less than'.

$>$ means 'greater than'.

$H_1$ means the experimental hypothesis.

$H_0$ means the null hypothesis.

# Appendix C
# Statistical probability tables

**Table C.1** Critical values of $\chi^2$ at various levels of probability. (For your $\chi^2$ value to be significant at a particular probability level, it should be *equal to* or *larger than* the critical values associated with the d.f. in your study)

| d.f. | Level of significance for a two-tailed test | | | | |
|------|------|------|------|------|------|
|      | 0.10 | 0.05 | 0.02 | 0.01 | 0.001 |
| 1  | 2.71  | 3.84  | 5.41  | 6.64  | 10.83 |
| 2  | 4.60  | 5.99  | 7.82  | 9.21  | 13.82 |
| 3  | 6.25  | 7.82  | 9.84  | 11.34 | 16.27 |
| 4  | 7.78  | 9.49  | 11.67 | 13.28 | 18.46 |
| 5  | 9.24  | 11.07 | 13.39 | 15.09 | 20.52 |
| 6  | 10.64 | 12.59 | 15.03 | 16.81 | 22.46 |
| 7  | 12.02 | 14.07 | 16.62 | 18.48 | 24.32 |
| 8  | 13.36 | 15.51 | 18.17 | 20.09 | 26.12 |
| 9  | 14.68 | 16.92 | 19.68 | 21.67 | 27.88 |
| 10 | 15.99 | 18.31 | 21.16 | 23.21 | 29.59 |
| 11 | 17.28 | 19.68 | 22.62 | 24.72 | 31.26 |
| 12 | 18.55 | 21.03 | 24.05 | 26.22 | 32.91 |
| 13 | 19.81 | 22.36 | 25.47 | 27.69 | 34.53 |
| 14 | 21.06 | 23.68 | 26.87 | 29.14 | 36.12 |
| 15 | 22.31 | 25.00 | 28.26 | 30.58 | 37.70 |
| 16 | 23.54 | 26.30 | 29.63 | 32.00 | 39.29 |
| 17 | 24.77 | 27.59 | 31.00 | 33.41 | 40.75 |
| 18 | 25.99 | 28.87 | 32.35 | 34.80 | 42.31 |
| 19 | 27.20 | 30.14 | 33.69 | 36.19 | 43.82 |
| 20 | 28.41 | 31.41 | 35.02 | 37.57 | 45.32 |
| 21 | 29.62 | 32.67 | 36.34 | 38.93 | 46.80 |
| 22 | 30.81 | 33.92 | 37.66 | 40.29 | 48.27 |
| 23 | 32.01 | 35.17 | 38.97 | 41.64 | 49.73 |
| 24 | 33.20 | 36.42 | 40.27 | 42.98 | 51.18 |
| 25 | 34.38 | 37.65 | 41.57 | 44.31 | 52.62 |
| 26 | 35.56 | 38.88 | 42.86 | 45.64 | 54.05 |
| 27 | 36.74 | 40.11 | 44.14 | 46.97 | 55.48 |
| 28 | 37.92 | 41.34 | 45.42 | 48.28 | 56.89 |
| 29 | 39.09 | 42.56 | 46.69 | 49.59 | 58.30 |
| 30 | 40.26 | 43.77 | 47.96 | 50.89 | 59.70 |

If you have a one-tailed hypothesis, look up your value as usual and simply *halve* the associated $p$ value shown for a two-tailed hypothesis.

**Table C.2** Critical values of *T* (Wilcoxon test) at various levels of probability. (For your *T* value to be significant at a particular probability level, it should be *equal to* or *less than* critical values associated with the *N* in your study)

| N | Level of significance for one-tailed test | | | | N | Level of significance for one-tailed test | | | |
|---|---|---|---|---|---|---|---|---|---|
| | 0.05 | 0.025 | 0.01 | 0.005 | | 0.05 | 0.025 | 0.01 | 0.005 |
| | Level of significance for two-tailed test | | | | | Level of significance for two-tailed test | | | |
| | 0.10 | 0.05 | 0.02 | 0.01 | | 0.10 | 0.05 | 0.02 | 0.01 |
| 5 | 1 | – | – | – | 28 | 130 | 117 | 102 | 92 |
| 6 | 2 | 1 | – | – | 29 | 141 | 127 | 111 | 100 |
| 7 | 4 | 2 | 0 | – | 30 | 152 | 137 | 120 | 109 |
| 8 | 6 | 4 | 2 | 0 | 31 | 163 | 148 | 130 | 118 |
| 9 | 8 | 6 | 3 | 2 | 32 | 175 | 159 | 141 | 128 |
| 10 | 11 | 8 | 5 | 3 | 33 | 188 | 171 | 151 | 138 |
| 11 | 14 | 11 | 7 | 5 | 34 | 201 | 183 | 162 | 149 |
| 12 | 17 | 14 | 10 | 7 | 35 | 214 | 195 | 174 | 160 |
| 13 | 21 | 17 | 13 | 10 | 36 | 228 | 208 | 186 | 171 |
| 14 | 26 | 21 | 16 | 13 | 37 | 242 | 222 | 198 | 183 |
| 15 | 30 | 25 | 20 | 16 | 38 | 256 | 235 | 211 | 195 |
| 16 | 36 | 30 | 24 | 19 | 39 | 271 | 250 | 224 | 208 |
| 17 | 41 | 35 | 28 | 23 | 40 | 287 | 264 | 238 | 221 |
| 18 | 47 | 40 | 33 | 28 | 41 | 303 | 279 | 252 | 234 |
| 19 | 54 | 46 | 38 | 32 | 42 | 319 | 295 | 267 | 248 |
| 20 | 60 | 52 | 43 | 37 | 43 | 336 | 311 | 281 | 262 |
| 21 | 68 | 59 | 49 | 43 | 44 | 353 | 327 | 297 | 277 |
| 22 | 75 | 66 | 56 | 49 | 45 | 371 | 344 | 313 | 292 |
| 23 | 83 | 73 | 62 | 55 | 46 | 389 | 361 | 329 | 307 |
| 24 | 92 | 81 | 69 | 61 | 47 | 408 | 379 | 345 | 323 |
| 25 | 101 | 90 | 77 | 68 | 48 | 427 | 397 | 362 | 339 |
| 26 | 110 | 98 | 85 | 76 | 49 | 446 | 415 | 380 | 356 |
| 27 | 120 | 107 | 93 | 84 | 50 | 466 | 434 | 398 | 373 |

Dashes in the table indicate that no decision is possible at the stated level of significance.

**Table C.3** Critical values of $t$ (related and unrelated $t$ tests) at various levels of probability. (For your $t$ value to be significant at a particular probability level, it should be *equal to* or *larger than* the critical values associated with the d.f. in your study)

| d.f. | 0.10 | 0.05 | 0.025 | 0.01 | 0.005 | 0.0005 |
|---|---|---|---|---|---|---|
| | | | Level of significance for one-tailed test | | | |
| | 0.20 | 0.10 | 0.05 | 0.02 | 0.01 | 0.001 |
| | | | Level of significance for two-tailed test | | | |
| 1 | 3.078 | 6.314 | 12.706 | 31.821 | 63.657 | 636.619 |
| 2 | 1.886 | 2.920 | 4.303 | 6.965 | 9.925 | 31.598 |
| 3 | 1.638 | 2.353 | 3.182 | 4.541 | 5.841 | 12.941 |
| 4 | 1.533 | 2.132 | 2.776 | 3.747 | 4.604 | 8.610 |
| 5 | 1.476 | 2.015 | 2.571 | 3.365 | 4.032 | 6.859 |
| 6 | 1.440 | 1.943 | 2.447 | 3.143 | 3.707 | 5.959 |
| 7 | 1.415 | 1.895 | 2.365 | 2.998 | 3.499 | 5.405 |
| 8 | 1.397 | 1.860 | 2.306 | 2.896 | 3.355 | 5.041 |
| 9 | 1.383 | 1.833 | 2.262 | 2.821 | 3.250 | 4.781 |
| 10 | 1.372 | 1.812 | 2.228 | 2.764 | 3.169 | 4.587 |
| 11 | 1.363 | 1.796 | 2.201 | 2.718 | 3.106 | 4.437 |
| 12 | 1.356 | 1.782 | 2.179 | 2.681 | 3.055 | 4.318 |
| 13 | 1.350 | 1.771 | 2.160 | 2.650 | 3.012 | 4.221 |
| 14 | 1.345 | 1.761 | 2.145 | 2.624 | 2.977 | 4.140 |
| 15 | 1.341 | 1.753 | 2.131 | 2.602 | 2.947 | 4.073 |
| 16 | 1.337 | 1.746 | 2.120 | 2.583 | 2.921 | 4.015 |
| 17 | 1.333 | 1.740 | 2.110 | 2.567 | 2.898 | 3.965 |
| 18 | 1.330 | 1.734 | 2.101 | 2.552 | 2.878 | 3.922 |
| 19 | 1.328 | 1.729 | 2.093 | 2.539 | 2.861 | 3.883 |
| 20 | 1.325 | 1.725 | 2.086 | 2.528 | 2.845 | 3.850 |
| 21 | 1.323 | 1.721 | 2.080 | 2.518 | 2.831 | 3.819 |
| 22 | 1.321 | 1.717 | 2.074 | 2.508 | 2.819 | 3.792 |
| 23 | 1.319 | 1.714 | 2.069 | 2.500 | 2.807 | 3.767 |
| 24 | 1.318 | 1.711 | 2.064 | 2.492 | 2.797 | 3.745 |
| 25 | 1.316 | 1.708 | 2.060 | 2.485 | 2.787 | 3.725 |
| 26 | 1.315 | 1.706 | 2.056 | 2.479 | 2.779 | 3.707 |
| 27 | 1.314 | 1.703 | 2.052 | 2.473 | 2.771 | 3.690 |
| 28 | 1.313 | 1.701 | 2.048 | 2.467 | 2.763 | 3.674 |
| 29 | 1.311 | 1.699 | 2.045 | 2.462 | 2.756 | 3.659 |
| 30 | 1.310 | 1.697 | 2.042 | 2.457 | 2.750 | 3.646 |
| 40 | 1.303 | 1.684 | 2.021 | 2.423 | 2.704 | 3.551 |
| 60 | 1.296 | 1.671 | 2.000 | 2.390 | 2.660 | 3.460 |
| 120 | 1.289 | 1.658 | 1.980 | 2.358 | 2.617 | 3.373 |
| ∞ | 1.282 | 1.645 | 1.960 | 2.326 | 2.576 | 3.291 |

When there is no exact d.f. use the next lowest number, except for very large d.f.s (well over 120), when you should use the infinity row. This is marked ∞.

**Table C.4(a)**  Critical values of $\chi_r^2$ (Friedman test) at various levels of probability for three conditions ($C = 3$). (For your $\chi_r^2$ value to be significant at a particular probability level, it should be *equal to* or *larger than* the critical values associated with the $C$ and $N$ in your study)

| N = 2 | | N = 3 | | N = 4 | | N = 5 | | N = 6 | | N = 7 | | N = 8 | | N = 9 | |
|---|---|---|---|---|---|---|---|---|---|---|---|---|---|---|---|
| $\chi_r^2$ | p | $\chi_r^2$ | p | $\chi_r^2$ | p | $\chi_r^2$ | p | $\chi_r^2$ | p | $\chi_r^2$ | p | $\chi_r^2$ | p | $\chi_r^2$ | p |
| 0 | 1.000 | 0.000 | 1.000 | 0.0 | 1.000 | 0.0 | 1.000 | 0.00 | 1.000 | 0.000 | 1.000 | 0.00 | 1.000 | 0.000 | 1.000 |
| 1 | 0.833 | 0.667 | 0.944 | 0.5 | 0.931 | 0.4 | 0.954 | 0.33 | 0.956 | 0.286 | 0.964 | 0.25 | 0.967 | 0.222 | 0.971 |
| 3 | 0.500 | 2.000 | 0.528 | 1.5 | 0.653 | 1.2 | 0.691 | 1.00 | 0.740 | 0.857 | 0.768 | 0.75 | 0.794 | 0.667 | 0.814 |
| 4 | 0.167 | 2.667 | 0.361 | 2.0 | 0.431 | 1.6 | 0.522 | 1.33 | 0.570 | 1.143 | 0.620 | 1.00 | 0.654 | 0.889 | 0.865 |
| | | 4.667 | 0.194 | 3.5 | 0.273 | 2.8 | 0.367 | 2.33 | 0.430 | 2.000 | 0.486 | 1.75 | 0.531 | 1.556 | 0.569 |
| | | 6.000 | 0.028 | 4.5 | 0.125 | 3.6 | 0.182 | 3.00 | 0.252 | 2.571 | 0.305 | 2.25 | 0.355 | 2.000 | 0.398 |
| | | | | 6.0 | 0.069 | 4.8 | 0.124 | 4.00 | 0.184 | 3.429 | 0.237 | 3.00 | 0.285 | 2.667 | 0.328 |
| | | | | 6.5 | 0.042 | 5.2 | 0.093 | 4.33 | 0.142 | 3.714 | 0.192 | 3.25 | 0.236 | 2.889 | 0.278 |
| | | | | 8.0 | 0.0046 | 6.4 | 0.039 | 5.33 | 0.072 | 4.571 | 0.112 | 4.00 | 0.149 | 3.556 | 0.187 |
| | | | | | | 7.6 | 0.024 | 6.33 | 0.052 | 5.429 | 0.085 | 4.75 | 0.120 | 4.222 | 0.154 |
| | | | | | | 8.4 | 0.0085 | 7.00 | 0.029 | 6.000 | 0.052 | 5.25 | 0.079 | 4.667 | 0.107 |
| | | | | | | 10.0 | 0.00077 | 8.33 | 0.012 | 7.143 | 0.027 | 6.25 | 0.047 | 5.556 | 0.069 |
| | | | | | | | | 9.00 | 0.0081 | 7.714 | 0.021 | 6.75 | 0.038 | 6.000 | 0.057 |
| | | | | | | | | 9.33 | 0.0055 | 8.000 | 0.016 | 7.00 | 0.030 | 6.222 | 0.048 |
| | | | | | | | | 10.33 | 0.0017 | 8.857 | 0.0084 | 7.75 | 0.018 | 6.889 | 0.031 |
| | | | | | | | | 12.00 | 0.00013 | 10.286 | 0.0036 | 9.00 | 0.0099 | 8.000 | 0.019 |
| | | | | | | | | | | 10.571 | 0.0027 | 9.25 | 0.0080 | 8.222 | 0.016 |
| | | | | | | | | | | 11.143 | 0.0012 | 9.75 | 0.0048 | 8.667 | 0.010 |
| | | | | | | | | | | 12.286 | 0.00032 | 10.75 | 0.0024 | 9.556 | 0.0060 |
| | | | | | | | | | | 14.000 | 0.000021 | 12.00 | 0.0011 | 10.667 | 0.0035 |
| | | | | | | | | | | | | 12.25 | 0.00086 | 10.889 | 0.0029 |
| | | | | | | | | | | | | 13.00 | 0.00026 | 11.556 | 0.0013 |
| | | | | | | | | | | | | 14.25 | 0.000061 | 12.667 | 0.00066 |
| | | | | | | | | | | | | 16.00 | 0.0000036 | 13.556 | 0.00035 |
| | | | | | | | | | | | | | | 14.000 | 0.00020 |
| | | | | | | | | | | | | | | 14.222 | 0.000097 |
| | | | | | | | | | | | | | | 14.889 | 0.000054 |
| | | | | | | | | | | | | | | 16.222 | 0.000011 |
| | | | | | | | | | | | | | | 18.000 | 0.0000006 |

These values are all for a two-tailed test only.

**Table C.4(b)** Critical values of $\chi_r^2$ (Friedman test) at various levels of probability for four conditions ($C = 4$). (For your $\chi_r^2$ value to be significant at a particular probability level, it should be *equal to* or *larger than* the critical values associated with the $C$ and $N$ in your study)

| N = 2 | | N = 3 | | N = 4 | | | |
|---|---|---|---|---|---|---|---|
| $\chi_r^2$ | $p$ | $\chi_r^2$ | $p$ | $\chi_r^2$ | $p$ | $\chi_r^2$ | $p$ |
| 0.0 | 1.000 | 0.0 | 1.000 | 0.0 | 1.000 | 5.7 | 0.141 |
| 0.6 | 0.958 | 0.6 | 0.958 | 0.3 | 0.992 | 6.0 | 0.105 |
| 1.2 | 0.834 | 1.0 | 0.910 | 0.6 | 0.928 | 6.3 | 0.094 |
| 1.8 | 0.792 | 1.8 | 0.727 | 0.9 | 0.900 | 6.6 | 0.077 |
| 2.4 | 0.625 | 2.2 | 0.608 | 1.2 | 0.800 | 6.9 | 0.068 |
| 3.0 | 0.542 | 2.6 | 0.524 | 1.5 | 0.754 | 7.2 | 0.054 |
| 3.6 | 0.458 | 3.4 | 0.446 | 1.8 | 0.677 | 7.5 | 0.052 |
| 4.2 | 0.375 | 3.8 | 0.342 | 2.1 | 0.649 | 7.8 | 0.036 |
| 4.8 | 0.208 | 4.2 | 0.300 | 2.4 | 0.524 | 8.1 | 0.033 |
| 5.4 | 0.167 | 5.0 | 0.207 | 2.7 | 0.508 | 8.4 | 0.019 |
| 6.0 | 0.042 | 5.4 | 0.175 | 3.0 | 0.432 | 8.7 | 0.014 |
| | | 5.8 | 0.148 | 3.3 | 0.389 | 9.3 | 0.012 |
| | | 6.6 | 0.075 | 3.6 | 0.355 | 9.6 | 0.006 9 |
| | | 7.0 | 0.054 | 3.9 | 0.324 | 9.9 | 0.006 2 |
| | | 7.4 | 0.033 | 4.5 | 0.242 | 10.2 | 0.002 7 |
| | | 8.2 | 0.017 | 4.8 | 0.200 | 10.8 | 0.001 6 |
| | | 9.0 | 0.001 7 | 5.1 | 0.190 | 11.1 | 0.000 94 |
| | | | | 5.4 | 0.158 | 12.0 | 0.000 072 |

These values are all for a two-tailed test only.

**Table C.5** Critical values of $L$ (Page's $L$ trend test) at various levels of probability. (For your $L$ value to be significant at a particular probability level, it should be *equal to* or *larger than* the critical values associated with the $C$ and $N$ in your study)

| $N$ | \multicolumn{4}{c}{$C$ (no. of conditions)} | $p <$ |
| --- | --- | --- | --- | --- | --- |
|  | 3 | 4 | 5 | 6 |  |
| 2 | – | – | 109 | 178 | 0.001 |
|  | – | 60 | 106 | 173 | 0.01 |
|  | 28 | 58 | 103 | 166 | 0.05 |
| 3 | – | 89 | 160 | 260 | 0.001 |
|  | 42 | 87 | 155 | 252 | 0.01 |
|  | 41 | 84 | 150 | 244 | 0.05 |
| 4 | 56 | 117 | 210 | 341 | 0.001 |
|  | 55 | 114 | 204 | 331 | 0.01 |
|  | 54 | 111 | 197 | 321 | 0.05 |
| 5 | 70 | 145 | 259 | 420 | 0.001 |
|  | 68 | 141 | 251 | 409 | 0.01 |
|  | 66 | 137 | 244 | 397 | 0.05 |
| 6 | 83 | 172 | 307 | 499 | 0.001 |
|  | 81 | 167 | 299 | 486 | 0.01 |
|  | 79 | 163 | 291 | 474 | 0.05 |
| 7 | 96 | 198 | 355 | 577 | 0.001 |
|  | 93 | 193 | 346 | 563 | 0.01 |
|  | 91 | 189 | 338 | 550 | 0.05 |
| 8 | 109 | 225 | 403 | 655 | 0.001 |
|  | 106 | 220 | 393 | 640 | 0.01 |
|  | 104 | 214 | 384 | 625 | 0.05 |
| 9 | 121 | 252 | 451 | 733 | 0.001 |
|  | 119 | 246 | 441 | 717 | 0.01 |
|  | 116 | 240 | 431 | 701 | 0.05 |
| 10 | 134 | 278 | 499 | 811 | 0.001 |
|  | 131 | 272 | 487 | 793 | 0.01 |
|  | 128 | 266 | 477 | 777 | 0.05 |
| 11 | 147 | 305 | 546 | 888 | 0.001 |
|  | 144 | 298 | 534 | 869 | 0.01 |
|  | 141 | 292 | 523 | 852 | 0.05 |
| 12 | 160 | 331 | 593 | 965 | 0.001 |
|  | 156 | 324 | 581 | 946 | 0.01 |
|  | 153 | 317 | 570 | 928 | 0.05 |

These values are all for a one-tailed test only.

**Table C.6** Critical values of $F$ (anovas) at various levels of probability. (For your $F$ value to be significant at a particular probability level, it should be *equal to or larger than* the critical values associated with $v_1$ and $v_2$ in your study)

*(a) Critical values of F at p < 0.05*

| $v_2$ | $v_1$ 1 | 2 | 3 | 4 | 5 | 6 | 7 | 8 | 10 | 12 | 24 | ∞ |
|---|---|---|---|---|---|---|---|---|---|---|---|---|
| 1 | 161.4 | 199.5 | 215.7 | 224.6 | 230.2 | 234.0 | 236.8 | 238.9 | 241.9 | 243.9 | 249.0 | 254.3 |
| 2 | 18.5 | 19.0 | 19.2 | 19.2 | 19.3 | 19.3 | 19.4 | 19.4 | 19.4 | 19.4 | 19.5 | 19.5 |
| 3 | 10.13 | 9.55 | 9.28 | 9.12 | 9.01 | 8.94 | 8.89 | 8.85 | 8.79 | 8.74 | 8.64 | 8.53 |
| 4 | 7.71 | 6.94 | 6.59 | 6.39 | 6.26 | 6.16 | 6.09 | 6.04 | 5.96 | 5.91 | 5.77 | 5.63 |
| 5 | 6.61 | 5.79 | 5.41 | 5.19 | 5.05 | 4.95 | 4.88 | 4.82 | 4.74 | 4.68 | 4.53 | 4.36 |
| 6 | 5.99 | 5.14 | 4.76 | 4.53 | 4.39 | 4.28 | 4.21 | 4.15 | 4.06 | 4.00 | 3.84 | 3.67 |
| 7 | 5.59 | 4.74 | 4.35 | 4.12 | 3.97 | 3.87 | 3.79 | 3.73 | 3.64 | 3.57 | 3.41 | 3.23 |
| 8 | 5.32 | 4.46 | 4.07 | 3.84 | 3.69 | 3.58 | 3.50 | 3.44 | 3.35 | 3.28 | 3.12 | 2.93 |
| 9 | 5.12 | 4.26 | 3.86 | 3.63 | 3.48 | 3.37 | 3.29 | 3.23 | 3.14 | 3.07 | 2.90 | 2.71 |
| 10 | 4.96 | 4.10 | 3.71 | 3.48 | 3.33 | 3.22 | 3.14 | 3.07 | 2.98 | 2.91 | 2.74 | 2.54 |
| 11 | 4.84 | 3.98 | 3.59 | 3.36 | 3.20 | 3.09 | 3.01 | 2.95 | 2.85 | 2.79 | 2.61 | 2.40 |
| 12 | 4.75 | 3.89 | 3.49 | 3.26 | 3.11 | 3.00 | 2.91 | 2.85 | 2.75 | 2.69 | 2.51 | 2.30 |
| 13 | 4.67 | 3.81 | 3.41 | 3.18 | 3.03 | 2.92 | 2.83 | 2.77 | 2.67 | 2.60 | 2.42 | 2.21 |
| 14 | 4.60 | 3.74 | 3.34 | 3.11 | 2.96 | 2.85 | 2.76 | 2.70 | 2.60 | 2.53 | 2.35 | 2.13 |
| 15 | 4.54 | 3.68 | 3.29 | 3.06 | 2.90 | 2.79 | 2.71 | 2.64 | 2.54 | 2.48 | 2.29 | 2.07 |
| 16 | 4.49 | 3.63 | 3.24 | 3.01 | 2.85 | 2.74 | 2.66 | 2.59 | 2.49 | 2.42 | 2.24 | 2.01 |
| 17 | 4.45 | 3.59 | 3.20 | 2.96 | 2.81 | 2.70 | 2.61 | 2.55 | 2.45 | 2.38 | 2.19 | 1.96 |
| 18 | 4.41 | 3.55 | 3.16 | 2.93 | 2.77 | 2.66 | 2.58 | 2.51 | 2.41 | 2.34 | 2.15 | 1.92 |
| 19 | 4.38 | 3.52 | 3.13 | 2.90 | 2.74 | 2.63 | 2.54 | 2.48 | 2.38 | 2.31 | 2.11 | 1.88 |

| | | | | | | | | | | | | |
|---|---|---|---|---|---|---|---|---|---|---|---|---|
| 20 | 4.35 | 3.49 | 3.10 | 2.87 | 2.71 | 2.60 | 2.51 | 2.45 | 2.35 | 2.28 | 2.08 | 1.84 |
| 21 | 4.32 | 3.47 | 3.07 | 2.84 | 2.68 | 2.57 | 2.49 | 2.42 | 2.32 | 2.25 | 2.05 | 1.81 |
| 22 | 4.30 | 3.44 | 3.05 | 2.82 | 2.66 | 2.55 | 2.46 | 2.40 | 2.30 | 2.23 | 2.03 | 1.78 |
| 23 | 4.28 | 3.42 | 3.03 | 2.80 | 2.64 | 2.53 | 2.44 | 2.37 | 2.27 | 2.20 | 2.00 | 1.76 |
| 24 | 4.26 | 3.40 | 3.01 | 2.78 | 2.62 | 2.51 | 2.42 | 2.36 | 2.25 | 2.18 | 1.98 | 1.73 |
| 25 | 4.24 | 3.39 | 2.99 | 2.76 | 2.60 | 2.49 | 2.40 | 2.34 | 2.24 | 2.16 | 1.96 | 1.71 |
| 26 | 4.23 | 3.37 | 2.98 | 2.74 | 2.59 | 2.47 | 2.39 | 2.32 | 2.22 | 2.15 | 1.95 | 1.69 |
| 27 | 4.21 | 3.35 | 2.96 | 2.73 | 2.57 | 2.46 | 2.37 | 2.31 | 2.20 | 2.13 | 1.93 | 1.67 |
| 28 | 4.20 | 3.34 | 2.95 | 2.71 | 2.56 | 2.45 | 2.36 | 2.29 | 2.19 | 2.12 | 1.91 | 1.65 |
| 29 | 4.18 | 3.33 | 2.93 | 2.70 | 2.55 | 2.43 | 2.35 | 2.28 | 2.18 | 2.10 | 1.90 | 1.64 |
| 30 | 4.17 | 3.32 | 2.92 | 2.69 | 2.53 | 2.42 | 2.33 | 2.27 | 2.16 | 2.09 | 1.89 | 1.62 |
| 32 | 4.15 | 3.29 | 2.90 | 2.67 | 2.51 | 2.40 | 2.31 | 2.24 | 2.14 | 2.07 | 1.86 | 1.59 |
| 34 | 4.13 | 3.28 | 2.88 | 2.65 | 2.49 | 2.38 | 2.29 | 2.23 | 2.12 | 2.05 | 1.84 | 1.57 |
| 36 | 4.11 | 3.26 | 2.87 | 2.63 | 2.48 | 2.36 | 2.28 | 2.21 | 2.11 | 2.03 | 1.82 | 1.55 |
| 38 | 4.10 | 3.24 | 2.85 | 2.62 | 2.46 | 2.35 | 2.26 | 2.19 | 2.09 | 2.02 | 1.81 | 1.53 |
| 40 | 4.08 | 3.23 | 2.84 | 2.61 | 2.45 | 2.34 | 2.25 | 2.18 | 2.08 | 2.00 | 1.79 | 1.51 |
| 60 | 4.00 | 3.15 | 2.76 | 2.53 | 2.37 | 2.25 | 2.17 | 2.10 | 1.99 | 1.92 | 1.70 | 1.39 |
| 120 | 3.92 | 3.07 | 2.68 | 2.45 | 2.29 | 2.18 | 2.09 | 2.02 | 1.91 | 1.83 | 1.61 | 1.25 |
| ∞ | 3.84 | 3.00 | 2.60 | 2.37 | 2.21 | 2.10 | 2.01 | 1.94 | 1.83 | 1.75 | 1.52 | 1.00 |

When there is no exact number for the d.f., use the next lowest number. For very large d.f.s (i.e. well over 120) you should use the row for infinity, marked ∞.

These values are all for a two-tailed test only.

**Table C.6** Critical values of $F$ (anovas) at various levels of probability. (For your $F$ value to be significant at a particular probability level, it should be *equal to or larger than* the critical values associated with $v_1$ and $v_2$ in your study)

(b) *Critical values of $F$ at $p < 0.025$*

| $v_2$ | \( v_1 \) | | | | | | | | | | | |
|---|---|---|---|---|---|---|---|---|---|---|---|---|
| | *1* | *2* | *3* | *4* | *5* | *6* | *7* | *8* | *10* | *12* | *24* | *∞* |
| 1 | 648 | 800 | 864 | 900 | 922 | 937 | 948 | 957 | 969 | 977 | 997 | 1018 |
| 2 | 38.5 | 39.0 | 39.2 | 39.2 | 39.3 | 39.3 | 39.4 | 39.4 | 39.4 | 39.4 | 39.5 | 39.5 |
| 3 | 17.4 | 16.0 | 15.4 | 15.1 | 14.9 | 14.7 | 14.6 | 14.5 | 14.4 | 14.3 | 14.1 | 13.9 |
| 4 | 12.22 | 10.65 | 9.98 | 9.60 | 9.36 | 9.20 | 9.07 | 8.98 | 8.84 | 8.75 | 8.51 | 8.26 |
| 5 | 10.01 | 8.43 | 7.76 | 7.39 | 7.15 | 6.98 | 6.85 | 6.76 | 6.62 | 6.52 | 6.28 | 6.02 |
| 6 | 8.81 | 7.26 | 6.60 | 6.23 | 5.99 | 5.82 | 5.70 | 5.60 | 5.46 | 5.37 | 5.12 | 4.85 |
| 7 | 8.07 | 6.54 | 5.89 | 5.52 | 5.29 | 5.12 | 4.99 | 4.90 | 4.76 | 4.67 | 4.42 | 4.14 |
| 8 | 7.57 | 6.06 | 5.42 | 5.05 | 4.82 | 4.65 | 4.53 | 4.43 | 4.30 | 4.20 | 3.95 | 3.67 |
| 9 | 7.21 | 5.71 | 5.08 | 4.72 | 4.48 | 4.32 | 4.20 | 4.10 | 3.96 | 3.87 | 3.61 | 3.33 |
| 10 | 6.94 | 5.46 | 4.83 | 4.47 | 4.24 | 4.07 | 3.95 | 3.85 | 3.72 | 3.62 | 3.37 | 3.08 |
| 11 | 6.72 | 5.26 | 4.63 | 4.28 | 4.04 | 3.88 | 3.76 | 3.66 | 3.53 | 3.43 | 3.17 | 2.88 |
| 12 | 6.55 | 5.10 | 4.47 | 4.12 | 3.89 | 3.73 | 3.61 | 3.51 | 3.37 | 3.28 | 3.02 | 2.72 |
| 13 | 6.41 | 4.97 | 4.35 | 4.00 | 3.77 | 3.60 | 3.48 | 3.39 | 3.25 | 3.15 | 2.89 | 2.60 |
| 14 | 6.30 | 4.86 | 4.24 | 3.89 | 3.66 | 3.50 | 3.38 | 3.29 | 3.15 | 3.05 | 2.79 | 2.49 |
| 15 | 6.20 | 4.76 | 4.15 | 3.80 | 3.58 | 3.41 | 3.29 | 3.20 | 3.06 | 2.96 | 2.70 | 2.40 |
| 16 | 6.12 | 4.69 | 4.08 | 3.73 | 3.50 | 3.34 | 3.22 | 3.12 | 2.99 | 2.89 | 2.63 | 2.32 |
| 17 | 6.04 | 4.62 | 4.01 | 3.66 | 3.44 | 3.28 | 3.16 | 3.06 | 2.92 | 2.82 | 2.56 | 2.25 |
| 18 | 5.98 | 4.56 | 3.95 | 3.61 | 3.38 | 3.22 | 3.10 | 3.01 | 2.87 | 2.77 | 2.50 | 2.19 |
| 19 | 5.92 | 4.51 | 3.90 | 3.56 | 3.33 | 3.17 | 3.05 | 2.96 | 2.82 | 2.72 | 2.45 | 2.13 |

| | | | | | | | | | | | | |
|---|---|---|---|---|---|---|---|---|---|---|---|---|
| 20 | 5.87 | 4.46 | 3.86 | 3.51 | 3.29 | 3.13 | 3.01 | 2.91 | 2.77 | 2.68 | 2.41 | 2.09 |
| 21 | 5.83 | 4.42 | 3.82 | 3.48 | 3.25 | 3.09 | 2.97 | 2.87 | 2.73 | 2.64 | 2.37 | 2.04 |
| 22 | 5.79 | 4.38 | 3.78 | 3.44 | 3.22 | 3.05 | 2.93 | 2.84 | 2.70 | 2.60 | 2.33 | 2.00 |
| 23 | 5.75 | 4.35 | 3.75 | 3.41 | 3.18 | 3.02 | 2.90 | 2.81 | 2.67 | 2.57 | 2.30 | 1.97 |
| 24 | 5.72 | 4.32 | 3.72 | 3.38 | 3.15 | 2.99 | 2.87 | 2.78 | 2.64 | 2.54 | 2.27 | 1.94 |
| 25 | 5.69 | 4.29 | 3.69 | 3.35 | 3.13 | 2.97 | 2.85 | 2.75 | 2.61 | 2.51 | 2.24 | 1.91 |
| 26 | 5.66 | 4.27 | 3.67 | 3.33 | 3.10 | 2.94 | 2.82 | 2.73 | 2.59 | 2.49 | 2.22 | 1.88 |
| 27 | 5.63 | 4.24 | 3.65 | 3.31 | 3.08 | 2.92 | 2.80 | 2.71 | 2.57 | 2.47 | 2.19 | 1.85 |
| 28 | 5.61 | 4.22 | 3.63 | 3.29 | 3.06 | 2.90 | 2.78 | 2.69 | 2.55 | 2.45 | 2.17 | 1.83 |
| 29 | 5.59 | 4.20 | 3.61 | 3.27 | 3.04 | 2.88 | 2.76 | 2.67 | 2.53 | 2.43 | 2.15 | 1.81 |
| 30 | 5.57 | 4.18 | 3.59 | 3.25 | 3.03 | 2.87 | 2.75 | 2.65 | 2.51 | 2.41 | 2.14 | 1.79 |
| 32 | 5.53 | 4.15 | 3.56 | 3.22 | 3.00 | 2.84 | 2.72 | 2.62 | 2.48 | 2.38 | 2.10 | 1.75 |
| 34 | 5.50 | 4.12 | 3.53 | 3.19 | 2.97 | 2.81 | 2.69 | 2.59 | 2.45 | 2.35 | 2.08 | 1.72 |
| 36 | 5.47 | 4.09 | 3.51 | 3.17 | 2.94 | 2.79 | 2.66 | 2.57 | 2.43 | 2.33 | 2.05 | 1.69 |
| 38 | 5.45 | 4.07 | 3.48 | 3.15 | 2.92 | 2.76 | 2.64 | 2.55 | 2.41 | 2.31 | 2.03 | 1.66 |
| 40 | 5.42 | 4.05 | 3.46 | 3.13 | 2.90 | 2.74 | 2.62 | 2.53 | 2.39 | 2.29 | 2.01 | 1.64 |
| 60 | 5.29 | 3.93 | 3.34 | 3.01 | 2.79 | 2.63 | 2.51 | 2.41 | 2.27 | 2.17 | 1.88 | 1.48 |
| 120 | 5.15 | 3.80 | 3.23 | 2.89 | 2.67 | 2.52 | 2.39 | 2.30 | 2.16 | 2.05 | 1.76 | 1.31 |
| ∞ | 5.02 | 3.69 | 3.12 | 2.79 | 2.57 | 2.41 | 2.29 | 2.19 | 2.05 | 1.94 | 1.64 | 1.00 |

When there is no exact number for the d.f., use the next lowest number. For very large d.f.s (i.e. well over 120) you should use the row for infinity, marked ∞.

These values are all for a two-tailed test only.

**Table C.6** Critical values of $F$ (anovas) at various levels of probability. (For your $F$ value to be significant at a particular probability level, it should be *equal to* or *larger than* the critical values associated with $v_1$ and $v_2$ in your study)

*(c) Critical values of $F$ at $p < 0.01$*

| $v_2$ | | | | | | $v_1$ | | | | | | |
|---|---|---|---|---|---|---|---|---|---|---|---|---|
| | 1 | 2 | 3 | 4 | 5 | 6 | 7 | 8 | 10 | 12 | 24 | ∞ |
| 1 | 4052 | 5000 | 5403 | 5625 | 5764 | 5859 | 5928 | 5981 | 6056 | 6106 | 6235 | 6366 |
| 2 | 98.5 | 99.0 | 99.2 | 99.2 | 99.3 | 99.3 | 99.4 | 99.4 | 99.4 | 99.4 | 99.5 | 99.5 |
| 3 | 34.1 | 30.8 | 29.5 | 28.7 | 28.2 | 27.9 | 27.7 | 27.5 | 27.2 | 27.1 | 26.6 | 26.1 |
| 4 | 21.2 | 18.0 | 16.7 | 16.0 | 15.5 | 15.2 | 15.0 | 14.8 | 14.5 | 14.4 | 13.9 | 13.5 |
| 5 | 16.26 | 13.27 | 12.06 | 11.39 | 10.97 | 10.67 | 10.46 | 10.29 | 10.05 | 9.89 | 9.47 | 9.02 |
| 6 | 13.74 | 10.92 | 9.78 | 9.15 | 8.75 | 8.47 | 8.26 | 8.10 | 7.87 | 7.72 | 7.31 | 6.88 |
| 7 | 12.25 | 9.55 | 8.45 | 7.85 | 7.46 | 7.19 | 6.99 | 6.84 | 6.62 | 6.47 | 6.07 | 5.65 |
| 8 | 11.26 | 8.65 | 7.59 | 7.01 | 6.63 | 6.37 | 6.18 | 6.03 | 5.81 | 5.67 | 5.28 | 4.86 |
| 9 | 10.56 | 8.02 | 6.99 | 6.42 | 6.06 | 5.80 | 5.61 | 5.47 | 5.26 | 5.11 | 4.73 | 4.31 |
| 10 | 10.04 | 7.56 | 6.55 | 5.99 | 5.64 | 5.39 | 5.20 | 5.06 | 4.85 | 4.71 | 4.33 | 3.91 |
| 11 | 9.65 | 7.21 | 6.22 | 5.67 | 5.32 | 5.07 | 4.89 | 4.74 | 4.54 | 4.40 | 4.02 | 3.60 |
| 12 | 9.33 | 6.93 | 5.95 | 5.41 | 5.06 | 4.82 | 4.64 | 4.50 | 4.30 | 4.16 | 3.78 | 3.36 |
| 13 | 9.07 | 6.70 | 5.74 | 5.21 | 4.86 | 4.62 | 4.44 | 4.30 | 4.10 | 3.96 | 3.59 | 3.17 |
| 14 | 8.86 | 6.51 | 5.56 | 5.04 | 4.70 | 4.46 | 4.28 | 4.14 | 3.94 | 3.80 | 3.43 | 3.00 |
| 15 | 8.68 | 6.36 | 5.42 | 4.89 | 4.56 | 4.32 | 4.14 | 4.00 | 3.80 | 3.67 | 3.29 | 2.87 |
| 16 | 8.53 | 6.23 | 5.29 | 4.77 | 4.44 | 4.20 | 4.03 | 3.89 | 3.69 | 3.55 | 3.18 | 2.75 |
| 17 | 8.40 | 6.11 | 5.18 | 4.67 | 4.34 | 4.10 | 3.93 | 3.79 | 3.59 | 3.46 | 3.08 | 2.65 |
| 18 | 8.29 | 6.01 | 5.09 | 4.58 | 4.25 | 4.01 | 3.84 | 3.71 | 3.51 | 3.37 | 3.00 | 2.57 |
| 19 | 8.18 | 5.93 | 5.01 | 4.50 | 4.17 | 3.94 | 3.77 | 3.63 | 3.43 | 3.30 | 2.92 | 2.49 |

| | | | | | | | | | | | |
|---|---|---|---|---|---|---|---|---|---|---|---|
| 20 | 2.42 | 2.86 | 3.23 | 3.37 | 3.56 | 3.70 | 3.87 | 4.10 | 4.43 | 4.94 | 5.85 | 8.10 |
| 21 | 2.36 | 2.80 | 3.17 | 3.31 | 3.51 | 3.64 | 3.81 | 4.04 | 4.37 | 4.87 | 5.78 | 8.02 |
| 22 | 2.31 | 2.75 | 3.12 | 3.26 | 3.45 | 3.59 | 3.76 | 3.99 | 4.31 | 4.82 | 5.72 | 7.95 |
| 23 | 2.26 | 2.70 | 3.07 | 3.21 | 3.41 | 3.54 | 3.71 | 3.94 | 4.26 | 4.76 | 5.66 | 7.88 |
| 24 | 2.21 | 2.66 | 3.03 | 3.17 | 3.36 | 3.50 | 3.67 | 3.90 | 4.22 | 4.72 | 5.61 | 7.82 |
| 25 | 2.17 | 2.62 | 2.99 | 3.13 | 3.32 | 3.46 | 3.63 | 3.86 | 4.18 | 4.68 | 5.57 | 7.77 |
| 26 | 2.13 | 2.58 | 2.96 | 3.09 | 3.29 | 3.42 | 3.59 | 3.82 | 4.14 | 4.64 | 5.53 | 7.72 |
| 27 | 2.10 | 2.55 | 2.93 | 3.06 | 3.26 | 3.39 | 3.56 | 3.78 | 4.11 | 4.60 | 5.49 | 7.68 |
| 28 | 2.06 | 2.52 | 2.90 | 3.03 | 3.23 | 3.36 | 3.53 | 3.75 | 4.07 | 4.57 | 5.45 | 7.64 |
| 29 | 2.03 | 2.49 | 2.87 | 3.00 | 3.20 | 3.33 | 3.50 | 3.73 | 4.04 | 4.54 | 5.42 | 7.60 |
| 30 | 2.01 | 2.47 | 2.84 | 2.98 | 3.17 | 3.30 | 3.47 | 3.70 | 4.02 | 4.51 | 5.39 | 7.56 |
| 32 | 1.96 | 2.42 | 2.80 | 2.93 | 3.13 | 3.26 | 3.43 | 3.65 | 3.97 | 4.46 | 5.34 | 7.50 |
| 34 | 1.91 | 2.38 | 2.76 | 2.90 | 3.09 | 3.22 | 3.39 | 3.61 | 3.93 | 4.42 | 5.29 | 7.45 |
| 36 | 1.87 | 2.35 | 2.72 | 2.86 | 3.05 | 3.18 | 3.35 | 3.58 | 3.89 | 4.38 | 5.25 | 7.40 |
| 38 | 1.84 | 2.32 | 2.69 | 2.83 | 3.02 | 3.15 | 3.32 | 3.54 | 3.86 | 4.34 | 5.21 | 7.35 |
| 40 | 1.80 | 2.29 | 2.66 | 2.80 | 2.99 | 3.12 | 3.29 | 3.51 | 3.83 | 4.31 | 5.18 | 7.31 |
| 60 | 1.60 | 2.12 | 2.50 | 2.63 | 2.82 | 2.95 | 3.12 | 3.34 | 3.65 | 4.13 | 4.98 | 7.08 |
| 120 | 1.38 | 1.95 | 2.34 | 2.47 | 2.66 | 2.79 | 2.96 | 3.17 | 3.48 | 3.95 | 4.79 | 6.85 |
| ∞ | 1.00 | 1.79 | 2.18 | 2.32 | 2.51 | 2.64 | 2.80 | 3.02 | 3.32 | 3.78 | 4.61 | 6.63 |

When there is no exact number for the d.f., use the next lowest number. For very large d.f.s (i.e. well over 120) you should use the row for infinity, marked ∞.

These values are all for a two-tailed test only.

**Table C.6** Critical values of $F$ (anovas) at various levels of probability. (For your $F$ value to be significant at a particular probability level, it should be *equal to or larger than* the critical values associated with $v_1$ and $v_2$ in your study)

*(d) Critical values of F at p < 0.001*

| $v_2$ | $v_1$ | | | | | | | | | | | |
|---|---|---|---|---|---|---|---|---|---|---|---|---|
| | *1* | *2* | *3* | *4* | *5* | *6* | *7* | *8* | *10* | *12* | *24* | *∞* |
| 1* | 4053 | 5000 | 5404 | 5625 | 5764 | 5859 | 5929 | 5981 | 6056 | 6107 | 6235 | 6366* |
| 2 | 998.5 | 999.0 | 999.2 | 999.2 | 999.3 | 999.3 | 999.4 | 999.4 | 999.4 | 999.4 | 999.5 | 999.5 |
| 3 | 167.0 | 148.5 | 141.1 | 137.1 | 134.6 | 132.8 | 131.5 | 130.6 | 129.2 | 128.3 | 125.9 | 123.5 |
| 4 | 74.14 | 61.25 | 56.18 | 53.44 | 51.71 | 50.53 | 49.66 | 49.00 | 48.05 | 47.41 | 45.77 | 44.05 |
| 5 | 47.18 | 37.12 | 33.20 | 31.09 | 29.75 | 28.83 | 28.16 | 27.65 | 26.92 | 26.42 | 25.14 | 23.79 |
| 6 | 35.51 | 27.00 | 23.70 | 21.92 | 20.80 | 20.03 | 19.46 | 19.03 | 18.41 | 17.99 | 16.90 | 15.75 |
| 7 | 29.25 | 21.69 | 18.77 | 17.20 | 16.21 | 15.52 | 15.02 | 14.63 | 14.08 | 13.71 | 12.73 | 11.70 |
| 8 | 25.42 | 18.49 | 15.83 | 14.39 | 13.48 | 12.86 | 12.40 | 12.05 | 11.54 | 11.19 | 10.30 | 9.34 |
| 9 | 22.86 | 16.39 | 13.90 | 12.56 | 11.71 | 11.13 | 10.69 | 10.37 | 9.87 | 9.57 | 8.72 | 7.81 |
| 10 | 21.04 | 14.91 | 12.55 | 11.28 | 10.48 | 9.93 | 9.52 | 9.20 | 8.74 | 8.44 | 7.64 | 6.76 |
| 11 | 19.69 | 13.81 | 11.56 | 10.35 | 9.58 | 9.05 | 8.66 | 8.35 | 7.92 | 7.63 | 6.85 | 6.00 |
| 12 | 18.64 | 12.97 | 10.80 | 9.63 | 8.89 | 8.38 | 8.00 | 7.71 | 7.29 | 7.00 | 6.25 | 5.42 |
| 13 | 17.82 | 12.31 | 10.21 | 9.07 | 8.35 | 7.86 | 7.49 | 7.21 | 6.80 | 6.52 | 5.78 | 4.97 |
| 14 | 17.14 | 11.78 | 9.73 | 8.62 | 7.92 | 7.44 | 7.08 | 6.80 | 6.40 | 6.13 | 5.41 | 4.60 |
| 15 | 16.59 | 11.34 | 9.34 | 8.25 | 7.57 | 7.09 | 6.74 | 6.47 | 6.08 | 5.81 | 5.10 | 4.31 |
| 16 | 16.12 | 10.97 | 9.01 | 7.94 | 7.27 | 6.80 | 6.46 | 6.19 | 5.81 | 5.55 | 4.85 | 4.06 |
| 17 | 15.72 | 10.66 | 8.73 | 7.68 | 7.02 | 6.56 | 6.22 | 5.96 | 5.58 | 5.32 | 4.63 | 3.85 |
| 18 | 15.38 | 10.39 | 8.49 | 7.46 | 6.81 | 6.35 | 6.02 | 5.76 | 5.39 | 5.13 | 4.45 | 3.67 |
| 19 | 15.08 | 10.16 | 8.28 | 7.27 | 6.62 | 6.18 | 5.85 | 5.59 | 5.22 | 4.97 | 4.29 | 3.51 |

| | | | | | | | | | | | | |
|---|---|---|---|---|---|---|---|---|---|---|---|---|
| 20 | 14.82 | 9.95 | 8.10 | 7.10 | 6.46 | 6.02 | 5.69 | 5.44 | 5.08 | 4.82 | 4.15 | 3.38 |
| 21 | 14.59 | 9.77 | 7.94 | 6.95 | 6.32 | 5.88 | 5.56 | 5.31 | 4.95 | 4.70 | 4.03 | 3.26 |
| 22 | 14.38 | 9.61 | 7.80 | 6.81 | 6.19 | 5.76 | 5.44 | 5.19 | 4.83 | 4.58 | 3.92 | 3.15 |
| 23 | 14.19 | 9.47 | 7.67 | 6.70 | 6.08 | 5.65 | 5.33 | 5.09 | 4.73 | 4.48 | 3.82 | 3.05 |
| 24 | 14.03 | 9.34 | 7.55 | 6.59 | 5.98 | 5.55 | 5.23 | 4.99 | 4.64 | 4.39 | 3.74 | 2.97 |
| 25 | 13.88 | 9.22 | 7.45 | 6.49 | 5.89 | 5.46 | 5.15 | 4.91 | 4.56 | 4.31 | 3.66 | 2.89 |
| 26 | 13.74 | 9.12 | 7.36 | 6.41 | 5.80 | 5.38 | 5.07 | 4.83 | 4.48 | 4.24 | 3.59 | 2.82 |
| 27 | 13.61 | 9.02 | 7.27 | 6.33 | 5.73 | 5.31 | 5.00 | 4.76 | 4.41 | 4.17 | 3.52 | 2.75 |
| 28 | 13.50 | 8.93 | 7.19 | 6.25 | 5.66 | 5.24 | 4.93 | 4.69 | 4.35 | 4.11 | 3.46 | 2.69 |
| 29 | 13.39 | 8.85 | 7.12 | 6.19 | 5.59 | 5.18 | 4.87 | 4.64 | 4.29 | 4.05 | 3.41 | 2.64 |
| 30 | 13.29 | 8.77 | 7.05 | 6.12 | 5.53 | 5.12 | 4.82 | 4.58 | 4.24 | 4.00 | 3.36 | 2.59 |
| 32 | 13.12 | 8.64 | 6.94 | 6.01 | 5.43 | 5.02 | 4.72 | 4.48 | 4.14 | 3.91 | 3.27 | 2.50 |
| 34 | 12.97 | 8.52 | 6.83 | 5.92 | 5.34 | 4.93 | 4.63 | 4.40 | 4.06 | 3.83 | 3.19 | 2.42 |
| 36 | 12.83 | 8.42 | 6.74 | 5.84 | 5.26 | 4.86 | 4.56 | 4.33 | 3.99 | 3.76 | 3.12 | 2.35 |
| 38 | 12.71 | 8.33 | 6.66 | 5.76 | 5.19 | 4.79 | 4.49 | 4.26 | 3.93 | 3.70 | 3.06 | 2.29 |
| 40 | 12.61 | 8.25 | 6.59 | 5.70 | 5.13 | 4.73 | 4.44 | 4.21 | 3.87 | 3.64 | 3.01 | 2.23 |
| 60 | 11.97 | 7.77 | 6.17 | 5.31 | 4.76 | 4.37 | 4.09 | 3.86 | 3.54 | 3.32 | 2.69 | 1.89 |
| 120 | 11.38 | 7.32 | 5.78 | 4.95 | 4.42 | 4.04 | 3.77 | 3.55 | 3.24 | 3.02 | 2.40 | 1.54 |
| ∞ | 10.83 | 6.91 | 5.42 | 4.62 | 4.10 | 3.74 | 3.47 | 3.27 | 2.96 | 2.74 | 2.13 | 1.00 |

* Critical values to the right of $v_2 = 1$ should all be multiplied by 100, i.e. 4053 should be 405 300.

When there is no exact number for the d.f., use the next lowest number. For very large d.f.s (i.e. well over 120) you should use the row for infinity, marked ∞.

These values are all for a two-tailed test only.

**Table C.7** Critical values of $U$ (Mann-Whitney $U$ test)) at various levels of probability. (For your $U$ value to be significant at a particular probability level, it should be *equal to* or *less than* the critical value associated with $n_1$ and $n_2$ in your study)

| $n_2$ | $n_1$ | | | | | | | | | | | | | | | | | | | |
|---|---|---|---|---|---|---|---|---|---|---|---|---|---|---|---|---|---|---|---|---|
| | *1* | *2* | *3* | *4* | *5* | *6* | *7* | *8* | *9* | *10* | *11* | *12* | *13* | *14* | *15* | *16* | *17* | *18* | *19* | *20* |

*(a) Critical values of U for a one-tailed test at 0.005; two-tailed test at 0.01*

| $n_2$ | 1 | 2 | 3 | 4 | 5 | 6 | 7 | 8 | 9 | 10 | 11 | 12 | 13 | 14 | 15 | 16 | 17 | 18 | 19 | 20 |
|---|---|---|---|---|---|---|---|---|---|---|---|---|---|---|---|---|---|---|---|---|
| 1 | – | – | – | – | – | – | – | – | – | – | – | – | – | – | – | – | – | – | – | – |
| 2 | – | – | – | – | – | – | – | – | – | – | – | – | – | – | – | – | – | – | 0 | 0 |
| 3 | – | – | – | – | – | – | – | – | 0 | 0 | 0 | 1 | 1 | 1 | 2 | 2 | 2 | 2 | 3 | 3 |
| 4 | – | – | – | – | – | 0 | 0 | 1 | 1 | 2 | 2 | 3 | 3 | 4 | 5 | 5 | 6 | 6 | 7 | 8 |
| 5 | – | – | – | – | 0 | 1 | 1 | 2 | 3 | 4 | 5 | 6 | 7 | 7 | 8 | 9 | 10 | 11 | 12 | 13 |
| 6 | – | – | – | 0 | 1 | 2 | 3 | 4 | 5 | 6 | 7 | 9 | 10 | 11 | 12 | 13 | 15 | 16 | 17 | 18 |
| 7 | – | – | – | 0 | 1 | 3 | 4 | 6 | 7 | 9 | 10 | 12 | 13 | 15 | 16 | 18 | 19 | 21 | 22 | 24 |
| 8 | – | – | – | 1 | 2 | 4 | 6 | 7 | 9 | 11 | 13 | 15 | 17 | 18 | 20 | 22 | 24 | 26 | 28 | 30 |
| 9 | – | – | 0 | 1 | 3 | 5 | 7 | 9 | 11 | 13 | 16 | 18 | 20 | 22 | 24 | 27 | 29 | 31 | 33 | 36 |
| 10 | – | – | 0 | 2 | 4 | 6 | 9 | 11 | 13 | 16 | 18 | 21 | 24 | 26 | 29 | 31 | 34 | 37 | 39 | 42 |
| 11 | – | – | 0 | 2 | 5 | 7 | 10 | 13 | 16 | 18 | 21 | 24 | 27 | 30 | 33 | 36 | 39 | 42 | 45 | 48 |
| 12 | – | – | 1 | 3 | 6 | 9 | 12 | 15 | 18 | 21 | 24 | 27 | 31 | 34 | 37 | 41 | 44 | 47 | 51 | 54 |
| 13 | – | – | 1 | 3 | 7 | 10 | 13 | 17 | 20 | 24 | 27 | 31 | 34 | 38 | 42 | 45 | 49 | 53 | 56 | 60 |
| 14 | – | – | 1 | 4 | 7 | 11 | 15 | 18 | 22 | 26 | 30 | 34 | 38 | 42 | 46 | 50 | 54 | 58 | 63 | 67 |
| 15 | – | – | 2 | 5 | 8 | 12 | 16 | 20 | 24 | 29 | 33 | 37 | 42 | 46 | 51 | 55 | 60 | 64 | 69 | 73 |
| 16 | – | – | 2 | 5 | 9 | 13 | 18 | 22 | 27 | 31 | 36 | 41 | 45 | 50 | 55 | 60 | 65 | 70 | 74 | 79 |
| 17 | – | – | 2 | 6 | 10 | 15 | 19 | 24 | 29 | 34 | 39 | 44 | 49 | 54 | 60 | 65 | 70 | 75 | 81 | 86 |
| 18 | – | – | 2 | 6 | 11 | 16 | 21 | 26 | 31 | 37 | 42 | 47 | 53 | 58 | 64 | 70 | 75 | 81 | 87 | 92 |
| 19 | – | 0 | 3 | 7 | 12 | 17 | 22 | 28 | 33 | 39 | 45 | 51 | 56 | 63 | 69 | 74 | 81 | 87 | 93 | 99 |
| 20 | – | 0 | 3 | 8 | 13 | 18 | 24 | 30 | 36 | 42 | 48 | 54 | 60 | 67 | 73 | 79 | 86 | 92 | 99 | 105 |

*(b) Critical values of U for a one-tailed test at 0.01; two-tailed test at 0.02*

| $n_2$ | 1 | 2 | 3 | 4 | 5 | 6 | 7 | 8 | 9 | 10 | 11 | 12 | 13 | 14 | 15 | 16 | 17 | 18 | 19 | 20 |
|---|---|---|---|---|---|---|---|---|---|---|---|---|---|---|---|---|---|---|---|---|
| 1 | – | – | – | – | – | – | – | – | – | – | – | – | – | – | – | – | – | – | – | – |
| 2 | – | – | – | – | – | – | – | – | – | – | – | – | – | 0 | 0 | 0 | 0 | 0 | 1 | 1 |
| 3 | – | – | – | – | – | – | 0 | 0 | 1 | 1 | 1 | 2 | 2 | 2 | 3 | 3 | 4 | 4 | 4 | 5 |
| 4 | – | – | – | – | 0 | 1 | 1 | 2 | 3 | 3 | 4 | 5 | 5 | 6 | 7 | 7 | 8 | 9 | 9 | 10 |
| 5 | – | – | – | 0 | 1 | 2 | 3 | 4 | 5 | 6 | 7 | 8 | 9 | 10 | 11 | 12 | 13 | 14 | 15 | 16 |
| 6 | – | – | – | 1 | 2 | 3 | 4 | 6 | 7 | 8 | 9 | 11 | 12 | 13 | 15 | 16 | 18 | 19 | 20 | 22 |
| 7 | – | – | 0 | 1 | 3 | 4 | 6 | 7 | 9 | 11 | 12 | 14 | 16 | 17 | 19 | 21 | 23 | 24 | 26 | 28 |
| 8 | – | – | 0 | 2 | 4 | 6 | 7 | 9 | 11 | 13 | 15 | 17 | 20 | 22 | 24 | 26 | 28 | 30 | 32 | 34 |
| 9 | – | – | 1 | 3 | 5 | 7 | 9 | 11 | 14 | 16 | 18 | 21 | 23 | 26 | 28 | 31 | 33 | 36 | 38 | 40 |
| 10 | – | – | 1 | 3 | 6 | 8 | 11 | 13 | 16 | 19 | 22 | 24 | 27 | 30 | 33 | 36 | 38 | 41 | 44 | 47 |
| 11 | – | – | 1 | 4 | 7 | 9 | 12 | 15 | 18 | 22 | 25 | 28 | 31 | 34 | 37 | 41 | 44 | 47 | 50 | 53 |
| 12 | – | – | 2 | 5 | 8 | 11 | 14 | 17 | 21 | 24 | 28 | 31 | 35 | 38 | 42 | 46 | 49 | 53 | 56 | 60 |
| 13 | – | 0 | 2 | 5 | 9 | 12 | 16 | 20 | 23 | 27 | 31 | 35 | 39 | 43 | 47 | 51 | 55 | 59 | 63 | 67 |
| 14 | – | 0 | 2 | 6 | 10 | 13 | 17 | 22 | 26 | 30 | 34 | 38 | 43 | 47 | 51 | 56 | 60 | 65 | 69 | 73 |
| 15 | – | 0 | 3 | 7 | 11 | 15 | 19 | 24 | 28 | 33 | 37 | 42 | 47 | 51 | 56 | 61 | 66 | 70 | 75 | 80 |
| 16 | – | 0 | 3 | 7 | 12 | 16 | 21 | 26 | 31 | 36 | 41 | 46 | 51 | 56 | 61 | 66 | 71 | 76 | 82 | 87 |
| 17 | – | 0 | 4 | 8 | 13 | 18 | 23 | 28 | 33 | 38 | 44 | 49 | 55 | 60 | 66 | 71 | 77 | 82 | 88 | 93 |
| 18 | – | 0 | 4 | 9 | 14 | 19 | 24 | 30 | 36 | 41 | 47 | 53 | 59 | 65 | 70 | 76 | 82 | 88 | 94 | 100 |
| 19 | – | 1 | 4 | 9 | 15 | 20 | 26 | 32 | 38 | 44 | 50 | 56 | 63 | 69 | 75 | 82 | 88 | 94 | 101 | 107 |
| 20 | – | 1 | 5 | 10 | 16 | 22 | 28 | 34 | 40 | 47 | 53 | 60 | 67 | 73 | 80 | 87 | 93 | 100 | 107 | 114 |

**Table C.7** (con't)

| $n_2$ | $n_1$ | | | | | | | | | | | | | | | | | | | |
|---|---|---|---|---|---|---|---|---|---|---|---|---|---|---|---|---|---|---|---|---|
| | *1* | *2* | *3* | *4* | *5* | *6* | *7* | *8* | *9* | *10* | *11* | *12* | *13* | *14* | *15* | *16* | *17* | *18* | *19* | *20* |

*(c) Critical values of U for a one-tailed test at 0.025; two-tailed test at 0.05*

| $n_2$ | 1 | 2 | 3 | 4 | 5 | 6 | 7 | 8 | 9 | 10 | 11 | 12 | 13 | 14 | 15 | 16 | 17 | 18 | 19 | 20 |
|---|---|---|---|---|---|---|---|---|---|---|---|---|---|---|---|---|---|---|---|---|
| 1 | – | – | – | – | – | – | – | – | – | – | – | – | – | – | – | – | – | – | – | – |
| 2 | – | – | – | – | – | – | – | 0 | 0 | 0 | 0 | 1 | 1 | 1 | 1 | 1 | 2 | 2 | 2 | 2 |
| 3 | – | – | – | – | 0 | 1 | 1 | 2 | 2 | 3 | 3 | 4 | 4 | 5 | 5 | 6 | 6 | 7 | 7 | 8 |
| 4 | – | – | – | 0 | 1 | 2 | 3 | 4 | 4 | 5 | 6 | 7 | 8 | 9 | 10 | 11 | 11 | 12 | 13 | 13 |
| 5 | – | – | 0 | 1 | 2 | 3 | 5 | 6 | 7 | 8 | 9 | 11 | 12 | 13 | 14 | 15 | 17 | 18 | 19 | 20 |
| 6 | – | – | 1 | 2 | 3 | 5 | 6 | 8 | 10 | 11 | 13 | 14 | 16 | 17 | 19 | 21 | 22 | 24 | 25 | 27 |
| 7 | – | – | 1 | 3 | 5 | 6 | 8 | 10 | 12 | 14 | 16 | 18 | 20 | 22 | 24 | 26 | 28 | 30 | 32 | 34 |
| 8 | – | 0 | 2 | 4 | 6 | 8 | 10 | 13 | 15 | 17 | 19 | 22 | 24 | 26 | 29 | 31 | 34 | 36 | 38 | 41 |
| 9 | – | 0 | 2 | 4 | 7 | 10 | 12 | 15 | 17 | 20 | 23 | 26 | 28 | 31 | 34 | 37 | 39 | 42 | 45 | 48 |
| 10 | – | 0 | 3 | 5 | 8 | 11 | 14 | 17 | 20 | 23 | 26 | 29 | 33 | 36 | 39 | 42 | 45 | 48 | 52 | 55 |
| 11 | – | 0 | 3 | 6 | 9 | 13 | 16 | 19 | 23 | 26 | 30 | 33 | 37 | 40 | 44 | 47 | 51 | 55 | 58 | 62 |
| 12 | – | 1 | 4 | 7 | 11 | 14 | 18 | 22 | 26 | 29 | 33 | 37 | 41 | 45 | 49 | 53 | 57 | 61 | 65 | 69 |
| 13 | – | 1 | 4 | 8 | 12 | 16 | 20 | 24 | 28 | 33 | 37 | 41 | 45 | 50 | 54 | 59 | 63 | 67 | 72 | 76 |
| 14 | – | 1 | 5 | 9 | 13 | 17 | 22 | 26 | 31 | 36 | 40 | 45 | 50 | 55 | 59 | 64 | 67 | 74 | 78 | 83 |
| 15 | – | 1 | 5 | 10 | 14 | 19 | 24 | 29 | 34 | 39 | 44 | 49 | 54 | 59 | 64 | 70 | 75 | 80 | 85 | 90 |
| 16 | – | 1 | 6 | 11 | 15 | 21 | 26 | 31 | 37 | 42 | 47 | 53 | 59 | 64 | 70 | 75 | 81 | 86 | 92 | 98 |
| 17 | – | 2 | 6 | 11 | 17 | 22 | 28 | 34 | 39 | 45 | 51 | 57 | 63 | 67 | 75 | 81 | 87 | 93 | 99 | 105 |
| 18 | – | 2 | 7 | 12 | 18 | 24 | 30 | 36 | 42 | 48 | 55 | 61 | 67 | 74 | 80 | 86 | 93 | 99 | 106 | 112 |
| 19 | – | 2 | 7 | 13 | 19 | 25 | 32 | 38 | 45 | 52 | 58 | 65 | 72 | 78 | 85 | 92 | 99 | 106 | 113 | 119 |
| 20 | – | 2 | 8 | 13 | 20 | 27 | 34 | 41 | 48 | 55 | 62 | 69 | 76 | 83 | 90 | 98 | 105 | 112 | 119 | 127 |

*(d) Critical values of U for a one-tailed test at 0.05; two-tailed test at 0.10*

| $n_2$ | 1 | 2 | 3 | 4 | 5 | 6 | 7 | 8 | 9 | 10 | 11 | 12 | 13 | 14 | 15 | 16 | 17 | 18 | 19 | 20 |
|---|---|---|---|---|---|---|---|---|---|---|---|---|---|---|---|---|---|---|---|---|
| 1 | – | – | – | – | – | – | – | – | – | – | – | – | – | – | – | – | – | – | 0 | 0 |
| 2 | – | – | – | – | 0 | 0 | 0 | 1 | 1 | 1 | 1 | 2 | 2 | 2 | 3 | 3 | 3 | 4 | 4 | 4 |
| 3 | – | – | 0 | 0 | 1 | 2 | 2 | 3 | 3 | 4 | 5 | 5 | 6 | 7 | 7 | 8 | 9 | 9 | 10 | 11 |
| 4 | – | – | 0 | 1 | 2 | 3 | 4 | 5 | 6 | 7 | 8 | 9 | 10 | 11 | 12 | 14 | 15 | 16 | 17 | 18 |
| 5 | – | 0 | 1 | 2 | 4 | 5 | 6 | 8 | 9 | 11 | 12 | 13 | 15 | 16 | 18 | 19 | 20 | 22 | 23 | 25 |
| 6 | – | 0 | 2 | 3 | 5 | 7 | 8 | 10 | 12 | 14 | 16 | 17 | 19 | 21 | 23 | 25 | 26 | 28 | 30 | 32 |
| 7 | – | 0 | 2 | 4 | 6 | 8 | 11 | 13 | 15 | 17 | 19 | 21 | 24 | 26 | 28 | 30 | 33 | 35 | 37 | 39 |
| 8 | – | 1 | 3 | 5 | 8 | 10 | 13 | 15 | 18 | 20 | 23 | 26 | 28 | 31 | 33 | 36 | 39 | 41 | 44 | 47 |
| 9 | – | 1 | 3 | 6 | 9 | 12 | 15 | 18 | 21 | 24 | 27 | 30 | 33 | 36 | 39 | 42 | 45 | 48 | 51 | 54 |
| 10 | – | 1 | 4 | 7 | 11 | 14 | 17 | 20 | 24 | 27 | 31 | 34 | 37 | 41 | 44 | 48 | 51 | 55 | 58 | 62 |
| 11 | – | 1 | 5 | 8 | 12 | 16 | 19 | 23 | 27 | 31 | 34 | 38 | 42 | 46 | 50 | 54 | 57 | 61 | 65 | 69 |
| 12 | – | 2 | 5 | 9 | 13 | 17 | 21 | 26 | 30 | 34 | 38 | 42 | 47 | 51 | 55 | 60 | 64 | 68 | 72 | 77 |
| 13 | – | 2 | 6 | 10 | 15 | 19 | 24 | 28 | 33 | 37 | 42 | 47 | 51 | 56 | 61 | 65 | 70 | 75 | 80 | 84 |
| 14 | – | 2 | 7 | 11 | 16 | 21 | 26 | 31 | 36 | 41 | 46 | 51 | 56 | 61 | 66 | 71 | 77 | 82 | 87 | 92 |
| 15 | – | 3 | 7 | 12 | 18 | 23 | 28 | 33 | 39 | 44 | 50 | 55 | 61 | 66 | 72 | 77 | 83 | 88 | 94 | 100 |
| 16 | – | 3 | 8 | 14 | 19 | 25 | 30 | 36 | 42 | 48 | 54 | 60 | 65 | 71 | 77 | 83 | 89 | 95 | 101 | 107 |
| 17 | – | 3 | 9 | 15 | 20 | 26 | 33 | 39 | 45 | 51 | 57 | 64 | 70 | 77 | 83 | 89 | 96 | 102 | 109 | 115 |
| 18 | – | 4 | 9 | 16 | 22 | 28 | 35 | 41 | 48 | 55 | 61 | 68 | 75 | 82 | 88 | 95 | 102 | 109 | 116 | 123 |
| 19 | 0 | 4 | 10 | 17 | 23 | 30 | 37 | 44 | 51 | 58 | 65 | 72 | 80 | 87 | 94 | 101 | 109 | 116 | 123 | 130 |
| 20 | 0 | 4 | 11 | 18 | 25 | 32 | 39 | 47 | 54 | 62 | 69 | 77 | 84 | 92 | 100 | 107 | 115 | 123 | 130 | 138 |

Dashes in the table mean that no decision is possible for those *n* values at the given level of significance.

**Table C.8** Critical values of $H$ (Kruskal-Wallis test) at various levels of probability. (For your $H$ value to be significant at a particular probability level, it should be *equal to* or *larger than* the critical values associated with the $n$s in your study)

| $n_1$ | $n_2$ | $n_3$ | $H$ | $p$ | $n_1$ | $n_2$ | $n_3$ | $H$ | $p$ |
|---|---|---|---|---|---|---|---|---|---|
| 2 | 1 | 1 | 2.7000 | 0.500 | 4 | 3 | 2 | 6.4444 | 0.008 |
| 2 | 2 | 1 | 3.6000 | 0.200 | | | | 6.3000 | 0.011 |
| | | | | | | | | 5.4444 | 0.046 |
| 2 | 2 | 2 | 4.5714 | 0.067 | | | | 5.4000 | 0.051 |
| | | | 3.7143 | 0.200 | | | | 4.5111 | 0.098 |
| | | | | | | | | 4.4444 | 0.102 |
| 3 | 1 | 1 | 3.2000 | 0.300 | | | | | |
| 3 | 2 | 1 | 4.2857 | 0.100 | 4 | 3 | 3 | 6.7455 | 0.010 |
| | | | 3.8571 | 0.133 | | | | 6.7091 | 0.013 |
| | | | | | | | | 5.7909 | 0.046 |
| 3 | 2 | 2 | 5.3572 | 0.029 | | | | 5.7273 | 0.050 |
| | | | 4.7143 | 0.048 | | | | 4.7091 | 0.092 |
| | | | 4.5000 | 0.067 | | | | 4.7000 | 0.101 |
| | | | 4.4643 | 0.105 | | | | | |
| | | | | | 4 | 4 | 1 | 6.6667 | 0.010 |
| 3 | 3 | 1 | 5.1429 | 0.043 | | | | 6.1667 | 0.022 |
| | | | 4.5714 | 0.100 | | | | 4.9667 | 0.048 |
| | | | 4.0000 | 0.129 | | | | 4.8667 | 0.054 |
| | | | | | | | | 4.1667 | 0.082 |
| 3 | 3 | 2 | 6.2500 | 0.011 | | | | 4.0667 | 0.102 |
| | | | 5.3611 | 0.032 | | | | | |
| | | | 5.1389 | 0.061 | 4 | 4 | 2 | 7.0364 | 0.006 |
| | | | 4.5556 | 0.100 | | | | 6.8727 | 0.011 |
| | | | 4.2500 | 0.121 | | | | 5.4545 | 0.046 |
| | | | | | | | | 5.2364 | 0.052 |
| 3 | 3 | 3 | 7.2000 | 0.004 | | | | 4.5545 | 0.098 |
| | | | 6.4889 | 0.011 | | | | 4.4455 | 0.103 |
| | | | 5.6889 | 0.029 | | | | | |
| | | | 5.6000 | 0.050 | 4 | 4 | 3 | 7.1439 | 0.010 |
| | | | 5.0667 | 0.086 | | | | 7.1364 | 0.011 |
| | | | 4.6222 | 0.100 | | | | 5.5985 | 0.049 |
| | | | | | | | | 5.5758 | 0.051 |
| 4 | 1 | 1 | 3.5714 | 0.200 | | | | 4.5455 | 0.099 |
| 4 | 2 | 1 | 4.8214 | 0.057 | | | | 4.4773 | 0.102 |
| | | | 4.5000 | 0.076 | | | | | |
| | | | 4.0179 | 0.114 | 4 | 4 | 4 | 7.6538 | 0.008 |
| | | | | | | | | 7.5385 | 0.011 |
| 4 | 2 | 2 | 6.0000 | 0.014 | | | | 5.6923 | 0.049 |
| | | | 5.3333 | 0.033 | | | | 5.6538 | 0.054 |
| | | | 5.1250 | 0.052 | | | | 4.6539 | 0.097 |
| | | | 4.4583 | 0.100 | | | | 4.5001 | 0.104 |
| | | | 4.1667 | 0.105 | | | | | |
| | | | | | 5 | 1 | 1 | 3.8571 | 0.143 |
| 4 | 3 | 1 | 5.8333 | 0.021 | 5 | 2 | 1 | 5.2500 | 0.036 |
| | | | 5.2083 | 0.050 | | | | 5.0000 | 0.048 |
| | | | 5.0000 | 0.057 | | | | 4.4500 | 0.071 |
| | | | 4.0556 | 0.093 | | | | 4.2000 | 0.095 |
| | | | 3.8889 | 0.129 | | | | 4.0500 | 0.119 |

**Table C.8** (con't)

| $n_1$ | $n_2$ | $n_3$ | $H$ | $p$ | $n_1$ | $n_2$ | $n_3$ | $H$ | $p$ |
|---|---|---|---|---|---|---|---|---|---|
| *Size of groups* | | | | | *Size of groups* | | | | |
| 5 | 2 | 2 | 6.5333 | 0.008 | 5 | 4 | 4 | 7.7604 | 0.009 |
|   |   |   | 6.1333 | 0.013 |   |   |   | 7.7440 | 0.011 |
|   |   |   | 5.1600 | 0.034 |   |   |   | 5.6571 | 0.049 |
|   |   |   | 5.0400 | 0.056 |   |   |   | 5.6176 | 0.050 |
|   |   |   | 4.3733 | 0.090 |   |   |   | 4.6187 | 0.100 |
|   |   |   | 4.2933 | 0.122 |   |   |   | 4.5527 | 0.102 |
| 5 | 3 | 1 | 6.4000 | 0.012 | 5 | 5 | 1 | 7.3091 | 0.009 |
|   |   |   | 4.9600 | 0.048 |   |   |   | 6.8364 | 0.011 |
|   |   |   | 4.8711 | 0.052 |   |   |   | 5.1273 | 0.046 |
|   |   |   | 4.0178 | 0.095 |   |   |   | 4.9091 | 0.053 |
|   |   |   | 3.8400 | 0.123 |   |   |   | 4.1091 | 0.086 |
| 5 | 3 | 2 | 6.9091 | 0.009 |   |   |   | 4.0364 | 0.105 |
|   |   |   | 6.8218 | 0.010 | 5 | 5 | 2 | 7.3385 | 0.010 |
|   |   |   | 5.2509 | 0.049 |   |   |   | 7.2692 | 0.010 |
|   |   |   | 5.1055 | 0.052 |   |   |   | 5.3385 | 0.047 |
|   |   |   | 4.6509 | 0.091 |   |   |   | 5.2462 | 0.051 |
|   |   |   | 4.4945 | 0.101 |   |   |   | 4.6231 | 0.097 |
| 5 | 3 | 3 | 7.0788 | 0.009 |   |   |   | 4.5077 | 0.100 |
|   |   |   | 6.9818 | 0.011 | 5 | 5 | 3 | 7.5780 | 0.010 |
|   |   |   | 5.6485 | 0.049 |   |   |   | 7.5429 | 0.010 |
|   |   |   | 5.5152 | 0.051 |   |   |   | 5.7055 | 0.046 |
|   |   |   | 4.5333 | 0.097 |   |   |   | 5.6264 | 0.051 |
|   |   |   | 4.4121 | 0.109 |   |   |   | 4.5451 | 0.100 |
| 5 | 4 | 1 | 6.9545 | 0.008 |   |   |   | 4.5363 | 0.102 |
|   |   |   | 6.8400 | 0.011 | 5 | 5 | 4 | 7.8229 | 0.010 |
|   |   |   | 4.9855 | 0.044 |   |   |   | 7.7914 | 0.010 |
|   |   |   | 4.8600 | 0.056 |   |   |   | 5.6657 | 0.049 |
|   |   |   | 3.9873 | 0.098 |   |   |   | 5.6429 | 0.050 |
|   |   |   | 3.9600 | 0.102 |   |   |   | 4.5229 | 0.099 |
| 5 | 4 | 2 | 7.2045 | 0.009 |   |   |   | 4.5200 | 0.101 |
|   |   |   | 7.1182 | 0.010 | 5 | 5 | 5 | 8.0000 | 0.009 |
|   |   |   | 5.2727 | 0.049 |   |   |   | 7.9800 | 0.010 |
|   |   |   | 5.2682 | 0.050 |   |   |   | 7.7800 | 0.049 |
|   |   |   | 4.5409 | 0.098 |   |   |   | 5.6600 | 0.051 |
|   |   |   | 4.5182 | 0.101 |   |   |   | 4.5600 | 0.100 |
| 5 | 4 | 3 | 7.4449 | 0.010 |   |   |   | 4.5000 | 0.102 |
|   |   |   | 7.3949 | 0.011 |   |   |   |   |   |
|   |   |   | 5.6564 | 0.049 |   |   |   |   |   |
|   |   |   | 5.6308 | 0.050 |   |   |   |   |   |
|   |   |   | 4.5487 | 0.099 |   |   |   |   |   |
|   |   |   | 4.5231 | 0.103 |   |   |   |   |   |

These values are all for a two-tailed test only.

**Table C.9** Critical values of $S$ (Jonckheere trend test) at various levels of probability. (For your $S$ value to be significant at a particular probability level, it should be *equal to* or *larger than* the critical values associated with $C$ and $n$ in your study)

| $C$ | | | | | $n$ | | | | |
| | 2 | 3 | 4 | 5 | 6 | 7 | 8 | 9 | 10 |
|---|---|---|---|---|---|---|---|---|---|
| *(a) Significance level $p < 0.05$* | | | | | | | | | |
| 3 | 10 | 17 | 24 | 33 | 42 | 53 | 64 | 76 | 88 |
| 4 | 14 | 26 | 38 | 51 | 66 | 82 | 100 | 118 | 138 |
| 5 | 20 | 34 | 51 | 71 | 92 | 115 | 140 | 166 | 194 |
| 6 | 26 | 44 | 67 | 93 | 121 | 151 | 184 | 219 | 256 |
| *(b) Significance level $p < 0.01$* | | | | | | | | | |
| 3 | – | 23 | 32 | 45 | 59 | 74 | 90 | 106 | 124 |
| 4 | 20 | 34 | 50 | 71 | 92 | 115 | 140 | 167 | 195 |
| 5 | 26 | 48 | 72 | 99 | 129 | 162 | 197 | 234 | 274 |
| 6 | 34 | 62 | 94 | 130 | 170 | 213 | 260 | 309 | 361 |

These values are all for a one-tailed test only.

**Table C.10** Critical values of $r_s$ (Spearman test) at various levels of probability. (For your $r_s$ value to be significant at a particular probability level, it should be *equal to* or *larger than* the critical values associated with $N$ in your study)

| $N$ (number of subjects) | Level of significance for one-tailed test | | | |
| | 0.05 | 0.025 | 0.01 | 0.005 |
| | Level of significance for two-tailed test | | | |
| | 0.10 | 0.05 | 0.02 | 0.01 |
|---|---|---|---|---|
| 5 | 0.900 | 1.000 | 1.000 | – |
| 6 | 0.829 | 0.886 | 0.943 | 1.000 |
| 7 | 0.714 | 0.786 | 0.893 | 0.929 |
| 8 | 0.643 | 0.738 | 0.833 | 0.881 |
| 9 | 0.600 | 0.683 | 0.783 | 0.833 |
| 10 | 0.564 | 0.648 | 0.746 | 0.794 |
| 12 | 0.506 | 0.591 | 0.712 | 0.777 |
| 14 | 0.456 | 0.544 | 0.645 | 0.715 |
| 16 | 0.425 | 0.506 | 0.601 | 0.665 |
| 18 | 0.399 | 0.475 | 0.564 | 0.625 |
| 20 | 0.377 | 0.450 | 0.534 | 0.591 |
| 22 | 0.359 | 0.428 | 0.508 | 0.562 |
| 24 | 0.343 | 0.409 | 0.485 | 0.537 |
| 26 | 0.329 | 0.392 | 0.465 | 0.515 |
| 28 | 0.317 | 0.377 | 0.448 | 0.496 |
| 30 | 0.306 | 0.364 | 0.432 | 0.478 |

When there is no exact number of subjects use the next lowest number.

**Table C.11** Critical values of $r$ (Pearson test) at various levels of probability. (For your $r$ value to be significant at a particular probability level, it should be *equal to* or *larger than* the critical values associated with the d.f. in your study.)

| d.f. = N − 2 | Level of significance for one-tailed test | | | | |
| | 0.05 | 0.025 | 0.01 | 0.005 | 0.0005 |
| | Level of significance for two-tailed test | | | | |
| | 0.10 | 0.05 | 0.02 | 0.01 | 0.001 |
|---|---|---|---|---|---|
| 1 | 0.9877 | 0.9969 | 0.9995 | 0.9999 | 1.0000 |
| 2 | 0.9000 | 0.9500 | 0.9800 | 0.9900 | 0.9990 |
| 3 | 0.8054 | 0.8783 | 0.9343 | 0.9587 | 0.9912 |
| 4 | 0.7293 | 0.8114 | 0.8822 | 0.9172 | 0.9741 |
| 5 | 0.6694 | 0.7545 | 0.8329 | 0.8745 | 0.9507 |
| 6 | 0.6215 | 0.7067 | 0.7887 | 0.8343 | 0.9249 |
| 7 | 0.5822 | 0.6664 | 0.7498 | 0.7977 | 0.8982 |
| 8 | 0.5494 | 0.6319 | 0.7155 | 0.7646 | 0.8721 |
| 9 | 0.5214 | 0.6021 | 0.6851 | 0.7348 | 0.8471 |
| 10 | 0.4973 | 0.5760 | 0.6581 | 0.7079 | 0.8233 |
| 11 | 0.4762 | 0.5529 | 0.6339 | 0.6835 | 0.8010 |
| 12 | 0.4575 | 0.5324 | 0.6120 | 0.6614 | 0.7800 |
| 13 | 0.4409 | 0.5139 | 0.5923 | 0.6411 | 0.7603 |
| 14 | 0.4259 | 0.4973 | 0.5742 | 0.6226 | 0.7420 |
| 15 | 0.4124 | 0.4821 | 0.5577 | 0.6055 | 0.7246 |
| 16 | 0.4000 | 0.4683 | 0.5425 | 0.5897 | 0.7084 |
| 17 | 0.3887 | 0.4555 | 0.5285 | 0.5751 | 0.6932 |
| 18 | 0.3783 | 0.4438 | 0.5155 | 0.5614 | 0.6787 |
| 19 | 0.3687 | 0.4329 | 0.5034 | 0.5487 | 0.6652 |
| 20 | 0.3598 | 0.4227 | 0.4921 | 0.5368 | 0.6524 |
| 25 | 0.3233 | 0.3809 | 0.4451 | 0.4869 | 0.5974 |
| 30 | 0.2960 | 0.3494 | 0.4093 | 0.4487 | 0.5541 |
| 35 | 0.2746 | 0.3246 | 0.3810 | 0.4182 | 0.5189 |
| 40 | 0.2573 | 0.3044 | 0.3578 | 0.3932 | 0.4896 |
| 45 | 0.2428 | 0.2875 | 0.3384 | 0.3721 | 0.4648 |
| 50 | 0.2306 | 0.2732 | 0.3218 | 0.3541 | 0.4433 |
| 60 | 0.2108 | 0.2500 | 0.2948 | 0.3248 | 0.4078 |
| 70 | 0.1954 | 0.2319 | 0.2737 | 0.3017 | 0.3799 |
| 80 | 0.1829 | 0.2172 | 0.2565 | 0.2830 | 0.3568 |
| 90 | 0.1726 | 0.2050 | 0.2422 | 0.2673 | 0.3375 |
| 100 | 0.1638 | 0.1946 | 0.2301 | 0.2540 | 0.3211 |

When there is no exact d.f. use the next lowest number.

**Table C.12** Critical values of *s* (Kendall's coefficient of concordance) at various levels of probability. (For your *s* value to be significant at a particular probability level, it should be *equal to* or *larger than* the critical values associated with *n* and *N* in your study.)

| n | N = 3 | N = 4 | N = 5 | N = 6 | N = 7 |
|---|---|---|---|---|---|
| *(a) Critical values of s at p = 0.05* | | | | | |
| 3 | – | – | 64.4 | 103.9 | 157.3 |
| 4 | – | 49.5 | 88.4 | 143.3 | 217.0 |
| 5 | – | 62.6 | 112.3 | 182.4 | 276.2 |
| 6 | – | 75.7 | 136.1 | 221.4 | 335.2 |
| 8 | 48.1 | 101.7 | 183.7 | 299.0 | 453.1 |
| 10 | 60.0 | 127.8 | 231.2 | 376.7 | 571.0 |
| 15 | 89.8 | 192.9 | 349.8 | 570.5 | 864.9 |
| 20 | 119.7 | 258.0 | 468.5 | 764.4 | 1158.7 |
| *(b) Critical values of s at p = 0.01* | | | | | |
| 3 | – | – | 75.6 | 122.8 | 185.6 |
| 4 | – | 61.4 | 109.3 | 176.2 | 265.0 |
| 5 | – | 80.5 | 142.8 | 229.4 | 343.8 |
| 6 | – | 99.5 | 176.1 | 282.4 | 422.6 |
| 8 | 66.8 | 137.4 | 242.7 | 388.3 | 579.9 |
| 10 | 85.1 | 175.3 | 309.1 | 494.0 | 737.0 |
| 15 | 131.0 | 269.8 | 475.2 | 758.2 | 1129.5 |
| 20 | 177.0 | 364.2 | 641.2 | 1022.2 | 1521.9 |

The values are all for a one-tailed test only.

A dash in the table means that no decision can be made at this level.

Where there is no exact *n* value, use the next *smallest* number.

## Appendix D

# Answers to exercises in the text

## D.1 Chapter 3

1. (i) (a) Where the patient is nursed and degree of satisfaction with nursing care. (b) 'There is no relationship between where a patient is nursed and degree of satisfaction with nursing care.' (c) The IV is where the patient is nursed; the DV is satisfaction with nursing care.
   (ii) (a) Ethnic origin and responsiveness to pain. (b) 'There is no relationship between ethnic origins and responsiveness to pain.' (c) The IV is ethnic origin; the DV is responsiveness to pain.
   (iii) (a) Type of bed-rest and rate of healing. (b) 'There is no relationship between type of bed-rest and rate of healing of varicose ulcers.' (c) The IV is type of bed-rest; the DV is rate of healing.
   (iv) (a) Type of training and likelihood of staying in ward-based jobs. (b) 'There is no relationship between type of training and the likelihood of staying in ward-based jobs.' (c) The IV is type of training; the DV is likelihood of staying in a ward-based job.
2. One experimental design for this hypothesis is given in Table D.1. Alternatively, you could have a second experimental condition, where the women had used an alternative form of contraception for the same length of time. The length of time taken to become pregnant for each group would be compared to see if there are differences between them. (It would be very difficult in this study to obtain a pre-test measure of fertility.)

**Table D.1** Sample experimental design

|  | IV | Post-test measure of DV |
|---|---|---|
| Experimental condition | Oral contraception for, say, five years | Length of time taken to become pregnant |
| Control condition | No oral contraception for equivalent length of time | Length of time taken to become pregnant |

**Table D.2** Sample correlational design

| Subject | Variable 1: length of time on oral contraception (months) | Variable 2: time taken to become pregnant |
|---|---|---|
| 1 | 0 | |
| 2 | 6 | |
| 3 | 12 | |
| 4 | 18 | |
| 5 | 24 | |
| 12 | 54 | |

A correlational design for this hypothesis might be to select a group of women who had taken oral contraception for varying periods of time and measure how long it took each one to become pregnant (see, e.g., Table D.2).

The time taken to become pregnant would be assessed and its relationship to the length of time on oral contraception assessed for a link or association.

3.  (i) (a) Negative correlation (*low* scores on temperature related to *high* scores on incidence of hypothermia). (b) The correlation coefficient would be around the $-1.0$ end.

  (ii)  (a) A positive correlation (*high* scores on AFP related to *high* scores on severity scale). (b) The correlation coefficient would be around the $+1.0$ end.

  (iii) (a) A negative correlation (*high* scores on waiting time related to *low* scores on satisfaction). (b) The correlation coefficient would be around the $-1.0$ end.

  (iv) (a) A positive correlation (*high* scores on amount of jogging related to *high* scores on joint damage). (b) The correlation coefficient would be around the $+1.0$ end.

4. Strongest relationship is $-0.803$; weakest is $+0.113$.

## D.2  Chapter 5

1. To test this hypothesis you would select a group of patients with burns caused by wet heat and a second group of patients with burns caused by dry heat.

  The constant errors are as follows:

  (i) Severity of burn. This could be controlled by ensuring that both groups of patients have burns of comparable severity.

  (ii) Site of burn. This could be controlled by ensuring that both groups of patients' burns were in similar areas (e.g. on a joint, or pressure area).

  (iii) Age of patient. This could be controlled by selecting patients for each group from within a particular age group.

  (iv) General physical condition of the patient. This could be controlled by

eliminating any patient with diabetes, circulatory problems or other medical condition which would predispose to slow healing.
You may have thought of additional constant errors besides these, which should also be controlled for.

The random errors are factors such as individual physical reactions to treatment procedures, temporary health states of the patients, etc. These could be dealt with by ensuring that patients in each group are randomly selected.

2. To test this hypothesis you would select a group of patients nursed in side wards and a second group nursed in main wards and compare the groups for post-operative depression.

The constant errors that would need to be controlled are as follows:

(i) Type of operation. This could be controlled by ensuring both groups of patients had the same operation.

(ii) Sex. This could be controlled by selecting either *all* males or *all* females or equal number of each in both groups.

(iii) Age. This could be controlled by selecting patients from a particular age range.

(iv) Family history of psychiatric problem. This could be controlled by eliminating any subject with this type of family history.

(v) Predisposing factors to depression. This could be controlled by eliminating any subjects with such a predisposition.

You may have listed other constant errors which would need to be controlled. The random errors include temporary mood states, idiosyncratic reactions to drugs, anaesthesia, etc., personality, etc. These errors could be dealt with by *random* selection of patients for each group.

3. (a) $p = 3\%$; (b) $p = 7\%$; (c) $p = 0.1\%$.
4. (a) $p = 1\%$; (b) $p = 5\%$; (c) $p = 0.1\%$; (d) $p = 50\%$; (e) $p = 10\%$; (f) $p = 0.05$; (g) $p = 0.03$; (h) $p = 0.01$; (i) $p = 0.15$; (j) $p = 0.02$.
5. (a) 1% chance of random error accounting for the results.
   (b) 7% chance of random error accounting for the results.
   (c) 3% chance of random error accounting for the results.
   (d) 5% chance of random error accounting for the results.
   (e) 10% chance of random error accounting for the results.
6. (a), (c) and (d) would be classified as significant using the standard 5% significance level.
7. $p = 0.01, 0.03, 0.05, 0.07, 0.10$.

# D.3  Chapter 6

1. (a) Interval/ratio; (b) nominal; (c) ordinal; (d) nominal; (e) interval/ratio.
2. You could use the following measures (or a variation of them) to monitor food intake.
   (a) Nominal. Did the patient take in any food?    Yes/No.

(b) Ordinal. How much food did the patient take in?

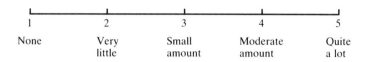

(c) Interval/ratio. How much food did the patient take in (e.g. number of ccs or number of ounces)?
3. Measuring urinary incontinence could be carried out in the following ways (or a variation)
   (a) Nominal. Was the patient continent/incontinent?
   (b) Ordinal. Was the patient

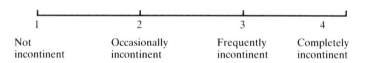

(c) Interval/ratio. How many times was the patient incontinent?

## D.4  Chapter 7

1. (a) $\chi^2$ test.
   (b) Wilcoxon or related $t$ test.
   (c) Spearman test.
   (d) Wilcoxon test.
2. (a) One-tailed (*more* physical contact is expected to *speed up* recovery).
   (b) One-tailed (women are *more* likely to be prescribed tranquillisers by male GPs).
   (c) Two-tailed (information is just expected to *alter* fear levels).
   (d) One-tailed (ototoxic drugs *increase* the likelihood of deafness).
   (e) Two-tailed (the grafts are predicted to be *differentially* successful).
3. (a) Different degrees of physical contact affect the speed of recovery in sick children.
   (b) There is a relationship between the sex of the GP and the likelihood of women patients being prescribed tranquillisers.
   (c) Orthopaedic patients who are given more information about their treatment experience less (or more) fear.
   (d) Ototoxic drugs taken during pregnancy affect the likelihood of deafness in the child.
   (e) Direct flap skin grafts are more (or less) successful than tubed pedicle grafts.

## D.5 Chapter 8

1. (a) $p < 0.10$; not significant.
   (b) $p = 0.025$; significant.
   (c) $p < 0.01$; significant.
   (d) $p < 0.05$; significant.
   (e) $p = 0.05$; significant.
2. $\chi^2 = 1.46$, d.f. $= 1$, not significant. The results of this study suggest that informing poorly controlled diabetics in writing of the potential problems of poor control has no effect on their compliance with a strict diet. The null hypothesis has
   been supported and the experimental hypothesis must be rejected.
3. (a) $p < 0.05$; significant.
   (b) $p = 0.01$; significant.
   (c) $p$ is greater than 5%; not significant.
   (d) $p < 0.10$; not significant.
   (e) $p < 0.005$; significant.
4. $T = 11.5$, $N = 13$, $p < 0.01$. The results of this study suggest that asthmatic
   patients suffer less severe asthma attacks when artificial E additives are eliminated from the diet (mean severity score before altering diet $= 4.73$, and
   after altering diet $= 3.2$). The experimental hypothesis can be accepted and the
   null hypothesis rejected.
5. (a) $p < 0.05$; significant.
   (b) $p < 0.10$; not significant.
   (c) $p = 0.01$; significant.
   (d) $p < 0.005$; significant.
   (e) $p < 0.10$; not significant.
6. $t = 2.69$, d.f. $= 11$, $p < 0.05$. The results of this study suggest that relaxation
   techniques have a significant effect on the blood pressure of hypertensive
   patients. Inspection of the mean scores for each condition indicates that systolic
   pressures are lower following a period of relaxation exercises (mean systolic
   score before relaxation $= 97.92$; mean score after relaxation $= 94.17$).

## D.6 Chapter 9

1. (a) $p < 0.10$; not significant.
   (b) $p < 0.02$; significant.
   (c) $p = 0.05$; significant.
   (d) $p < 0.10$; not significant.
   (e) $p < 0.01$; significant.
2. $Q = 8.02$, d.f. $= 3$, $p < 0.05$. The results of this study suggest that there is a
   significant relationship between quality of student nurse performance and type
   of clinical placement. Therefore the experimental hypothesis can be accepted
   and the null hypothesis rejected.

3. (a) $p < 0.042$; significant.
   (b) $p$ is greater than 0.10; not significant.
   (c) $p$ is greater than 0.10; not significant.
   (d) $p < 0.10$; not significant.
   (e) $p = 0.036$; significant.
4. $\chi_r^2 = 6.29$, $N = 6$, $C = 3$, $p < 0.072$, not significant. The results from this study suggest that there is no relationship between the progress of stroke victims and the nature of their home support structure. The null hypothesis must be accepted and the experimental hypothesis rejected.
5. (a) $p = 0.01$; significant.
   (b) $p$ is greater than 0.05; not significant.
   (c) $p < 0.001$; significant.
   (d) $p < 0.01$; significant.
   (e) $p < 0.05$; significant.
6. $L = 162.5$, $C = 3$, $N = 12$, $p < 0.001$. The results from this study suggest that student nurses are most likely to take sick leave during their psychiatric ward experience, followed by geriatric and paediatric ward experience respectively.
7. (a) $F_{bet}$: $p < 0.05$; significant. $F_{subj}$: $p$ is greater than 0.05; not significant.
   (b) $F_{bet}$: $p < 0.001$; significant. $F_{subj}$: $p < 0.001$; significant.
   (c) $F_{bet}$: $p < 0.001$; significant. $F_{subj}$: $p$ is greater than 0.05; not significant.
   (d) $F_{bet}$: $p$ is greater than 0.05; not significant. $F_{subj}$: $p < 0.05$; significant.
   (e) $F_{bet}$: $p < 0.025$; significant. $F_{subj}$: $p < 0.025$; significant.
8. The results of this study (Table D.3) suggest that student nurses perform differently on the nursing studies examination, according to their stage of training. These results cannot be attributed to individual differences among the sub-

**Table D.3**

| Sources of variation in scores | Sums of squares (SS) | Degrees of freedom (d.f.) | Mean squares (MS) | F ratios (F) |
|---|---|---|---|---|
| Variation in scores between conditions | 34.07 | 2 | 17.04 | 19.15 |
| Variation in scores between subjects | 18.17 | 9 | 2.02 | 2.27 |
| Variation due to random error | 15.93 | 18 | 0.89 | |
| Total | 68.17 | 29 | | |

For $F_{bet}$ (19.15) $p < 0.001$.
For $F_{subj}$ (2.27) $p$ is greater than 0.05 and is therefore not significant.

ject sample. Therefore we can conclude that the experimental hypothesis has been supported; the null hypothesis must be rejected.

9. Condition 1 × condition 2

$$F = 6.72, F' = 8.52, F^0 = 4.26; \text{ not significant.}$$

Condition 1 × condition 3

$$F = 37.56, F' = 8.52, F^0 = 4.26, p < 0.05; \text{ significant.}$$

Condition 2 × condition 3

$$F = 12.5, F' = 8.52, F^0 = 4.26; \text{ significant.}$$

These results suggest that student nurses' marks on their first and third year, and second and third year nursing studies exams differ significantly. There are, however, no significant differences between their year 1 and 2 marks.

# D.7 Chapter 10

1. (a) $p < 0.025$; significant.
   (b) $p < 0.05$; significant.
   (c) $p = 0.02$; significant.
   (d) $p$ is greater than 0.1; not significant.
   (e) $p < 0.001$; significant.
2. $\chi^2 = 6.59$, d.f. $= 1$, $p < 0.01$, significant. The results suggest that women who have a lumpectomy for carcinoma are less likely to suffer from severe depression in the six months following surgery, when compared with women who have a radical mastectomy. The experimental hypothesis has been supported; the null hypothesis must be rejected.
3. (a) $p < 0.005$; significant.
   (b) $p = 0.10$; not significant.
   (c) $p < 0.05$; significant.
   (d) $p < 0.02$; significant.
   (e) $p < 0.05$; significant.
4. $U = 86$; $U' = 124$; $n_1 = 15$; $n_2 = 14$; not significant. These results suggest there is no relationship between the level of aluminium in the water supply and the severity of Alzheimer's disease. The experimental hypothesis must be rejected and the null hypothesis accepted.
5. (a) $p < 0.05$; significant.
   (b) $p = 0.02$; significant.
   (c) $p < 0.10$; not significant.
   (d) $p < 0.01$; significant.
   (e) $p < 0.0005$; significant.
6. $t = 2.63$, d.f. $= 22$, $p < 0.01$; significant. These results indicate that women who have pre-hysterectomy counselling recover more quickly than those who do not

(mean score for the counselling group = 6.25; mean score for the non-counselling group = 7.33). The experimental hypothesis has therefore been supported; the null hypothesis can be rejected.

## D.8  Chapter 11

1. (a) $p$ is greater than 0.05; not significant.
   (b) $p < 0.02$; significant.
   (c) $p < 0.01$; significant.
   (d) $p = 0.05$; significant.
   (e) $p < 0.001$; significant.
2. $\chi^2 = 17.22$, d.f. = 4, $p < 0.01$. The results suggest that there is a significant relationship between the degree of maternal involvement in caring for a child in hospital and the stress experienced. The experimental hypothesis has been supported and the null hypothesis can be rejected.
3. (a) $p = 0.02$; significant.
   (b) $p$ is greater than 0.114; not significant.
   (c) $p = 0.01$; significant.
   (d) $p < 0.01$; significant.
   (e) $p < 0.001$; significant.
4. $H = 11.68$, d.f. = 2, $p < 0.009$. The results suggest that there is a significant difference in the evaluations by patients, nurses and doctors of the quality of student nurses trained on the pilot scheme. The experimental hypothesis has been supported and the null hypothesis can be rejected.
5. (a) $p < 0.05$; significant.
   (b) $p < 0.01$; significant.
   (c) $p$ is greater than 0.05; not significant.
   (d) $p < 0.05$; significant.
   (e) $p < 0.01$; significant.
6. $A = 51$, $B = 75$, $S = 27$, $C = 3$, $n = 5$, not significant. The results of this study suggest that there is no relationship between severity of heart attack and smoking behaviour of the patient (i.e. smokers having the most severe heart attacks, followed by ex-smokers and non-smokers). Therefore, the experimental hypothesis has not been supported and the null hypothesis must be accepted.
7. (a) $p < 0.025$; significant.
   (b) $p < 0.01$; significant.
   (c) $p$ is greater than 0.05; not significant.
   (d) $p < 0.05$; significant.
   (e) $p < 0.001$; significant.
8. The results (Table D.4) suggest that there is a significant relationship between the method of administering morphine to patients and the subsequent amount of

**Table D.4**

| Source of variance | SS | d.f. | MS | F ratio |
|---|---|---|---|---|
| Between conditions variance | 38.11 | 2 | 19.06 | 19.06 |
| Error variance | 15 | 15 | 1 | |
| Total | 53.11 | 17 | | |

$F = 19.06, p < 0.001$

pain experienced. The experimental hypothesis has been supported and the null hypothesis can be rejected.

9. Comparison 1 × comparison 2 (morphine pump v. regular injections)

$$F = 36.03, \ F^0 = 3.68, \ F' = 7.36$$

These results suggest that the pain relief derived from morphine administered by a morphine pump is significantly greater than when it is administered at regular intervals.

Comparison 1 × comparison 3 (morphine pump v. morphine on request)

$$F = 3.97, \ F^0 = 3.68, \ F' = 7.36$$

These results suggest that there is no significant difference in the pain relief derived from a morphine pump and morphine on request.

Comparison 2 × comparison 3 (regular injections v. morphine on request)

$$F = 16.12, \ F^0 = 3.68, \ F' = 7.36$$

These results suggest that morphine on request produces significantly more pain relief than when the morphine is given by regular injections.

## D.9 Chapter 12

1. (a) $p < 0.05$; significant.
   (b) $p$ is greater than 0.10; not significant.
   (c) $p = 0.05$; significant.
   (d) $p < 0.01$; significant.
   (e) $p < 0.025$; significant.
2. $r_s = +0.22$, $N = 12$, not significant. These results suggest that there is no significant relationship between the time it takes to get to work and the degree of stress experienced. The null hypothesis must be accepted and the experimental hypothesis rejected.
3. (a) $p = 0.005$; significant.

(b) $p$ is greater than 0.05; not significant.
(c) $p < 0.0005$; significant.
(d) $p$ is greater than 0.10; not significant.
(e) $p = 0.05$; significant.
4. $r = +0.16$, d.f. $= 10$, not significant. The results indicate that there is no significant relationship between length of time on oral contraception and length of time to conceive. The null hypothesis must be accepted and the experimental hypothesis rejected.
5. (a) $p < 0.05$; significant.
(b) $p$ is greater than 0.05; not significant.
(c) $p < 0.01$; significant.
(d) $p = 0.01$; significant.
(e) $p < 0.05$; significant.
6. $W = 0.72$, $s = 688$, $n = 10$, $N = 5$, $p < 0.01$. The results suggest that there is significant agreement amongst the nurses as to how the sum of money should be spent. The experimental hypothesis has been supported and the null hypothesis can be rejected.
7. $a = 1.91$    $b = 0.58$
Patient I $= 3.65$ on the adaptation scale.
Patient II $= 3.07$ on the adaptation scale.
Patient III $= 4.23$ on the adaptation scale.

## D.10 Chapter 13

1. The 95% confidence limits in this example are 0.29–0.61 or 29–61%.
2. The 90% confidence limits are: 4.2–5.4 months.
The 95% confidence limits are: 4.08–5.52 months.
The 99% confidence limits are: 3.82–5.78 months.
3. The 95% confidence limits in this example are 0.08–0.24 or 8–24%.
The 99% confidence limits are 0.06–0.26 or 6–26%.

## D.11 Appendix A

Table D.5 Appendix A

| Score | Rank |
| --- | --- |
| 2 | 1 |
| 3 | 3 |
| 4 | 5 |
| 3 | 3 |
| 5 | 6 |
| 3 | 3 |
| 7 | 7 |
| 8 | 8.5 |
| 8 | 8.5 |

# References

Bradley, J.I. and McClelland, J.N. (1978) *Basic Statistical Concepts*, Glenview, Illinois: Scott, Foresman.

Clifford, C. and Gough, S. (1990) *Nursing Research: A Skills Based Introduction*, Hemel Hempstead: Prentice Hall.

Darbyshire, P. (1986) 'When the face doesn't fit', *Nursing Times*, **82**(39), 28–30.

Downie, R.S. and Calman, K.C. (1987) *Healthy Respect: Ethics in Health Care*, London: Faber & Faber.

Ferguson, G.A. (1976) *Statistical Analysis in Psychology and Education*, Tokyo: McGraw-Hill Kogakusha.

Fisher, R.A. and Yates, F. (1963) *Statistical Tables for Biological, Agricultural and Medical Research*, New York: Hafner.

Greene, J. and D'Oliveira, M. (1982) *Learning to Use Statistical Tests in Psychology*, Milton Keynes: Open University Press.

Jaeger, R.M. (1983) *Statistics: A Spectator Sport*, Beverly Hills, CA: Sage.

Nunnally, J.C. (1975) *Introduction to Statistics for Psychology and Education*, New York: McGraw-Hill.

Oppenheim, A.N. (1966) *Questionnaire Design and Attitude Measurement*, London: Heinemann.

Reid, N.G. and Boore, J.R.P. (1987) *Research Methods and Statistics in Health Care*, London: Edward Arnold.

Robson, C. (1974) *Experiment, Design and Statistics in Psychology*, Harmondsworth: Penguin.

Royal College of Nursing (1977) *Ethics Related to Research in Nursing: Guidance for Nurses Involved in Research or any Study/Project Concerning Human Subjects*, Royal College of Nursing Research Society, London: Royal College of Nursing of the UK.

Royal College of Nursing Advisory Service (1981) *Directory of Nursing. Funds and Trusts.* Published, on behalf of the Royal College of Nursing, by and obtainable from Gazelle Book Services, Falcon House, Queen Square, Lancaster LA1 1RN.

Siegel, S. (1956) *Nonparametric Statistics for the Behavioural Sciences*, Tokyo: McGraw-Hill Kogakusha.

# Index

abbreviations, 279
abstract or summary, evaluation, 271
abstracting, 243–4
acknowledgements, 267
*AORN* (Association of Operating Room
    Nurses), 243
apparatus used
    describing in research report, 263
    evaluation, 274
authors of research paper, listing for
    publication, 258–9

*British Humanities Index*, 244
*British Journal of Hospital Medicine*, 243
*British Reports, Translations and Theses*, 244

calculations, rules of order, 280–1
*Cancer Nursing*, 243
card index file of information sources, 244–5
characteristic of population, 8
chi-squared test, 76, 141, 142–8
    calculation procedure, 144–5
    conditions required for use, 142, 143
    probability tables, 143, 146, 290
    worked example, 143–6
Cochran $Q$ test, 76, 109–14
    calculation procedure, 111–12
    conditions required for use, 108, 109
    probability tables, using, 109, 112
    worked example, 109–13
confidence, level of, 234
confidence interval, definition 227–8
confidence limits
    estimating population mean, 235–6, 238–9
    estimating proportions of the population,
        237–8, 239
    upper and lower, 8, 277, 235
constant errors, 41, 59
    correlational designs, 56
    experimental designs, 54–6
control conditions, 20, 75
correlation coefficient, 29–31
correlational designs, 10, 24–31
    collection of data, 25–6
    compared with experimental designs, 24–6,
        28

constant errors, 56
correlation coefficient, 29–31
    description, 24, 74
    ethical issues, 27
    experimenter bias effects and source of error,
        53–4
    main features, 194–5
    order effects and sources of error, 51–2, 53
    positive and negative correlations, 27–9, 30
    procedure, 195–6
    random errors, 58–9
    selection of subjects, 38
    statistical tests *see under* statistical tests
counterbalancing, 45, 51
critical values, 112
*Cumulative Index of Nursing and Allied Health
    Literature*, 244
*Current Research in Britain*, 244
current study, 25

decimal places, rounding, 284–5
degrees of freedom (d.f.), 79
Department of Health, Nursing Research
    Fellowship Scheme, 251
dependent variables (DV), 17–19
descriptive statistics, 5–6
design of research
    describing in research report, 262
    evaluation, 273
different subject designs, 38, 39, 40, 42–3, 47,
    75, 141, 161
    statistical tests *see under* statistical tests
discussion section of report, evaluation, 277–8
double-blind technique, 54, 59

error sources
    constant errors, 41
        correlational designs, 56
        experimental designs, 54–6
    description, 49
    experimenter bias effects, 53–4
    order effects and counterbalancing, 49–53
        correlational designs, 51–2, 53
        experimental designs, 49–51, 52–3
    random errors, 41
        correlational designs, 58–9

error sources (*continued*)
  experimental designs, 56–8
estimation, 7, 10, 225–40
  description, 225
  interval estimation, 8, 9, 226–8
  point estimation, 8, 9, 226–8
  procedure, 225–6
ethical issues, 21, 27, 248
experimental conditions, 20, 75
experimental designs, 10, 17–24
  compared with correlational designs, 24–6, 28
  considerations when constructing, 19–23
  constant errors, 54–6
  decision chart for selecting statistical test, 76
  description, 74
  different subject designs, 38, 39, 40, 42–3, 47
  ethical issues, 21
  example, 19–20
  experimenter bias effects and souce of error, 53–4
  five-condition, 23
  matched subject designs, 39, 40, 45–6, 47, 48
  order effects and sources of error, 49–51, 52–3
  random errors, 56–8
  same subject designs, 48
  statistical tests *see under* statistical tests, different subject designs *and* same and matched subject designs
  three-condition, 22, 23
  two-condition, 21, 22
  use of subjects, 38–48
    same subject designs, 38, 39, 43, 44–5
    variables, independent and dependent, 17–19
experimental hypothesis *see* hypothesis, experimental
experimenter bias effect, 53–4, 59
extended chi-squared test, 76, 162–70
  calculation procedure, 166–8
  conditions required for use, 162, 163–5
  probability tables, using, 164, 168
  worked example, 165–9

fatigue effects, 50
frequency polygon, 228, 229
Friedman test, 76, 114–20
  calculation procedure, 116–18
  conditions required for use, 108, 114
  probability tables, 114, 118, 293–4
  worked example, 115–18
funding research, 248, 250–2
  list of funding bodies, 250–1

*Geriatric Medicine*, 243
graphical presentation of results, 276

HO *see* null hypothesis
H1 *see* hypothesis, experimental

*Health Service Abstracts*, 244
*Hospital Abstracts*, 244
hypothesis, experimental
  defining, 241–2, 271
  description, 13–14, 16
  formulating, 9
  one-tailed, examples, 80–1, 82
  two-tailed, examples, 81, 83
hypothesis testing, 8, 9–10
  role in nursing research, 12
  *see also* correlational designs; experimental designs; project design

independent variables (IV), 17–19
*Index Medicus*, 244
indexing systems, 243–4
inferences, 5
inferential statistics, 3, 5–6, 10
information sources, recording, 244–5
*International Journal of Nursing Studies*, 243
*International Nursing Index*, 244
interval estimation, 8, 9, 235
  description, 226–7
  samples larger than thirty, 235–8
    estimating confidence limits, population mean, 235–6
  samples less than thirty, 238–9
    estimating confidence limits, population mean, 238–9
interval level of measurement, 65, 69–70, 71, 77
introduction to research report, 260–1, 271–2

Jonckheere trend test, 76, 176–81
  calculation procedure, 179–80
  conditions required for use, 162, 176–7
  probability tables, 177, 180, 308
  worked example, 177–80
*Journal of Advanced Nursing*, 243
*Journal of Paediatric Nursing*, 243
journals containing research articles, 243

Kendall's coefficient of concordance, 76, 197, 209–15
  calculation procedure, 212
  conditions required for use, 196, 209–10
  probability tables, 210, 213, 310
  worked example, 210–13
Kruskal-Wallis test, 76, 170–6
  calculation procedure, 172–4
  conditions required for use, 162, 170
  probability tables, 170, 174–5, 306–7
  worked example, 171–5

level of confidence, 234
libraries, using 243
linear regression, 197, 215–24
  calculation procedure, 220–2
  conditions required for use, 196, 216
  formula, 219

linear regression (*continued*)
  uses, 215–16
  worked example, 219–23
literature search, 242–4
lower confidence limit, definition 227

McNemar test, 76, 86, 87–94
  calculation procedure, 89–91
  conditions required for use, 87, 88
  probability tables, using, 91–2
  worked example, 88–93
Mann-Whitney *U* test, 76, 141, 148–54
  calculation procedure, 150–2
  conditions required for use, 142, 148
  probability tables, 148, 152, 304–5
  worked example, 148–53
matched subject designs, 39, 40, 45–6, 47, 48, 75
  description, 86, 87
  statistical tests *see* statistical tests, same and
      matched subject designs
  three or more conditions, description 107, 108
materials used
  describing in research report, 263
  evaluation, 274–5
measurement, levels, 75
  interval level, 65, 69–70, 71, 77
  nominal level, 65–7, 71, 77
  ordinal level, 65, 67–9, 71, 77
  ratio level, 65, 70–1, 77
method section of research report, 272–6

negative correlation, 28–9, 30
negative and positive numbers, 282–3
nominal level of measurement, 65–7, 71, 77
normal distribution, 228–9
  relationship with standard deviation, 232,
      233–4
normal distribution curve, 78, 229
null hypothesis, 14–16, 241, 242
*Nursing Bibliography*, 244
*Nursing Research*, 243
*Nursing Research Abstracts*, 244
*Nursing Times*, 243

one-way analysis of variance for related designs
      *see* one-way anova for related designs
one-way analysis of variance for unrelated
      designs *see* one-way anova for unrelated
      designs
one-way anova for related designs, 76, 126–35
  calculation procedure, 129–32
  conditions required for use, 108, 126
  probability tables, using, 126, 132–3
  worked example, 126–34
one-way anova for unrelated designs, 76, 181–9
  calculation procedure, 184–6
  conditions required for use, 162, 181
  probability tables, using, 182, 186–7
  worked example, 182–7

order effects, 43
  fatigue effects, 50
  practice effects, 50
  sources of error
    correlational designs, 51–2, 53
    experimental designs, 49–51, 52–3
ordinal level of measurement, 65, 67–9, 71, 77

Page's *L* trend test, 76, 120–6
  calculation procedure, 122–4
  conditions required for use, 108, 120
  probability tables, 121, 124, 295
  worked example, 121–4
parameters, 8, 225
Pearson test, 76, 197, 203–9
  calculation procedure, 205–6
  conditions required for use, 196, 203
  probability tables, 203, 206–7, 309
  worked example, 203–7
pilot trials, 253
point estimation, 8, 9, 226–7, 235
population, 6, 7, 12, 225
positive correlation, 27–8, 30
positive and negative numbers, 282–3
practice effects, 50
pre-test measure, 19
previous studies, reviewing, 242–4
probabilities, random errors, 59–60, 61
probability tables
  critical values of
    chi-squared, 290
    chi-squared (Friedman test), 293–4
    *F* (anovas), 296–303
    *H* (Kruskal-Wallis test), 306–7
    *L* (Page's *L* trend test), 295
    *r* (Pearson test), 309
    *r*ₛ (Spearman test), 308
    *S* (Jonckheere trend test), 308
    *s* (Kendall's coefficient of concordance),
        310
    *T* (Wilcoxon test), 291
    *t* (related and unrelated *t* tests), 292
    *U* (Mann-Whitney *U* test), 304–5
  description, 79–80
  using
    chi-squared test, 143, 146
    Cochran *Q* test, 109, 112
    extended chi-squared test, 164, 168
    Friedman test, 114, 118
    Jonckheere trend test, 177, 180
    Kendall's coefficient of concordance, 210,
        213
    Kruskal-Wallis test, 170, 174–5
    McNemar test, 91–2
    Mann-Whitney *U* test, 148, 152
    one-way anova for related designs, 126,
        132–3
    one-way anova for unrelated designs, 182,
        186–7

probability tables (*continued*)
   Page's *L* trend test, 121, 124
   Pearson test, 203, 206–7
   related *t* test, 104
   Spearman test, 197–8, 200–1
   unrelated *t* test, 154, 158–9
   Wilcoxon test, 98
project design, 16–31
   correlational designs, 24–31
   experimental designs, 17–24
publishing research findings
   choice of journal, 257
   format required, 257
   preparation of article, 257–68
      abstract or summary, 259–60
      acknowledgements, 267
      authors, listing, 258–9
      discussion section, 265–7
      introduction, 260–1
      method section, 261–4: apparatus used,
         263; design, 262; materials used, 263;
         research procedure, 263–4; subjects,
         262–3
      references, 267–8
      results section, 264–5
      title, 258

*Quality Assurance Abstracts*, 244
questionnaire design, 274–5

random errors, 41
   correlational designs, 58–9
   description, 56, 59
   experimental designs, 56–8
   probabilities, 59–60, 61
   significance levels, 60–1
random number tables, 37
random sampling, 58, 226
random selection of subjects, 34, 36–8
rank ordering data, 285–7
ratio level of measurement, 65, 70–1, 77
references, 267–8, 278
related *t* test, 76, 77, 86, 100–6
   calculation procedure, 102–4
   conditions required for use, 87, 100
   probability tables, 104, 292
   worked example, 101–5
representative sampling, 7
research
   definition, 1
   ethical issues, 21, 27, 248
   function in nursing, 2
   funding *see* funding research
   obtaining permission, 247
   pilot trials, 253
   planning and preparation, 241–7
      data analysis, 247
      defining area of interest, 241–2
      recording information sources, 244–5

reviewing previous studies, 242–4
   subjects, selecting, 246
research findings
   presentation, 253–4, 255–69
      evaluation, 276–7
      general guidelines, 255–7
      graphical presentation, 257
   publishing *see* publishing research findings
   support of hypothesis, 79–80
research procedure
   describing in research report, 263–4
   evaluation, 275–6
   summary of details, 252–3
research project, costing 248
research proposal
   functions, 247–9
   information included, 249–50
   layout, 250
   presentation, 251
   writing, 247–50
research reports
   evaluation, 270–9
      abstract or summary, 271
      discussion section, 277–8
      introduction, 271–2
      method section, 272–6: apparatus used,
         274; design, 273; materials used, 274–5;
         research procedure, 275–6; subjects,
         273–4;
      presentation of results, 276–7
      references, 278
      title, 270–1
results of research *see* research findings
retrospective study, 25, 52
Royal College of Nursing library, 243
Royal College of Nursing Personal Advisory
   Service, 251

same subject designs, 38, 39, 43, 44–5, 48, 75
   description, 86
   statistical tests *see* statistical tests, same and
      matched subject designs
   three or more conditions, description 107
sample, 6, 7, 12, 225
sample size, subjects, 35–6
sampling error, 36
scales of measurement *see* measurement
Scheffe multiple range test, 136–40
   calculation procedure, 137–9
   conditions required for use, 108, 136
   worked example, 136–9
Scheffe multiple range test (unrelated designs),
   189–93
   calculation procedure, 190–2
   conditions required for use, 162, 189
   worked example, 189–92
significance levels, 60–1
   comparison of one-tailed and two-tailed test,
      81–2
sources of error *see* error

Spearman test, 76, 197, 197–203
    calculation procedure, 199–200
    conditions required for use, 196, 197
    probability tables, 197–8, 200–1, 308
    worked example, 198–201
square roots, 283–4
squared numbers, 283–4
standard deviation, 228
    formula, 230–1
    relationship with normal distribution, 232,
        233–4
statistical probability tables *see* probability
    tables statistical tests
choice of, 247
    correlational designs, 194–224
        linear regression, 196, 197, 215–24
        non-parametric: Kendall's coefficient of
            concordance, (three or more sets of
            data), 196, 197, 209–15; Spearman test
            (two sets of data), 196, 197–203
        parametric, Pearson test (two sets of data),
            196, 197, 203–9
    different subject designs
        three or more conditions 161–94: extended
            chi-squared test, 161, 162–70;
            Jonckheere trend test, 161, 162, 176–81;
            Kruskal-Wallis test, 161, 162, 170–6;
            one-way anova for unrelated designs,
            161, 162, 181–9; Scheffe multiple range
            test (unrelated designs), 161, 162, 189–93
        two conditions only, 141–60: chi-squared
            test, 141, 142–8; Mann-Whitney *U* test,
            141, 142, 148–54; unrelated *t* test, 141,
            142, 154–60
    experimental designs *see* different subject
        designs; same and matched subject
        designs
    parametric tests comparison with non-
        parametric tests, 77–8, 83
    same and matched subject designs
        three or more conditions, 107–140:
            Cochran *Q* test, 108, 109–114; Friedman
            test, 108, 114–20; one-way anova for
            related designs, 108, 126–35; Page's *L*
            trend test, 108, 120–6; Scheffe multiple

        range test, 108, 136–40
        two conditions only, 86–106: McNemar test
            87–94; related *t* test, 100–6; Wilcoxon
            test, 94–9
    selection, 73–5, 76, 77, 83
statistics, role in nursing, 1
subject sample, evaluation 273–4
subjects
    definition, 33
    describing in research report, 262–3
    experimental design
        different subject designs, 38, 39, 40, 42–3,
            47
        matched subject designs, 39, 40, 45–6, 47,
            48
        same subject designs, 38, 39, 43, 44–5, 48
    random selection, 34, 36–8
    sample size, 35–6
    selection, 33–8, 246
        correlation design, 38
    use in experimental design, 38–48
symbols used in statistical formulae, 289

title of research paper, 258, 270–1
topic of research, defining, 241–2

unrelated *t* test, 76, 141, 154–60
    calculation procedure, 156–8
    conditions required for use, 142, 154
    probability tables, 154, 158–9, 292
    worked example, 155–9
upper confidence limit, definition, 227

variability, 230
variables, 13
    examples, 14
    identifying, 242
    independent and dependent, 17–19
    manipulating, 17

Wilcoxon test, 76, 77, 86, 94–9
    calculation procedure, 97–8
    conditions required for use, 87, 94
    probability tables, using, 98
    worked example, 95–9